The Rationality Quotient

The Rationality Quotient

Toward a Test of Rational Thinking

Keith E. Stanovich, Richard F. West, and Maggie E. Toplak

The MIT Press
Cambridge, Massachusetts
London, England

© 2016 Massachusetts Institute of Technology

Appendix © 2015 Keith E. Stanovich and Richard F. West

All rights reserved. No part of this book may be reproduced in any form by any electronic or mechanical means (including photocopying, recording, or information storage and retrieval) without permission in writing from the publisher.

This book was set in Stone Serif and Stone Sans by Toppan Best-set Premedia Limited. Printed and bound in the United States of America.

Library of Congress Cataloging-in-Publication Data

Names: Stanovich, Keith E., 1950- author. | Toplak, Maggie E., author.
Title: The rationality quotient : toward a test of rational thinking / Keith Stanovich, Richard West, and Maggie E. Toplak.
Description: Cambridge, MA : MIT Press, 2016. | Includes bibliographical references and index.
Identifiers: LCCN 2016002051 | ISBN 9780262034845 (hardcover : alk. paper)
Subjects: LCSH: Reasoning--Ability testing. | Reasoning (Psychology) | Intelligence levels. | Cognition.
Classification: LCC BF442 .S728 2016 | DDC 153.4/30285--dc23 LC record available at https://lccn.loc.gov/2016002051

10 9 8 7 6 5 4 3 2 1

For Paula, since the last book, the best times of our lives, my love

For Josh, my son and friend

For my parents Joseph and Veronika, for instilling in me the pursuit of knowledge and rationality

Contents

Preface ix
Acknowledgments xv

I Theoretical Underpinnings 1

1 Definitions of Rationality in Philosophy, Cognitive Science, and Lay Discourse 3
2 Rationality, Intelligence, and the Functional Architecture of the Mind 15
3 Overcoming Miserly Processing: Detection, Override, and Mindware 39
4 A Framework for the Comprehensive Assessment of Rational Thinking (CART) 63

II The Components of Rational Thought Assessed by the CART 75

5 Probabilistic and Statistical Reasoning 77
6 Scientific Reasoning 97
7 Avoidance of Miserly Information Processing: Direct Tests 111
8 Avoidance of Miserly Information Processing: Indirect Effects 141
9 Probabilistic Numeracy, Financial Literacy, Sensitivity to Expected Value, and Risk Knowledge 177
10 Contaminated Mindware 191
11 The Dispositions and Attitudes of Rationality 207

III Comprehensive Rational Thinking Assessment: Data and Conclusions 217

12 Associations among the Subtests: A Short-Form CART 219
13 Associations among the Subtests: The Full-Form CART 233
14 The CART: Context, Caveats, and Questions 269
15 The Social and Practical Implications of a Rational Thinking Test 297

Appendix: Structure and Sample Items for the Subtests and Scales of the Comprehensive Assessment of Rational Thinking 331
Notes 369
References 379
Author Index 441
Subject Index 455

Preface

We get surprised when someone we consider to be smart acts foolishly. When someone we consider to be *not* so smart acts foolishly, we tend not to be so surprised. But why should we be so surprised in the first case? It seems that smart people do foolish things all the time. Wasn't the financial crisis of 2008 just littered with smart people doing dumb things—from the buyers and sellers of the toxic mortgage securities to the homebuyers who seemed to think their house price would double every three years?

So if it is not rare for smart people to act foolishly, then why the surprise? In fact, the confusion here derives from being caught up in the inconsistencies and incoherence of folk language. The folk terms being used in this discussion are in dire need of some unpacking. Consider the title of an edited book to which we contributed a chapter: *Why Smart People Can Be So Stupid* (Sternberg, 2002). A typical dictionary definition of the adjectival form of the word "smart" is "characterized by sharp quick thought; bright" or "having or showing quick intelligence or ready mental capacity." Thus, being smart seems a lot like being intelligent, according to the dictionary. Dictionaries also tell us that a stupid person is "slow to learn or understand; lacking or marked by lack of intelligence." Thus, if a smart person is intelligent and stupid means a lack of intelligence, then the "smart person being stupid" phrase seems to make no sense.

However, a secondary definition of the word "stupid" is "tending to make poor decisions or careless mistakes"—a phrase that attenuates the sense of contradiction. A similar thing happens if we analyze the word "dumb" to see if the phrase "smart but acting dumb" makes sense. The primary definition describes "dumb" as the antonym of "intelligent," again leading to a contradiction. But in phrases referring to decisions or actions such as "What a dumb thing to do!" we see a secondary definition, like that of "stupid": "tending to make poor decisions or careless mistakes." These phrases pick out a particular meaning of "stupid" or "dumb"—albeit not the primary one.

It is likewise with the word "foolish." A foolish person is a person "lacking good sense or judgment; showing a lack of sense; unwise; without judgment or discretion." This picks out the aspect of "stupid" and "dumb" that we wish to focus on here—the aspect that refers not to intelligence (general mental "brightness") but instead to the tendency to make judicious decisions (or, rather, injudicious ones).

However we phrase it—"smart but acting dumb," "smart but acting foolish," or whatever—we have finally specified the phenomenon: intelligent people taking injudicious actions or holding unjustified beliefs. Folk psychology is picking out two different traits: mental "brightness" (intelligence) and making injudicious decisions (*rational* thinking). If we were clear about the fact that the two traits are different, the sense of paradox or surprise at the "smart but acting foolish" phenomenon would vanish. What perpetuates the surprise is that we tend to think of the two traits as one, or at least that they should be strongly associated.

The confusion is fostered because psychology has a measurement device (the intelligence test) for the first but not the second. Psychology has a long and storied history (over one hundred years old) of measuring the intelligence trait. There has been psychological work on rational thinking, but this research started much later and was not focused on individual differences. Our research group has conducted one of the longest extant investigations of individual differences in rational thinking processes. In the present book, we will attempt to synthesize this work by presenting the first prototype of a rational thinking test (the Comprehensive Assessment of Rational Thinking).

A novice psychology student might be a bit confused at this point—thinking that somewhere along the line he or she has heard definitions of intelligence that included rationality. Such a student would be right. Many *theoretical* definitions of intelligence incorporate the rationality concept by alluding to judgment and decision making in the definition. Other definitions emphasize behavioral adaptiveness and thus also fold rationality into intelligence. The problem here is that *none* of these components of rationality—adaptive responding, good judgment, and decision making—are assessed by *actual tests* of intelligence.

Preface

Publishers and proponents of IQ tests have encouraged the view that you get everything you need in mental assessment from such tests. They have also encouraged the view that even when this is not the case, the correlation with intelligence will be so high that it will not even be worth worrying about measuring anything else in the cognitive domain. But in fact, by giving an intelligence test, one does not automatically get a measure of rational thinking. To get the latter, we need to actually construct a test of rational thinking. That is what this book is about. Because we now have conceptually grounded theories of rationality and because we have a prodigious number of tasks that measure the components of rational thinking (Baron, 2008; Kahneman, 2011; Stanovich, 2011), it is now possible to see what would happen if we began from the ground up to construct a rationality test around that concept only.

Synthesizing theoretical work and empirical research that began over two decades ago (Stanovich, 1993; Stanovich & West, 1997, 1998c), we present a prototype of such a test in this volume. We have proceeded by grounding our project in the empirical literature on the nature of human judgment and decision making (Kahneman, 2011; Manktelow, 2012) and theoretical discussions of rationality in cognitive science (Evans, 2014; Stanovich, 2011, 2012). For years, we have been examining how one would go about constructing the best test of rational thinking if the focus were solely on that construct. Thus, we have not structured our investigations around what was previously known about intelligence. This point deserves elaboration, because a common misinterpretation of our work is that we are trying to improve intelligence tests. Not only is this *not* our goal, but it is a serious misunderstanding of what we are trying to achieve.

Unlike some writers, we do not see the usefulness of labeling every human cognitive skill as intelligence—particularly when there are readily existing concepts (both scientific concepts and folk concepts) for some of those things (e.g., rationality, creativity, wisdom, critical thinking, open-minded thinking, reflectivity, sensitivity to evidence). Calling everything "intelligence" has been the strategy of broad theorists whose stance serves to *inflate* the concept of intelligence. By inflation, we mean putting into the term more than what IQ tests measure. For example, broad theorists such as Howard Gardner and Robert Sternberg define concepts such as practical intelligence, bodily kinesthetic intelligence, emotional intelligence. We would argue that these broad theorists are adopting a *permissive* conceptualization of intelligence rather than a *grounded* conceptualization. Permissive theories include in their definitions of intelligence aspects of mental functioning that are captured by the vernacular term "intelligence" (e.g., adaptation to the environment, showing wisdom, creativity, etc.) *whether or not* these aspects are measured by existing tests of intelligence. Grounded theories, in contrast, confine the concept of intelligence to the set of mental abilities actually tested by IQ tests. Adopting permissive definitions of the concept of intelligence only serves to obscure what is absent from existing IQ tests. Instead, to highlight the *missing* elements in IQ tests, we adopt a thoroughly grounded notion of intelligence.

Thus, it follows that we do not view our project as an attempt to improve existing intelligence tests, but rather as an attempt to look at what a comprehensive assessment of the rationality concept might look like. Which subcomponents of rationality correlate with intelligence and which do not will be revealed after the fact. That is, we did not use observed correlations with intelligence (or the lack of such correlations) as a criterion for our test. We instead used the theoretical foundations of epistemic and instrumental rationality in the literature of philosophy and decision science (Manktelow, 2004; Over, 2004), as well as empirical work in the heuristics and biases tradition, to structure our test.

Our goal has always been to give the concept of rationality a fair hearing—almost as if it had been proposed prior to intelligence. Had that happened, what would a comprehensive test of rationality have looked like? We are running, if you will, an experiment on theoretically grounding and operationalizing an aspect of mental life that is important in its own right. We are not trying to improve IQ measurement. Nor are we concerned with the issue of incremental validity over IQ (whether our test can predict variance in important behaviors beyond what the current IQ tests can predict), or with logistical questions such as the length of testing if one were given both a rational thinking test and an IQ test.

Regarding correlations between rational thinking (and its components) and intelligence, we are prepared to let the chips fall where they may. If it were found that, in an unselected sample, a latent rationality variable and a latent intelligence variable had an absolutely perfect correlation, then we would be the first to say "What an absolutely extraordinary instrument the IQ test is—we have not known it all these years, but it perfectly measures individual differences in rationality as well!" However, on the other hand, we would want to make this concession only after a research effort that did not give pride of place to current IQ tests. Whatever the outcome empirically, our research effort will at least provide a measure of the rational thinking concept. But again, we are not using relationships with intelligence (or the lack of such relationships) to decide what to include in our test. If a rational thinking component was conceptually grounded and researched in the psychological literature, then it was a candidate for our assessment device.

In our society, what gets measured gets valued. With our Comprehensive Assessment of Rational Thinking (the CART) we aim to draw more attention to the skills of rational thought by measuring them systematically. It was in part a historical contingency that the operational measurement of the concept of intelligence is older than the operational measurement of rationality. If we were to say that operationalized intelligence dates from Spearman or Binet, then we could say that operationalized rationality dates from the axiomatic approach to utility theory of von Neumann and Morgenstern (and Savage)—and here we are talking about the 1940s and 1950s. One way to look at our research program is to say that we are looking at how one would measure rationality, ignoring the historical contingency that intelligence was established earlier as a measure of cognition.

So one more time: We are not trying to make a better intelligence test. Nor are we trying to find mental tests to add to intelligence tests that will incrementally improve the validity of the IQ test. We are trying to show how one would go about measuring rational thinking as a psychological construct in its own right. We wish to accentuate the importance of a domain of thinking that has been obscured because of the prominence of intelligence tests and their proxies. It is long overdue that we had more systematic ways of measuring these components of cognition, which are important in their own right but are missing from IQ tests. Rational thinking has a unique history grounded in philosophy and psychology, and several of its subcomponents are firmly identified with well-studied paradigms. The story we will tell in this book is of how we have turned this literature into the first comprehensive device for the assessment of rational thinking—the CART.

Acknowledgments

This book, and the CART itself, has been inevitable since Stanovich coined the term _dysrationalia_ in 1993. But even an inevitable product doesn't appear automatically, as the interim theoretical and empirical work took over twenty years! Our early empirical work on individual differences in rational thought (Stanovich & West, 1997, 1998c, 1999) was first cashed out in terms of theoretical insights concerning dual-process theory and evolutionary psychology that were relevant to the Great Rational Debate in cognitive science (Stanovich, 1999, 2004; Stanovich & West, 2000). The next phase of our empirical work (see Stanovich & West, 2008b) led to the book *What Intelligence Tests Miss* (Stanovich, 2009). From that book, it was clear that the next logical step was following through on our claim that nothing was preventing the construction of a test of rational thinking. We outlined an early version of our framework for assessing rational thinking in Stanovich (2011, chapter 10) and in Stanovich, West, and Toplak (2011a), along with suggested tasks. Those familiar with our earlier framework will notice several quite radical changes when they read the present volume. These changes resulted from the new conceptualization of heuristics and biases tasks that we offer in chapters 3 and 4.

One thing that did immensely speed the appearance of this volume and the development of the CART was a grant from John Templeton Foundation (JTF), which allowed Stanovich and West to have an exclusive three-year focus on the project as well as allowing extra time for the participation of co-author Maggie Toplak. Susan Arellano was instrumental in drawing our attention to the JTF for possible support. Craig Joseph dealt initially with our grant at JTF, but for most of its term, our grant has been in the capable hands of Sarah Clement, Director of Character Virtue Development. Staff members Richard Bollinger and Caitlin Younce helped us with many administrative details. The opinions expressed in this book are those of the authors and do not necessarily reflect the views of the John Templeton Foundation.

We thank our editor at the MIT Press, Phil Laughlin, for his enthusiasm for the project from the very beginning and for his speed in getting us signed so efficiently that we did not lose any time on the book itself. We have always felt that we had strong editorial support for this complex and difficult project. Judy Feldmann was very helpful with manuscript editing.

Several conferences were seminal in allowing us to discuss these ideas at length: the Fourth International Thinking Conference in Durham, England; the Conference on Dual-Process Theories of Reasoning and Rationality in Cambridge, England, organized by Jonathan Evans and Keith Frankish; a workshop on dual-process theory at the University of Virginia organized by Tim Wilson and Jonathan Evans; the Sixth International Thinking Conference in Venice, Italy; the NSF Workshop on Higher Cognition in Adolescents and Young Adults, organized by Valerie Reyna; and the Seventh International Thinking Conference in London.

Our intellectual debts in writing this book are immense and are represented in the wide literature that we cite. Decades ago, the work of Daniel Kahneman and Amos Tversky inspired our interest in rational thinking tasks that were new to psychology at the time. The work of Jonathan Evans inspired our early (Stanovich, 1999; Stanovich & West, 2000) and later (Stanovich, 2011; this volume) contributions to dual-process theory. Jonathan Baron's concept of actively open-minded thinking was an early inspiration for our work on individual differences in rational thought. David Perkins coined the term *mindware*, which we have been using for many years now.

Jonathan Evans of the University of Plymouth read the entire manuscript and made many discerning suggestions for revision. His astute and meticulous reading provided us with much "big picture" direction, as well as numerous scientific suggestions based on his erudite knowledge of the vast literature that we cover in this book.

Prior to the grant from the John Templeton Foundation, our empirical research was supported by grants received from the Social Sciences and Humanities Research Council of Canada and by the Canada Research Chairs program to Keith E. Stanovich. Support was also received from the Social Sciences and Humanities Research Council of Canada to Maggie E. Toplak.

Acknowledgments

Most members of the Stanovich/West/Toplak lab (a joint lab linking the University of Toronto, James Madison University, and York University) in the past two decades have contributed in some way to the research of our own that is cited in this volume. However, we must single out Richard Hohn and Tristan Kirkman for their dedicated work in managing the James Madison University lab and its research assistants during the length of the Templeton grant. Before their tenure, Christopher Runyon, Rebecca Marsh Runyon, and Russell Meserve did exemplary work managing the James Madison University lab. Robin Anderson, Chair of the Department of Graduate Psychology, and administrative assistant Rosa Turner provided vital support for numerous aspects of our work on the John Templeton Foundation Grant. At the York University lab we have been aided expertly by Dr. David Flora (statistical consultant) and by trainees Geoff B. Sorge, Alexandra Basile, and Mohamed Al-Haj. Although work from the University of Toronto lab is older than that from our other two sites, critical lab administration was done there years ago by Caroline Ho, Robyn Macpherson, and Walter Sá.

I Theoretical Underpinnings

1 Definitions of Rationality in Philosophy, Cognitive Science, and Lay Discourse

"Rationality" is a torturous and tortured term in intellectual discourse. It is contentious and has a multitude of definitions. The term is claimed by many disciplines and parsed slightly differently by each discipline. Philosophy, economics, decision theory, psychology—all claim the term and have their own definitions. For example, animal behaviorists claim to measure degrees of rationality in animals (Kacelnik, 2006), yet by some of the definitions used in other disciplines, animals couldn't have rationality at all.

Across disciplines, scholars agree on only one thing: there are many definitions of rationality and they differ across scholarly domains. For example, economist Robert Frank (2004) acknowledges that "there are many conceptions of rationality, each with its own strengths and weaknesses. Much of the ambiguity concerning rationality stems from the simple fact that no single conception has managed to prevail over its competitors" (p. 45). Philosophers Susan Hurley and Matthew Nudds (2006) argue similarly that "in application to human beings, different disciplines use 'rationality' in different senses; some focus on rational behavior and others on rational processes, in line with their different assumptions and purposes. As a result, there is a danger of talking at cross purposes in interdisciplinary discussion" (p. 6). As if the cross-disciplinary problem was not bad enough, even *within* a discipline such as psychology there tends to be a proliferation of rationality-related terms and distinctions. We do not intend to comment on all of them, but simply wish to illustrate the need to demarcate how we are using the term in this book. We cannot address every philosophical issue concerning rationality here. Instead, our choice of terms and definitions is determined by our overall task: to present a prototype of what a comprehensive test of rational thinking would look like.

Our conception of rationality will be drawn from decision theory and cognitive science because, for an assessment device, we obviously needed a concept that was empirically based and operationally grounded. Philosophical conceptions not so grounded are essentially useless for the purpose of constructing a rational thinking test. Additionally, we needed a definition of rationality that allows for individual differences—one where rational thought and rational responding falls on a continuum from highly rational to less rational, and on which people differ. In contrast, many philosophical notions of rationality are crafted so as to equate all humans—thus, by fiat, defining away the very individual differences we wish to study.

Nevertheless, many people are used to philosophical definitions that equate people on rational thinking. These definitions have entered lay discourse in ways that may cause confusion about our usage. Regardless of one's prior familiarity with discussions of the term "rationality," we would urge our readers not to get hung up on the specific term. If you dislike our use of the term in any part of the book, then just substitute the phrase "good thinking" and absolutely nothing will be lost.[1]

The Strong Sense of "Rationality" Used in This Volume

The term "rationality" has a strong and a weak sense. Things often go wrong right at the beginning among laypeople and academics alike, because for them a weak sense of the term is common, whereas among empirical scientists a strong sense is common. The strong sense of the term is the one used in cognitive science, and it will be the one we use throughout this book. In contrast, dictionary definitions of rationality tend to be of the weak sort—often seeming quite lame and unspecific, for example, "the state or quality of being in accord with reason."

The weak definitions of rationality derive from a categorical notion of rationality tracing to Aristotle that posits humans as the only animals who base actions on reason. As de Sousa (2007) has pointed out, such a notion of rationality as "based on reason" has as its opposite not irrationality but *arationality*. Aristotle's characterization is categorical—an organism's behavior is either based on thought or it is not rational. In this conception, humans are rational, other animals are not. There is no room for individual differences in rational thinking *among* humans in this view.

Rationality in its stronger sense—the sense employed in cognitive science and in this book—is a *normative* notion. Normative models of optimal judgment and decision making define rationality in the noncategorical manner used in cognitive science. Rationality thus comes in degrees defined by the distance of the thought or behavior from the optimum defined by a normative model (Etzioni, 2014). Thus, when a cognitive scientist terms a behavior irrational, he or she means that the behavior departs from the optimum prescribed by a particular normative model. The scientist is not implying that no thought or reasoning whatsoever was behind the behavior. Some of the hostility that has been engendered by experimental claims of human irrationality no doubt derive from a (perhaps tacit) influence of the Aristotelian view—the assumption that cognitive psychologists are saying that certain people are somehow "less than human" when they are said to behave irrationally. Thus, the term "irrationality" becomes coded as a particularly egregious insult in folk language. Our point here is that psychologists are not using the term in this way. They are adopting a different definition of rationality, one in which people are all fully human but can nonetheless differ in their rational tendencies.

To a layperson, lingering associations with the Aristotelian categorical view may well make the term "irrationality" sound more cutting than it actually is. As cognitive scientists, we could do better to signal that it is the noncategorical, continuous sense of "rationality" and "irrationality" that we use in our science. When we find a behavioral pattern that is less than optimally rational, we could easily say that it is "less than perfectly rational" rather than that it is irrational—with no loss of the intended meaning. Perhaps if this had been the habit in the literature, the Great Rationality Debate in cognitive science (Bishop & Trout, 2005; Cohen, 1981; Kelman, 2011; Stanovich, 1999, 2004; Stein, 1996; Tetlock & Mellers, 2002) would not have become so heated. For this reason, we will use the term "irrationality" sparingly in this book. Instead, our focus is on continuous variation in rational responses, running—in degrees—from perfectly rational through increasingly less rational responses.

Another reason why psychologists do not adopt the categorical definitions of rationality is that such definitions provide no motivation for cognitive reform or cognitive change. Continuous definitions of rationality motivate cognitive reform, because most people are less than perfectly or optimally rational; most people can improve their rational thinking tendencies. But adopting the continuous definition of cognitive science—one that motivates cognitive reform—necessitates designating some people as more rational than others (at least on a task-by-task basis). This is no different from saying that some people are more intelligent than others. Variation in rationality simply partitions a doubtless overlapping, but nevertheless distinct, cognitive space. People do differ in intelligence, and because of the ability to measure such differences, psychologists have been able to devise ways to make people more intelligent. Likewise, people do differ in rationality, and cognitive science has discovered many ways to make people more rational and thus to achieve more positive outcomes in their lives. For example, we can teach people to make better financial decisions, teach doctors to make better medical decisions, and advise educators on how to make better educational decisions.

Rationality: Instrumental and Epistemic

As mentioned above, dictionary definitions of rationality ("the possession of reason") tend to be weak and not specific enough to be testable. Additionally, for various reasons some writers have promulgated a caricature of rationality because they wish to downplay its importance. A common ploy is to imply that rationality means little more than the ability to solve the syllogistic reasoning problems that are encountered in Philosophy 101. The meaning of rationality in modern cognitive science is, in contrast, much more comprehensive, robust, and important.

We follow many cognitive science theorists in recognizing two types of rationality: instrumental and epistemic (Manktelow, 2004; Over, 2004). The simplest definition of instrumental rationality is: Behaving in the world so that you get exactly what you most want, given the resources (physical and mental) available to you. Somewhat more technically, we could characterize instrumental rationality as the optimization of the individual's goal fulfillment. Economists and cognitive scientists have refined the notion of optimization of goal fulfillment into the technical notion of expected utility.

The other type of rationality, epistemic rationality, concerns how well our beliefs map onto the actual structure of the world. Epistemic rationality is sometimes termed "theoretical rationality" or "evidential rationality" by philosophers. Likewise, instrumental rationality is sometimes termed "practical rationality." These two types of rationality are related, of course; to take actions that fulfill our goals, we need to base those actions on beliefs that properly match up with the world.

When epistemic and instrumental rationality are properly defined, virtually no one wishes to eschew them. Most people want their beliefs to be in some correspondence with reality, and they also want to act in ways that help them achieve their goals. Manktelow (2004) has emphasized the practicality of both types of rationality by noting that they concern two critical things: what is true and what to do. Epistemic rationality is about what is true and instrumental rationality is about what to do. For our beliefs to be rational they must correspond to the way the world is—they must be true. For our actions to be rational, they must be the best means toward our goals—they must be the best things to do. Nothing could be more practical or useful for a person's life than the thinking processes that help them find out what is true and what is best to do.

More formally, economists and cognitive scientists define instrumental rationality as the maximization of expected utility. To be instrumentally rational, a person must choose among options based on which option has the greatest expected utility. Decision situations can be broken down into three components: (1) possible actions; (2) possible states of the world; and (3) evaluations of the consequences of possible actions in each possible state of the world. Expected utility is calculated by taking the utility of each outcome and multiplying it by the probability of that outcome occurring and then summing those products over all of the possible outcomes.

The Axiomatic Approach to Rational Choice

In practice, assessing rationality in this computational manner can be difficult because eliciting personal probabilities can be tricky. Also, measuring the utilities of various consequences can be experimentally difficult. Fortunately, there is another useful way to measure the rationality of decisions and deviations from rationality. It has been proven through several formal analyses that if people's preferences follow certain consistent patterns (the so-called axioms of choice), then they are behaving as if they are maximizing utility (Dawes, 1998; Edwards, 1954; Jeffrey, 1983; Luce & Raiffa, 1957; Savage, 1954; von Neumann & Morgenstern, 1944). These analyses have led to what has been termed the "axiomatic approach" to whether people are maximizing utility. This approach allows us to more easily measure people's degrees of rationality by the experimental methods of cognitive science. The deviation from the optimal choice pattern according to the axioms is an (inverse) measure of the degree of rationality—the further one deviates from the optimal choice pattern, the less rational that person is.

The axiomatic approach to choice defines instrumental rationality as adherence to certain types of consistency and coherence criteria. For example, one such axiom is that of transitivity: If you prefer A to B and B to C, then you should prefer A to C. Violating transitivity is a serious violation of rationality because it can lead to what decision theorists call a "money pump"—a situation where, if you acted on your intransitive preferences, you could be drained of all your wealth (Schick, 1986).

The axiomatic approach characterizes someone who maximizes his or her utility as having stable, underlying preferences for each of the options presented in a decision situation. It is assumed that an optimally rational person's preferences for the options available are complete, well ordered, and well behaved in terms of the axioms of choice (Thaler, 2015). All of the axioms of choice (independence of irrelevant alternatives, transitivity, independence, reduction of compound lotteries, etc.), in one way or another, ensure that decisions are not influenced by irrelevant context (Stanovich, 2013). Because the strength of each preference—the utility of that option—exists in the brain before the option is even presented, nothing about the context of the presentation should affect the preference. If our preferences are affected by irrelevant context, our preferences cannot be stable and we cannot be maximizing our utility. As a result, how much our thinking is independent of irrelevant context becomes an important measure of rational thinking and will have a prominent place in our assessment battery. Considerable empirical evidence in cognitive science indicates that people sometimes violate these axioms of utility theory (Kahneman & Tversky, 2000; Thaler, 2015). It is also known that people vary widely in their tendency to adhere to the basic axioms of choice that define instrumental rationality.

We can use the axiomatic approach to assess epistemic rationality as well. Recall that the expected utility of an action involves multiplying the probability of an outcome by its utility and summing across possible outcomes. Thus, determining the best action involves estimating the probabilities of various outcomes. These probabilities are not conscious calculations,

Definitions of Rationality

of course—they are our confidence estimates about states of the world. They are our beliefs and the confidence we have in them. If our probabilistic judgments about the states of the world are wrong, decision making will not maximize our utility—our actions will not result in our getting what we most want. Thus, instrumental and epistemic rationality become intertwined. If we are to determine what to do, we need to make sure that our actions are based on what is true. It is in this sense that rationality of belief—epistemic rationality—is one of the foundations for rationality of action.

Rationality of belief is assessed by looking at a variety of probabilistic reasoning skills, evidence evaluation skills, and hypothesis testing skills. For a person to be epistemically rational, his probability estimates must follow the rules of objective probabilities. That is, his estimates must follow the so-called probability calculus. Mathematically, probability values follow certain rules. These rules form one of the most important normative models for subjective probability estimates. As was the case with instrumental rationality, an important research tradition in cognitive psychology has shown that people sometimes violate many of the rules of epistemic rationality.

Rationality and the Heuristics and Biases Literature

In constructing our rational thinking assessment instrument, we have drawn on the vast literature that has demonstrated violations of the normative models of instrumental and epistemic rationality. A substantial research literature—one comprising literally hundreds of empirical studies conducted over several decades—has firmly established that people's responses sometimes deviate from the performance considered normative on many reasoning tasks. For example, people assess probabilities incorrectly, they test hypotheses inefficiently, they violate the axioms of utility theory, they do not properly calibrate degrees of belief, their choices are affected by irrelevant context, they ignore alternative hypotheses when evaluating data, and they display numerous other information processing biases (Baron, 2008, 2014; Evans, 2014; Kahneman, 2011; Kahneman & Tversky, 2000; Koehler & Harvey, 2004; Manktelow, 2012; Thaler, 2015). We shall draw heavily on this research, especially that of the so-called heuristics and biases tradition inaugurated by Kahneman and Tversky in the early 1970s (Kahneman &

Tversky, 1972, 1973; Tversky & Kahneman, 1974). The term "biases" refers to the systematic errors that people make in choosing actions and in estimating probabilities, and the term "heuristic" refers to *why* people often make these errors—because they use mental shortcuts (heuristics) to solve many problems. We shall discuss the psychological theory of these mental shortcuts in several chapters of this book. Table 1.1 lists some of the tasks, effects, and biases that we have studied in our lab and from which we selected in order to construct the Comprehensive Assessment of Rational Thinking (CART).

Table 1.1
Sampling of the individual differences in heuristics and biases tasks studied in the Stanovich/West/Toplak Lab

Tasks, Effects, and Biases	Studies of Individual Differences from the Work of Our Lab
Baserate Neglect	Kokis et al. (2002); Stanovich & West (1998c, 1998d, 1999, 2008b); West et al. (2008)
Conjunction Fallacy	Stanovich & West (1998b); Toplak et al. (2011); West et al. (2008)
Framing Effects	Stanovich & West (1998b, 1999, 2008b); Toplak et al. (2014a, 2014b)
Anchoring Effect	Stanovich & West (2008b)
Sample Size Awareness	Toplak et al. (2011); West et al. (2008)
Regression to the Mean	Toplak et al. (2007, 2011); West et al. (2008)
Control Group Reasoning	Stanovich & West (1998c); Toplak et al. (2011); West et al. (2008)
Disjunctive Reasoning	Toplak & Stanovich (2002); West et al. (2008)
Temporal Discounting	Toplak et al. (2014a)
Gambler's Fallacy	Toplak et al. (2007, 2011); (West et al. (2008)
Probability Matching	Stanovich & West (2008b); Toplak et al. (2007, 2011); West & Stanovich (2003)
Overconfidence Effect	Stanovich & West (1998c)
Outcome Bias	Stanovich & West (1998c, 2008b); Toplak et al. (2007, 2011)
Ratio Bias	Kokis et al. (2002); Stanovich & West (2008b); Toplak et al. (2014a, 2014b); West et al. (2008)
Four-Card Selection Task	Stanovich & West (1998a, 2008b); Toplak & Stanovich (2002); Toplak et al. (2014a); West et al. (2008)

Table 1.1 (continued)

Tasks, Effects, and Biases	Studies of Individual Differences from the Work of Our Lab
Ignoring P(D/~H)	Stanovich & West (1998d, 1999); West et al. (2008)
Sunk Cost Effect	Stanovich & West (2008b); Toplak et al. (2011)
Risk-Benefit Confounding	Stanovich & West (2008b)
Covariation Detection	Stanovich & West (1998c, 1998d); Sá et al. (1999); Toplak et al. (2011); West et al. (2008)
Belief Bias in Syllogistic Reasoning	Macpherson & Stanovich (2007); Stanovich & West (1998c, 2008b); Toplak et al. (2014a, 2014b)
Omission Bias	Stanovich & West (2008b)
Informal Argument Evaluation	Stanovich & West (1997, 2008b); Sá et al. (1999)
Unconfounded Hypothesis Testing	Stanovich & West (1998c); Toplak et al. (2011)
Myside Bias	Sá, Kelley, Ho, & Stanovich (2005); Stanovich & West (2007, 2008a, 2008b); Toplak & Stanovich (2003); Toplak et al. (2014a, 2014b)
Expected Value Maximization	Stanovich, Grunewald, & West (2003); Toplak et al. (2007)
Newcomb's Problem	Stanovich & West (1999); Toplak & Stanovich (2002)
Prisoner's Dilemma	Stanovich & West (1999); Toplak & Stanovich (2002)
Hindsight Bias	Stanovich & West (1998c)
One-side Bias	Stanovich & West (2008a)
Certainty Effect	Stanovich & West (2008b)
Willingness to pay/ Willingness to accept	Stanovich & West (2008b)
Bias Blind Spot	West, Meserve, & Stanovich (2012); Toplak et al. (2014a)
Evaluability: Less is More Effect	Stanovich & West (2008b)
Proportion Dominance Effect	Stanovich & West (2008b)

Virtually all research psychologists acknowledge the importance of the cognitive space marked out by the award-winning heuristics and biases work. The press release for the 2002 Nobel Prize awarded to Daniel Kahneman drew attention to the roots of his work in "the analysis of human judgment and decision-making by cognitive psychologists." This work was lauded for "inspiring a new generation of researchers in economics and finance to enrich economic theory using insights from cognitive psychology into intrinsic human motivation" (The Royal Swedish Academy of Sciences, 2002a, 2002b). One reason that this work was so influential was that it addressed deep issues concerning human rationality. As the Nobel announcement noted, "Kahneman and Tversky discovered how judgment under uncertainty systematically departs from the kind of rationality postulated in traditional economic theory" (The Royal Swedish Academy of Sciences, 2002a, 2002b). The thinking errors uncovered by Kahneman and Tversky are thus not trivial errors in a parlor game. Being rational means acting to achieve one's own life goals using the best means possible. To violate the thinking rules examined in this Nobel-winning work thus has the practical consequence that we are less satisfied with our lives than we might be.

Cognitive scientist Steven Pinker seconded the Nobel committee when he argued that when "trying to identify what any educated person should know in the entire expanse of knowledge … the work on human cognition and probabilistic reason should be up there as one of the first things any educated person should know. I am unqualified in my respect for how important this work is" (Creative Leadership Forum, 2011). We agree with Pinker that these domains of thinking are something that every educated person should know, and his statement provides one way of thinking about what we are creating in the CART. What Pinker thinks "any educated person should know in the entire expanse of knowledge" is exactly what we are attempting to assess!

The skills of judgment and decision making are cognitive skills that are the foundation of rational thought and action, and they are missing from intelligence tests. This book, then, and the assessment device we are beginning to construct, can be viewed as a partial remedy for one of the most profound historical ironies of the behavioral sciences: the Nobel Prize was awarded for studies of cognitive characteristics that are entirely *missing*

Definitions of Rationality

from the most well-known mental assessment device in the behavioral sciences—the intelligence test (Stanovich, 2009). We hope that the CART will at least in part rectify this strange anomaly in the attention that our discipline has paid to different cognitive characteristics.

Because much of the operationalization of our framework of rational thinking comes from the heuristics and biases tradition, it is important to explicate the logic of its tasks in terms of contemporary theories of the functional architecture of the human mind. In the next chapter, we will outline the functional cognitive theory that we will use to interpret the tasks in this literature. In chapter 3, we will unpack the logic of heuristics and biases tasks in terms of this architecture. Once the architecture is revealed, a key point from the preface will become clear. Specifically, we will show that the concept of rationality and the concept of intelligence are two different things. Even more specifically, we will show that rationality is actually a more encompassing mental construct than intelligence. Thus, as measures of rationality, the tasks in the heuristics and biases literature, while tapping intelligence in part, actually encompass more cognitive processes and knowledge than are assessed by IQ tests.

Summary

Because rationality is an issue across many disciplines, it has acquired many different definitions. Some of those definitions actually serve to thwart the goal of assessing individual differences in the important mental faculties related to judgment and decision making. Not surprisingly, we have chosen definitions of rationality from cognitive science that are amenable to our program of measuring individual differences. They are empirically grounded, and data have shown that there are substantial individual differences when we measure rationality in the way that most cognitive scientists do. Another way to state our logic would be to say that *if* you are going to measure individual differences in rational thought, you *have* to define it in the way that we have. Any investigator would be free to define rationality in some other way, but the point would be that *the individual differences that we are studying would still be there*. We would then just need another name for them. In one sense, this would not be objectionable to us, but with one important caveat. Anyone objecting to designating the large set of heuristics and biases tasks that we will present in the subsequent chapters by the term "rationality" should definitely not call these measures part of intelligence. They are not measured by extant IQ tests, and IQ tests are less than perfect predictors of individual differences in these reasoning processes. Thus, we prefer our own default of calling our measure a measure of rational thinking. As we will argue in chapter 2, go ahead and call it a measure of something else—just don't call it *intelligence*.

2 Rationality, Intelligence, and the Functional Architecture of the Mind

Discussions of the intelligence concept often implicitly privilege the mental abilities tapped by IQ tests in ways that are not immediately apparent to the nonscientist. One of the primary reasons why this happens is that discussions of intelligence often proceed without a prior explication of a full model of human cognition. In the absence of such a complete model, it is understandable that the layperson naturally assumes that IQ tests measure all of the important cognitive characteristics.

In short, discussions of intelligence often go off the rails at the very beginning by failing to set the concept within a general model of cognitive functioning, thus inviting the default assumption that intelligence tests tap all of the critical features of the mind. Thus, in this chapter, we will first sketch out a consensus view of intelligence. We will then make an effort to place intelligence within a comprehensive model of cognition. By placing both intelligence and rationality within this comprehensive model, we will illustrate why rationality is actually a more encompassing concept than intelligence, contrary to many popular discussions.

A Grounded Theory of Intelligence

In the preface, we introduced the distinction between *permissive* conceptualizations of intelligence and *grounded* conceptualizations. Permissive theories include in their definitions aspects of functioning that are captured by the vernacular term "intelligence" (adaptation to the environment, etc.) but that are not actually included in IQ tests. Grounded theories, in contrast, confine the concept of intelligence to the set of mental abilities actually tested by extant IQ tests. For this volume, we adopt a standard grounded definition that involves a statistical abstraction from performance on established tests and cognitive ability indicators (Deary, 2013).

Our default grounded theory will be the Cattell–Horn–Carroll (CHC) theory of intelligence (Carroll, 1993; Cattell, 1963, 1998; Horn & Cattell, 1967).[1] It yields a scientific concept of general intelligence, usually symbolized as "g," and a small number of broad factors, of which two are dominant. Fluid intelligence (Gf) reflects reasoning abilities operating across of variety of domains—including novel ones. It is measured by tests of abstract thinking such as figural analogies, Raven's Matrices, and series completion (e.g., what is the next number in the series 1, 4, 5, 8, 9, 12, __). Crystallized intelligence (Gc) reflects declarative knowledge acquired from acculturated learning experiences. It is measured by vocabulary tasks, verbal comprehension, and general knowledge measures. Although substantially correlated, Gf and Gc reflect a long history of considering two aspects of intelligence: intelligence-as-process and intelligence-as-knowledge (Ackerman, 1996, 2014; Duncan, 2010; Hunt, 2011; Nisbett et al., 2012). In addition to Gf and Gc, other broad factors represent things like memory and learning, auditory perception, and processing speed (see Carroll, 1993, for a full account).

There is a large literature on the CHC theory and on the correlates of Gf and Gc (see Duncan, 2010; Duncan et al., 2008; Geary, 2005; Gignac, 2005; Kane & Engle, 2002; Mackintosh & Bennett, 2003; McArdle et al., 2002; McGrew, 2009; Nisbett et al., 2012). The theory's constructs have been validated in studies of brain injury, educational attainment, cognitive neuroscience, developmental trends, and information processing. There are, of course, alternative models to the CHC conception (Deary, 2013; Hunt, 2011). For example, Hunt (2011) discusses Johnson and Bouchard's (2005) g-VPR model as an alternative model that is empirically differentiable from the CHC theory. However, for the purposes of the theoretical contrast with rationality, it makes no difference which of the currently viable grounded theories of intelligence we choose. All of them ignore a critical level of cognitive analysis that is important for rationality.

Dual-Process Theory: The First Step toward a Model of Cognitive Architecture

As discussed in the previous chapter, the multitude of tasks from the heuristics and biases literature of cognitive psychology largely provide the operational definition of rational thinking in cognitive science. Discussion of these tasks often leads to conceptualizing rationality within a dual-process framework, because most of the tasks in the heuristics and biases literature were deliberately designed to pit an automatically triggered response against a normative response, usually (but not always[2]) generated by more controlled types of mental processing (Kahneman, 2011). Since Kahneman and Tversky launched the heuristics and biases approach in the 1970s, a wealth of evidence has accumulated in support of the dual-process approach (Evans & Stanovich, 2013).

A wide variety of evidence has converged on the conclusion that some type of dual-process notion is needed in a diverse set of specialty areas, including but not limited to cognitive psychology, economics, social psychology, naturalistic philosophy, decision theory, and clinical psychology (Alós-Ferrer & Strack, 2014; Chein & Schneider, 2012; Evans, 2008, 2010, 2014; Evans & Frankish, 2009; Evans & Stanovich, 2013; Lieberman, 2009; McLaren et al., 2014; Schneider & Chein, 2003; Sherman, Gawronski, & Trope, 2014; Smith & DeCoster, 2000; Stanovich, 1999, 2004). Evolutionary theorizing and neurophysiological work also have supported a dual-process conception (Corser & Jasper, 2014; Frank, Cohen, & Sanfey, 2009; Lieberman, 2009; McClure & Bickel, 2014; McClure, Laibson, Loewenstein & Cohen, 2004; Prado & Noveck, 2007; Toates, 2005, 2006; Tomlin et al., 2015).

Because there is now a plethora of dual-process theories (see Stanovich, 2011, 2012, for a list of the numerous versions of such theories), researchers tend to vary a lot in their terms for the two processes. For the purposes of this volume, we will most often adopt the Type 1/Type 2 terminology discussed by Evans and Stanovich (2013), and occasionally use the similar System 1/System 2 terminology of Stanovich (1999) and Kahneman (2011).[3] The defining feature of Type 1 processing is its autonomy—Type 1 processes execute automatically upon encountering their triggering stimuli, and they are not dependent on input from high-level control systems. Autonomous processes have other correlated features—for example, their execution tends to be rapid, they do not put a heavy load on central processing capacity, they tend to be associative—but these are not defining features (Stanovich & Toplak, 2012). The category of autonomous processes includes: processes of emotional regulation; the encapsulated modules for solving specific adaptive problems that have been posited by evolutionary psychologists; processes of implicit learning; and the automatic firing of overlearned associations (see Barrett & Kurzban, 2006; Carruthers, 2006; Evans, 2008, 2009; Moors & De Houwer, 2006; Samuels, 2005, 2009; Shiffrin & Schneider, 1977).

That these categories are disparate makes it clear that Type 1 processing is a grab bag—encompassing *both* innately specified processing modules or procedures and experiential associations that have been learned to automaticity, or become so ingrained as to be automatic. The many kinds of Type 1 processing have in common the property of autonomy, but otherwise, their neurophysiology and etiology might be considerably different. It is important to emphasize that Type 1 processing is not limited to modular subprocesses that meet all of the classic Fodorian (1983) criteria (see Stanovich, 1999, for a discussion). Type 1 processing also encompasses unconscious implicit learning and conditioning. In addition, many rules, stimulus discriminations, and decision-making principles that have been practiced to automaticity (e.g., Kahneman & Klein, 2009; Shiffrin & Schneider, 1977) are processed in a Type 1 manner. This learned information can sometimes be just as much a threat to rational behavior as are evolutionary modules that fire inappropriately in a modern environment. Rules learned to automaticity can be overgeneralized—they can autonomously trigger behavior when the situation is an exception to the class of events they are meant to cover (Arkes & Ayton, 1999; Hsee & Hastie, 2006).

In contrast with Type 1 processing, Type 2 processing is relatively slow and computationally expensive. Many Type 1 processes can operate at once in parallel, but Type 2 processing is largely serial. One of the most critical functions of Type 2 processing is to override Type 1 processing. This is sometimes necessary because autonomous processing has heuristic qualities. Autonomous processing is designed to get the response into the right ballpark when solving a problem or making a decision, but it is not designed for the type of fine-grained analysis called for in situations of unusual importance (financial decisions, judgments of fairness, employment decisions, legal judgments, and the like).

Type 1 processing heuristics depend on benign environments to provide obvious cues that elicit adaptive behaviors. In hostile environments, reliance on heuristics can be costly (see Hilton, 2003; Over, 2000; Stanovich, 2004). A benign environment is one that contains useful (that is, diagnostic) cues that can be exploited by various heuristics (for example, affect-triggering cues, vivid and salient stimulus components, convenient and accurate anchors). Additionally, for an environment to be benign, it must also contain no other individuals who will adjust their behavior to exploit those relying only on Type 1 processing. In contrast, a hostile environment for heuristics is one in which there are few cues that are usable by autonomous processes or there are misleading cues (Kahneman & Klein, 2009). Another way that an environment can turn hostile for a user of Type 1 processing occurs when other agents discern the simple cues that are being used and arrange them for their own advantage (for example, in advertisements, or the strategic design of supermarket floor space in order to maximize revenue).

All of the different kinds of Type 1 processing (processes of emotional regulation, Darwinian modules, associative and implicit learning processes) can produce suboptimal responses in a particular context if not overridden. For example, humans often act as "cognitive misers" (see chapter 3 for an extended discussion of this concept) by engaging in attribute substitution (Kahneman & Frederick, 2002)—the substitution of an easy-to-evaluate characteristic for a more difficult one, even if the easier one is less accurate. For example, the cognitive miser will substitute the less effortful attributes of vividness or affect for the more effortful retrieval of relevant facts (Kahneman, 2003; Li & Chapman, 2009; Slovic & Peters, 2006; Wang, 2009). But when we are evaluating important risks—such as the risk of certain activities and environments for our children—we do not want to substitute vividness for careful thought about the situation. In such situations, we want to employ Type 2 override processing to block the attribute substitution of the cognitive miser. This is the important process of *detecting the need for decoupling*, which we will discuss extensively in the next chapter.

Once detection of the conflict between the normative response and the response triggered by System 1 has taken place, Type 2 processing must display at least two related capabilities in order to override Type 1 processing. One is the capability of interrupting Type 1 processing. Type 2 processing thus involves inhibitory mechanisms that have been the focus of work on executive functioning (Best, Miller, & Jones, 2009; Hasher, Lustig, & Zacks, 2007; Miyake & Friedman, 2012). But the ability to suppress Type 1 processing gets the job only half done. Suppressing one response is not helpful unless there is a better response available to substitute for it. Where do these better responses come from? One answer is that they can come from processes of hypothetical reasoning and cognitive simulation that are a unique aspect of Type 2 processing (Evans, 2007a, 2010; Evans & Stanovich, 2013). When we reason hypothetically, we create temporary models of the world and test out actions (or alternative causes) in that simulated world. To reason hypothetically we must, however, have one critical cognitive capability—we must be able to prevent our representations of the real world from becoming confused with representations of imaginary situations. The so-called cognitive decoupling operations (Stanovich, 2011) are the central feature of Type 2 processing that make this possible, and they have implications for how we conceptualize both intelligence and rationality, as we shall see. The important issue for our purposes is that decoupling secondary representations from the world and then maintaining the decoupling while simulation is carried out is a Type 2 processing operation. It is computationally taxing and greatly restricts the ability to conduct any other Type 2 operation simultaneously. In fact, decoupling operations might well be a major contributor to a distinctive Type 2 property—its seriality.

Figure 2.1 presents a preliminary dual-process model of mind based on what we have outlined thus far. The figure shows the Type 2 override function we have been discussing, as well as the Type 2 process of simulation. The arrow going from Type 1 to Type 2 processes indicates that Type 2 processes receive inputs from Type 1 computations. These so-called preattentive processes (Evans, 2008) establish the content of most Type 2 processing.

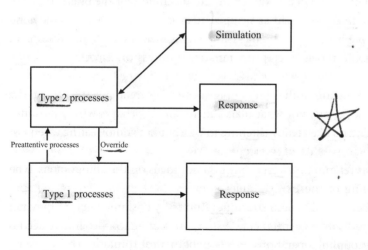

Figure 2.1
A preliminary dual-process model.

Fitting Intelligence into the Dual-Process Architecture

Where does intelligence fit into this model? First, we must mention an important part of the context for studying individual differences in intelligence. A process can be a critical component of cognition yet be a negligible source of individual differences. This occurs when people do not tend to vary much in that process. Such is the case with many Type 1 processes. They help us carry out a host of useful information-processing operations and adaptive behaviors (e.g., depth perception, face recognition, frequency estimation, syntactic processing, threat detection, emotive responses, color perception)—yet for the most part, people do not differ widely on many of these processes. This accounts for some of the confusion surrounding the use of the term "intelligence" in cognitive science.

If you pick up a magazine article or textbook on cognitive science, the author might describe the marvelous mechanisms we have for recognizing faces and refer to this as "a remarkable aspect of human intelligence." Likewise, a book on popular science might describe how we have mechanisms for parsing syntax when we process language and also refer to this as "a fascinating product of the evolution of the human intellect." Finally, a textbook on evolutionary psychology might describe the remarkably efficient mechanisms of kin recognition that operate in many animals, including humans. Such processes—face recognition, syntactic processing, kin recognition—are all parts of the machinery of the brain. They are also sometimes described as being part of human intelligence. Yet none of these processes is ever tapped on intelligence tests, for the reason we warned about above: Intelligence tests assess only those aspects of cognitive functioning on which people tend to differ widely. Intelligence tests are a bit like the listings for online dating sites—they are about the things that distinguish people, not what makes them similar. That is why the listings contain entries like "enjoy listening to U2" but not "enjoy eating when I'm hungry."

For this reason, intelligence tests do not focus on the autonomous Type 1 processing of the brain.[4] Intelligence tests, in their Gf (fluid) components, largely tap Type 2 processing. And to a substantial extent they tap the operation we have been emphasizing in this section—cognitive decoupling. Decoupling operations enable hypothetical thinking. They must be continually in force during any ongoing mental simulations, and the raw ability to sustain such simulations while keeping the relevant representations decoupled is one key aspect of the brain's computational power that is assessed by measures of fluid intelligence. This is becoming clear from converging work on executive function and working memory, which both display correlations with fluid intelligence that are quite high (Duncan et al., 2008; Gray, Chabris, & Braver, 2003; Hicks, Harrison, & Engle, 2015; Kane, 2003; Kane & Engle, 2002; Kane, Hambrick, & Conway, 2005; Salthouse, Atkinson, & Berish, 2003). This is because most measures of executive function, such as working memory, are direct or indirect indicators of a person's ability to sustain decoupling operations (Feldman Barrett, Tugade, & Engle, 2004). Thus, Type 2 processes are strongly associated with Gf (Burgess, Gray, Conway, & Braver, 2011; Chuderski, 2014; Engel de Abreu, Conway, & Gathercole, 2010; McVay & Kane, 2012; Mrazek et al., 2012). We shall work Gc (crystallized intelligence) into the model shortly, but first we turn to an even more critical complication.

Toward a Tripartite Model

In the preface, we argued that if we properly understand the difference between intelligence and rationality, there is nothing paradoxical at all about the idea that people might actually be smart but act dumb. In this section we will explain why. The first step in the explanation is to understand that rational thinking involves a level in the hierarchical control system of the brain that is only weakly tapped by IQ tests. This can be understood by unpacking the logic of the override of autonomous subsystems—specifically, in understanding that override needs to be conceptualized in terms of two levels of processing. To understand the two levels in a vernacular way, consider two imaginary stories.

Both stories involve a woman walking along a cliff. The stories are both sad—the woman dies at the end. The purpose of this exercise is to get us to think about how we explain the death in each story. In incident A, a woman is walking along a cliff by the ocean and goes to step on a large rock, but what appears to be a rock is not a rock at all. Instead, it is actually the side of a crevice, and she falls down the crevice and dies. In incident B, a woman commits suicide by jumping off an ocean cliff and dies when she is crushed on the rocks below.

In both cases, at the most basic level, when we ask ourselves for an explanation of why the woman died, we might say that the answer is the same. The same laws of physics that operate in incident A (the gravitational laws that describe why the woman will be crushed upon impact) also operate in incident B. However, we feel that the laws of gravity and force somehow do not provide a complete explanation of what has happened in either incident. Further, when we attempt a more fine-grained explanation, incidents A and B seem to call for a different level of explanation, if we wish to zero in on the *essential* cause of death.

In analyzing incident A, a psychologist would be prone to say that when processing a stimulus (the crevice that looked somewhat like a rock), the woman's information-processing system malfunctioned—sending the wrong information to response decision mechanisms, which then resulted in a disastrous motor response. We will refer to this level of analysis as the "algorithmic level."[5] In computer science, this would be the level of the instructions in the abstract language used to program the computer (BASIC, C, Java, etc.). The cognitive psychologist works largely at this level by showing that human performance can be explained by positing certain information-processing mechanisms in the brain (input-coding mechanisms, perceptual-registration mechanisms, short- and long-term-memory-storage systems, etc.). For example, a simple task of letter pronunciation might entail encoding the letter, storing it in short-term memory, comparing it with information stored in long-term memory, and, if a match occurs, making a response decision and then executing a motor response. In the case of the woman in incident A, the algorithmic level is the right level to explain her unfortunate demise. Her perceptual registration and classification mechanisms malfunctioned by providing incorrect information to response-decision mechanisms, causing her to step into the crevice.

Incident B, on the other hand, does not involve such an algorithmic-level information-processing error. The woman's perceptual apparatus accurately recognized the edge of the cliff, and her motor command centers quite accurately programmed her body to jump off the cliff. The computational processes posited at the algorithmic level of analysis executed quite perfectly. No error at this level of analysis explains why the woman is dead in incident B. Instead, this woman died because of her overall goals and how these goals interacted with her beliefs about the world in which she lived.

We will present our model of cognitive architecture (building on Stanovich, 2011) in the spirit of Dan Dennett's (1996) book *Kinds of Minds*, where he used that title to suggest that within the brain of humans are control systems of very different types—different kinds of minds. In our terms, the woman in incident A had a problem with the algorithmic mind, and the woman in incident B had a problem with the reflective mind.[6] This terminology captures the fact that we turn to an analysis of goals, desires, and beliefs to understand a case such as B. The algorithmic level provides an incomplete explanation of behavior in cases like incident B, because it provides an information-processing explanation of how the brain is carrying out a particular task (in this case, jumping off of a cliff), but no explanation of *why* the brain is carrying out this particular task. We turn to the level of the reflective mind when we ask questions about the *goals* of the system's computations (*what* the system is attempting to compute and *why*). In short, the reflective mind is concerned with the goals of the system, beliefs relevant to those goals, and the choice of action that is optimal given the system's goals and beliefs. All of these characteristics implicate the reflective mind in many issues of rationality. High computational efficiency in the algorithmic mind is not a sufficient condition for rationality.

Our attempt to differentiate the levels of control involved in Type 2 processing creates a kind of tripartite theory of mind as displayed in figure 2.2. On this theory, the mind has three levels: the autonomous, the algorithmic, and the reflective. The psychological literature provides much converging evidence and theory to support such a structure. First, psychometricians have long distinguished typical performance situations from optimal (sometimes termed "maximal") performance situations (Ackerman & Kanfer, 2004; Cronbach, 1949; Sternberg, Grigorenko, & Zhang, 2008). Typical performance situations are unconstrained in that no overt instructions to maximize performance are given, and the task interpretation is determined to some extent by the subject. The goals to be pursued in the task are left somewhat open. The issue is what a person would typically do in such a situation, given few constraints. Typical performance measures implicate, at least in part, the reflective mind—they assess goal prioritization and epistemic regulation. In contrast, optimal performance situations are those where the task interpretation is determined externally. The person performing the task is told the rules that maximize performance. Thus, optimal performance tasks assess questions of the *efficiency* of goal

Rationality, Intelligence, and Architecture of the Mind

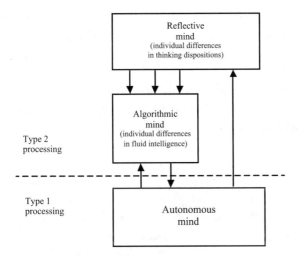

Figure 2.2
The tripartite structure of the mind and the locus of individual differences in cognition.

pursuit—they capture the processing efficiency of the algorithmic mind. All tests of intelligence or cognitive aptitude are optimal performance assessments, whereas measures of critical or rational thinking are often assessed under typical performance conditions.

The difference between the algorithmic mind and the reflective mind is captured in another well-established distinction in the measurement of individual differences—the distinction between cognitive ability and thinking dispositions. The former are, as just mentioned, measures of the efficiency of the algorithmic mind. The latter travel under a variety of names in psychology—"thinking dispositions" or "cognitive styles" being the two most popular. Many thinking dispositions concern beliefs, belief structure, and, importantly, attitudes toward forming and changing beliefs. Other thinking dispositions that have been identified concern a person's goals and goal hierarchy. Examples of some thinking dispositions that have been investigated by psychologists include: actively open-minded thinking, need for cognition, consideration of future consequences, need for closure, and dogmatism (Baron et al., 2015; Cacioppo et al., 1996; Kruglanski & Webster, 1996; Schommer-Aikins, 2004; Stanovich, 1999, 2011; Sternberg, 2003; Sternberg & Grigorenko, 1997; Strathman et al., 1994).

The types of cognitive propensities that these thinking disposition measures reflect include: the tendency to collect information before making up one's mind, the tendency to seek various points of view before coming to a conclusion, the disposition to think extensively about a problem before responding, the tendency to calibrate the degree of strength of one's opinion to the degree of evidence available, the tendency to think about future consequences before taking action, the tendency to explicitly weigh pluses and minuses of situations before making a decision, and the tendency to seek nuance and avoid absolutism. In short, individual differences in thinking dispositions reflect variation in people's goal management, epistemic values, and epistemic self-regulation—differences in the operations of the reflective mind.

The cognitive abilities assessed by intelligence tests are not of this type. They are not about high-level personal goals and their regulation, or about the tendency to change beliefs in the face of contrary evidence, or about how knowledge acquisition is internally regulated when not externally directed. People have indeed come up with *definitions* of intelligence that encompass such things. Such permissive theorists (see the preface of this volume) often define intelligence in ways that encompass rational action and belief, but, nevertheless, *the actual measures of intelligence in use assess only algorithmic-level cognitive capacity*.

Figure 2.2 represents the classification of individual differences in the tripartite view. The broken horizontal line represents the location of the key distinction in older, dual-process views. Figure 2.2 identifies variation in fluid intelligence (Gf) with individual differences in the efficiency of processing of the algorithmic mind. We have argued previously that individual differences in fluid intelligence are a key indicator of the variability across individuals in the ability to override and sustain decoupling operations. In contrast, the reflective mind is identified with individual differences in thinking dispositions related to beliefs and goals.

The Tripartite Model and Why Rationality Is a More Encompassing Construct Than Intelligence

Figure 2.2 highlights an important sense in which rationality is a more encompassing construct than intelligence. As previously discussed, to be rational, a person must have well-calibrated beliefs and must act appropriately on those beliefs to achieve goals—both properties of the reflective mind. The person must also, of course, have the algorithmic-level machinery that enables him or her to carry out the actions and to process the environment in a way that enables the correct beliefs to be fixed and the correct actions to be taken. Thus, individual differences in rational thought and action can arise because of individual differences in fluid intelligence (the algorithmic mind) or because of individual differences in thinking dispositions (the reflective mind).

To put it simply, the concept of rationality encompasses two things (thinking dispositions and algorithmic-level capacity), whereas the concept of intelligence—at least as it is commonly operationalized—is largely confined to algorithmic-level capacity. Intelligence tests do not attempt to measure aspects of epistemic or instrumental rationality, nor do they examine any thinking dispositions that relate to rationality. It is clear from figure 2.2 why rationality and intelligence can become dissociated. Rational thinking depends on our thinking dispositions as well as on our algorithmic efficiency. Thus, as long as variation in thinking dispositions is not perfectly correlated with variation in fluid intelligence, there is a statistical possibility that rationality and intelligence will diverge.

In fact, substantial empirical evidence indicates that individual differences in thinking dispositions and intelligence are far from perfectly correlated. Studies (e.g., Ackerman & Heggestad, 1997; Cacioppo et al., 1996; Kanazawa, 2004; Zeidner & Matthews, 2000) have indicated that measures of intelligence display only moderate to weak correlations (usually less than 0.30) with some thinking dispositions (e.g., actively open-minded thinking, need for cognition) and near zero correlations with others (e.g., conscientiousness, curiosity, diligence). Other important evidence supports the conceptual distinction made here between algorithmic cognitive capacity and thinking dispositions. For example, across a variety of tasks from the heuristics and biases literature, it has consistently been found that rational thinking dispositions will predict variance after the effects of general intelligence have been controlled (Bruine de Bruin, Parker, & Fischhoff, 2007; Finucane & Gullion, 2010; Klaczynski & Lavallee, 2005; Kokis et al., 2002; Macpherson & Stanovich, 2007; Parker & Fischhoff, 2005; Stanovich & West, 1997, 1998c; Toplak et al., 2007; Toplak & Stanovich, 2002; Toplak, West, & Stanovich, 2011, 2014a, 2014b).

There is one particularly important point to understand about thinking dispositions that will be important for how they are represented in our rational thinking assessment device (the CART—Comprehensive Assessment of Rational Thinking). It is that the thinking dispositions of the reflective mind are the psychological mechanisms that *underlie* rational thought. Maximizing these dispositions is *not* the criterion of rational thought itself. Rationality involves instead the maximization of goal achievement via judicious decision making and optimizing the fit of belief to evidence. The thinking dispositions of the reflective mind are a means to these ends of achieving our goals. Certainly, high levels of such commonly studied dispositions as deliberativeness and belief flexibility are needed for rational thought and action. But this does not necessarily mean that the maximal levels are optimal. Rather, there must be a balance. One does not maximize the deliberativeness dimension, for example, because such a person might get lost in interminable pondering and never make a decision. Likewise, one does not maximize the thinking disposition of belief flexibility either, because such a person might end up with a pathologically unstable personality. Deliberativeness and belief flexibility are "good" cognitive styles, in that most people are not high enough on these dimensions, so that "more" would be better (Baron, 1985, 2008). However, they are not meant to be maximized. For this reason (and others—see chapter 11) thinking disposition subscales are supplemental measures in the CART and not treated as direct measures of rational thinking themselves.

The Fleshed-Out Tripartite Model

Figure 2.3 illustrates more completely the functions of the various levels of cognition and how they control each other. There, it is clear that the override capacity itself is a property of the algorithmic mind, indicated by the arrow labeled A. However, previous dual-process theories have tended to ignore the higher-level cognitive operation that *initiates* the override function in the first place. This initiation of the override function is a dispositional property of the reflective mind that is related to rationality. In the model in figure 2.3, it corresponds to arrow B, which represents the instruction to the algorithmic mind to override the Type 1 response by taking it offline. This is a different mental function from the override function itself

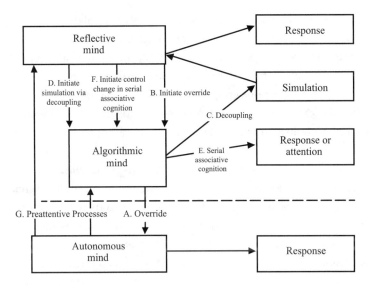

Figure 2.3
A more complete model of the tripartite structure of the mind.

(arrow A), and the evidence cited above indicates that the two functions are indexed by different types of individual differences.

The override function has loomed so large in dual-process theory that it has somewhat overshadowed the simulation process—the process that computes the alternative responses that makes the override worthwhile. Thus, figure 2.3 explicitly represents the simulation function as well as the fact that the instruction to initiate simulation originates in the reflective mind. The decoupling operation (indicated by arrow C)—the process of separating simulations of reality from reality itself—is also carried out by the algorithmic mind. But again, the instruction to *initiate* simulation (indicated by arrow D) is carried out by the reflective mind. Two different types of individual differences are associated with the initiation call and the decoupling operator—specifically, thinking dispositions with the initiation, and fluid intelligence with the decoupling itself. Also represented is the fact that the higher levels of control receive inputs from the autonomous mind (arrow G) via "preattentive processes" (Evans, 2006, 2009).

The arrows labeled E and F reflect the decoupling and higher-level control of a kind of Type 2 processing (serial associative cognition) that does not involve fully explicit cognitive simulation (see Stanovich, 2011). There are types of slow, serial cognition that do not involve simulating alternative worlds and exploring them exhaustively. Their existence points to an important fact: All hypothetical thinking involves Type 2 processing (Evans & Over, 2004), but not all Type 2 processing involves hypothetical thinking. Serial associative cognition represents this latter category.

Recall that the category of Type 1 processes is composed of: affective responses; previously learned responses that have been practiced to automaticity; conditioned responses; and adaptive modules that have been shaped by our evolutionary history. These cover many situations indeed, but modern life still creates many problems for which none of these mechanisms is suited. That is why the laboratory tasks of modern cognitive psychology are, contrary to what many of their detractors claim, actually *good* proxies for the reasoning novelties of the modern world (Stanovich, 2004).

Consider Wason's (1966) four-card selection task, which has generated a vast literature (e.g., Evans, 2014; Evans, Newstead, & Byrne, 1993; Stanovich, 1999; Stanovich & West, 1998a). It remains an excellent context for demonstrating the difference between serial associative cognition and an exhaustive exploration of alternative worlds via cognitive simulation. The abstract version of the problem is often presented as follows. Consider four cards lying on a table. Each one of the cards has a letter on one side and a number on the other side. Here is a rule: "If a card has a vowel on its letter side, then it has an even number on its number side." Two of the cards are letter-side up, and two of the cards are number-side up. Your task is to decide which card or cards must be turned over in order to find out whether the rule is true or false. Indicate which cards must be turned over. The four cards confronting the subject have the stimuli K, A, 8, and 5 showing.

The correct answer is A and 5, the only two cards that could show the rule to be false. However, the majority of subjects incorrectly answer A and 8, showing a so-called matching bias. Evans (2006) has pointed out that the previous emphasis on the matching bias (Evans, 1972, 1998; Evans & Lynch, 1973) might have led some investigators to infer that higher-level Type 2 processing is not occurring in the task. However, Evans (2006) presents evidence that Type 2 processing is occurring during the task—even in the majority of subjects who do not give the normatively correct response (A and 5) but instead give the A and 8 response.

First, in discussing the card inspection paradigm (Evans, 1996) that he pioneered (see also Lucas & Ball, 2005; Roberts & Newton, 2001), Evans

(2006) notes that although subjects look disproportionately at the cards they will choose (the finding leading to the inference that autonomous Type 1 processes were determining the responses), the lengthy amount of time they spend on those cards suggests that analytic Type 2-thought is occurring (if only to generate justification for the automatically triggered choices). Second, in verbal protocol studies, subjects can justify their responses (indeed, can justify *any* set of responses they are told are correct; see Evans & Wason, 1976) with analytic arguments—arguments that sometimes refer to the hidden side of cards chosen.

Our position is that Evans (2006) is correct that Type 2 cognition is occurring in the task, but it is not full-blown cognitive simulation of alternative world models. It is Type 2 processing of a shallower type. When think-aloud protocols are analyzed, it has seemed that most subjects are engaging in some slow, serial processing, but of a type that was simply incomplete. A typical protocol from a subject might go something like this: "Well, let's see, I'd turn the A to see if there is an even number on the back. Then I'd turn the 8 to make sure a vowel is on the back." Then the subject stops. Several things are apparent here. First, it makes sense that subjects are engaging in some kind of Type 2 processing. Most Type 1 processes would be of no help on this problem. Affective processing is not engaged, so processes of emotional regulation are no help. Unless the subject is a philosophy major, he or she has no highly practiced procedures (logic) that have become automatized that would be of any help. Finally, the problem is evolutionarily unprecedented, so no Darwinian modules would be helpful.

The subject is left to rely on Type 2 processing, but the processing is seriously incomplete in the example given. The subject has relied on serial associative cognition rather than exhaustive simulation of an alternative world—a world that includes situations in which the rule is false. The subject has not constructed the false case—a vowel with an odd number on the back. Nor has the subject gone systematically through the cards asking the question of whether that card could be a vowel/odd combination. Answer: K (no), A(yes), 8(no), 5(yes). Such a procedure yields the correct choice of A and 5. Instead, the subject with this protocol started from the model given—the rule as true—and then just worked through implications of what would be expected if the rule were true. A fully simulated world with all the possibilities—including the possibility of a false rule—was never constructed. The fact that the subject refers to hidden sides of the cards does not mean that any alternative model of the situation has been constructed beyond what was given by the experimenter and the subject's own assumption that the rule is true.

Thus, the kind of Type 2 processing that is occurring in this task is not full-blown cognitive simulation of alternative world models. It is thinking of a shallower type—cognition that is inflexibly locked into an associative mode that takes as its starting point a model of the world that is given to the subject. Serial associative cognition is not rapid and parallel in the manner of the systems contained in the autonomous mind, but it is nonetheless rather inflexibly locked into an associative mode that takes as its starting point the model of the world that is most easy to construct. In the selection task, subjects reason from a single focal model—systematically generating associations from this focal model but never constructing another model of the situation. In our model of cognitive architecture, this is what we term "serial associative cognition with a focal bias."

Serial Associative Cognition as Miserly Processing

One way to contextualize the idea of focal bias is as the second stage in a framework for thinking about human information processing that we will discuss repeatedly in this book and extensively in chapter 3—the idea of humans as cognitive misers. People are cognitive misers because their basic tendency is to default to processing mechanisms of low computational expense. Humorously, Hull (2001) has said that "the rule that human beings seem to follow is to engage the brain only when all else fails—and usually not even then" (p. 37). More seriously, Richerson and Boyd (2005) have put the same point in terms of its origins in evolution: "In effect, all animals are under stringent selection pressure to be as stupid as they can get away with" (p. 135).

There are in fact two aspects of cognitive miserliness. Dual-process theory has heretofore highlighted only Rule 1 of the cognitive miser: Default to Type 1 processing whenever possible. But defaulting to Type 1 processing is not always possible—particularly in novel situations where there are no stimuli available to signal domain-specific evolutionary modules, nor perhaps any information with which to run overlearned and well-compiled procedures that the autonomous mind has acquired through practice. Type 2 processing procedures will be necessary in such cases, but a cognitive miser default is operating even there. Rule 2 of the cognitive miser is: When Type 2 processing is necessary, default to serial associative cognition with a focal bias (*not* fully decoupled cognitive simulation).

The notion of a focal bias conjoins several closely related ideas in the literature—Evans, Over, and Handley's (2003) singularity principle; Johnson-Laird's (1999, 2005, 2006) principle of truth; Kahneman's (2011) WYSIATI tendency (what you see is all there is); focusing (Legrenzi, Girotto, & Johnson-Laird, 1993); the effect/effort issues discussed by Sperber, Cara, and Girotto (1995); the metacognitive myopia of Fiedler (2012); and finally, the focalism (Wilson, Wheatley, Meyers, Gilbert, & Axsom, 2000) and belief acceptance (Gilbert, 1991) issues that have been prominent in the social psychological literature. Focal bias combines all of these tendencies into the basic idea that the information processor is strongly disposed to deal with only the most easily constructed cognitive model.

As a result, the focal model that will dominate processing—the only model that serial associative cognition deals with—is the most easily constructed model. The focal model tends to represent only one state of affairs (the Evans et al., 2003, singularity idea); it accepts what is directly presented and models what is presented as true (e.g., Gilbert, 1991; Johnson-Laird, 1999); it is a model that minimizes effort (Sperber et al., 1995); it ignores moderating factors (as the social psychological literature has demonstrated, e.g., Wilson et al., 2000)—probably because taking account of those factors would necessitate modeling several alternative worlds, and this is just what focal processing allows us to avoid. And finally, given the voluminous literature in cognitive science on belief bias and the informal reasoning literature on myside bias, the easiest models to represent clearly appear to be those closest to what a person already believes and has modeled previously (e.g., Evans, 2002, 2014; Stanovich, West, & Toplak, 2013).

Thus, serial associative cognition is defined by its reliance on a single focal model that triggers all subsequent thought. Framing effects, for instance, are a clear example of serial associative cognition with a focal bias. As Kahneman (2003) notes, "The basic principle of framing is the passive acceptance of the formulation given" (p. 703). The frame presented to the subject is taken as focal, and all subsequent thought derives from it rather than from alternative framings because the latter would necessitate more computationally expensive simulation operations.

Having introduced the idea of serial associative cognition, we can now return to figure 2.3 and identify a third function of the reflective mind—initiating an interrupt of serial associative cognition (arrow F). This interrupt signal alters the next step in a serial associative sequence that would otherwise direct thought. This interrupt signal might stop serial associative cognition altogether in order to initiate a comprehensive simulation (arrow C). Alternatively, it might start a new serial associative chain (arrow E) from a different starting point by altering the temporary focal model that is the source of a current associative chain.

The Importance of Knowledge Structures (Mindware)

One aspect of dual-process theory that has been relatively neglected is that successful Type 2 override operations require both procedural and declarative knowledge. Although taking the Type 1 response priming offline might itself be procedural, the process of synthesizing an alternative response often utilizes stored knowledge of various types. During the simulation process, declarative knowledge and strategic rules (linguistically coded strategies) are used to transform a decoupled representation.

The knowledge, rules, procedures, and strategies that can be retrieved and used to transform decoupled representations have been referred to as "mindware," a term coined by David Perkins in a 1995 book (Clark, 2001, uses the term in a slightly different way from Perkins's original coinage). The mindware available for use during cognitive simulation is, in part, the product of past learning experiences. This means that individuals will differ in their abilities to simulate better alternatives to a Type 1 response based on variation in their available mindware.

In fact, each of the levels in the tripartite model has to access knowledge to carry out its operations. One's reflective mind accesses not only one's general knowledge structures but, importantly, one's opinions, beliefs, and reflectively acquired goal structure (considered preferences; see Gauthier, 1986). One's algorithmic mind accesses micro-strategies for cognitive operations and production system rules for sequencing behaviors and thoughts. Finally, one's autonomous mind not only accesses evolutionarily compiled encapsulated knowledge bases, but also retrieves information that has become tightly compiled and automatically activated as a result of overlearning and practice. Of course, the knowledge bases we have mentioned are those that are *unique* to each mind. Algorithmic- and reflective-level processes also receive inputs from the computations of the autonomous mind (see the G arrows in figure 2.3, which indicate the influence of preattentive processes).

Because the CHC theory of intelligence is one of the most comprehensively validated theories of intelligence available, it is important to see how two of its major components miss critical aspects of rational thought. Fluid intelligence will, of course, have some relation to rationality because it indicates the computational power of the algorithmic mind to sustain decoupling. Because override and simulation are important for rational thought, Gf will definitely facilitate rational action in some situations. Nevertheless, the tendency to initiate override (arrow B in figure 2.3) and to initiate simulation activities (arrow D in figure 2.3) are both aspects of the reflective mind not assessed by intelligence tests, so the tests will miss these components of rationality. Such propensities are instead indexed by measures of typical performance (cognitive styles and thinking dispositions) as opposed to measures of maximal performance such as IQ tests.

The situation with respect to Gc is a little different. Rational thought depends critically on the acquisition of certain types of knowledge. That knowledge would, in the abstract, be classified as crystallized intelligence. But is it the kind of crystallized knowledge that is assessed on actual tests of intelligence? The answer is "no." The knowledge structures that support rational thought are specialized. They cluster in the domains of probabilistic reasoning, causal reasoning, and scientific reasoning, as will be clear in subsequent chapters where we introduce the components of the CART. In contrast, the crystallized knowledge assessed on IQ tests is deliberately designed to be nonspecialized. The designers of the tests, to ensure that the sampling of vocabulary and knowledge is fair and unbiased, explicitly attempt to *broadly* sample vocabulary, verbal comprehension domains, and general knowledge. In short, Gc, as traditionally measured, does not assess individual differences in rationality.

The Behavioral Tendencies and Knowledge Bases That Support Rationality Are Based in All Minds

It may have seemed from our discussion so far that only the algorithmic and reflective minds are implicated in rational thought. Such an interpretation would be mistaken. In fact, the autonomous mind, as well as the algorithmic and reflective minds, often operates to support rational thought. There is one particular way that the autonomous mind supports rationality that we would like to emphasize. It is the point mentioned in the previous section, that the autonomous mind contains rational rules and normative strategies that have been tightly compiled and that are automatically activated as a result of overlearning and practice. This means that, for some people, in some instances, the normative response emanates directly from the autonomous mind rather than from the more costly Type 2 process of simulation.

Figure 2.4 illustrates more clearly the point we wish to make here. This figure has been simplified by the removal of all the arrow labels and the removal of the boxes representing serial associative cognition, as well as some response boxes. The type of accessing of mindware that is most discussed in the literature is represented in the upper right. In the case represented there, a nonnormative response from the autonomous mind has been interrupted and the computationally taxing process of simulating an

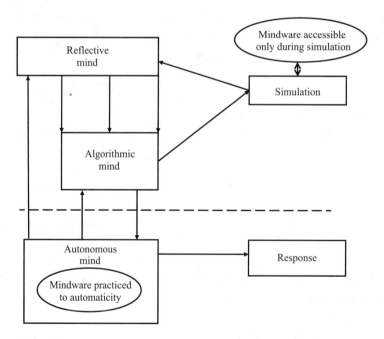

Figure 2.4
A simplified model showing both automatized mindware and mindware accessible during simulation.

alternative response is underway. That simulation involves the computationally expensive process of accessing mindware for the simulation.

In contrast to this type of normative mindware access, indicated in the lower left of the figure is a qualitatively different way that mindware can determine the normative response. The figure indicates the point we stressed earlier, that within the autonomous mind can reside normative rules and rational strategies that have been practiced to automaticity and that can *automatically* compete with (and often immediately defeat) any alternative nonnormative response that is also stored in the autonomous mind. So, as should be clear from figure 2.4, it does not follow from the output of a normative response that System 2 was necessarily the genesis of the rational responding. Neither does it necessarily follow (as has been wrongly inferred in much recent research on dual-process theory) that a rapid response should necessarily be an incorrect one. The main purpose of figure 2.4 is to concretize the idea that the normative mindware of rational responding is not exclusively retrieved during simulation activities, but can become implicated in performance directly and automatically from the autonomous mind if it has been practiced enough. As we will see in chapter 3, this complicates the interpretation of performance on heuristics and biases tasks in a dual-process context.

Our Model of Cognitive Architecture, Rationality, and Subsequent Chapters

The cognitive architecture that we have outlined is far from complete, but it is detailed enough to serve as a general framework for presenting our Comprehensive Assessment of Rational Thinking (CART). Stanovich (2011) discusses more detailed empirical evidence than we have presented here, particularly evidence regarding the separability of the algorithmic mind from the reflective mind. The present model provides enough of a framework for us to contextualize and classify heuristics and biases tasks in subsequent chapters.

The classification of those tasks that begins in chapter 3 derives from the fact that, according to the model just presented, rationality requires three mental characteristics. First, the reflective mind must be characterized by the tendency to initiate the override of suboptimal responses generated by the autonomous mind and to initiate simulation activities that will result in a better response. Second, algorithmic-level cognitive capacity (Gf) is needed for override and sustained simulation activities. Finally, the mindware that allows the computation of rational responses needs to be available and accessible during simulation activities or be accessible from the autonomous mind (see figure 2.4) because it has been highly practiced. Intelligence tests primarily assess only the second of these three characteristics that determine rational thought and action. This is why rationality requires more than just intelligence.

In chapter 3 we will unpack the logic of heuristics and biases tasks, showing how they are quite multifarious but nevertheless fall along some definable dimensions. We will describe the various dimensions of heuristics and biases tasks that allow us to develop a taxonomy of the types of cognitive errors that are made on such tasks. In chapter 4, we use that taxonomy of errors to define a positive framework for assessing rational thinking. We identify in that chapter the panoply of task types that make up the CART. In chapters 5 through 11 we will take up in turn each of the tasks that are used to operationalize rational thinking in our assessment device.

3 Overcoming Miserly Processing: Detection, Override, and Mindware

Heuristics and biases tasks were designed for human brains, not animal brains. What we mean by this is that heuristics and biases tasks were designed for brains that could at least *potentially* experience conscious conflict. This is why Kahneman (2000) stressed that "Tversky and I always thought of the heuristics and biases approach as a two-process theory" (p. 682). All multiple-process models of mind, including dual-process theories (Evans & Stanovich, 2013), capture a phenomenal aspect of human decision making that is profoundly important—that humans often feel alienated from their choices. We display what both folk psychology and philosophers term "weakness of will." For example, we continue to smoke even when we know that it is a harmful habit; or we order dessert after a large meal, merely an hour after pledging to ourselves that we would not. However, we also display alienation from our responses in situations that do not involve weakness of will—we may, for example, find ourselves recoiling from the sight of a disfigured person even after a lifetime of dedication to inclusion and social harmony.

This feeling of alienation, although emotionally discomfiting when it occurs, is actually a reflection of a unique aspect of human cognition, namely, the use of the metarepresentational abilities of Type 2 processing to enable a cognitive critique of our beliefs and our desires. Humans alone appear to be able to represent a model of an idealized (i.e., hypothesized) preference structure, while still maintaining a first-order model of current response tendencies.

Another way to put this would be to say that heuristics and biases were designed for Popperian and Gregorian creatures, not Darwinian and Skinnerian ones. Here, we are referring again to the terminology from Dennett's (1996) short but provocative book *Kinds of Minds*, which describes the overlapping short-leashed and long-leashed strategies embodied in our brains by labeling them as different "minds." These minds are all lodged within the same brain in the case of humans, and are all operating simultaneously to solve problems. The minds reflect increasingly powerful mechanisms for predicting the future world. The Darwinian mind uses prewired reflexes (metaphorically, "Do *this* when *x* happens because it is best"). The Skinnerian mind uses operant conditioning to shape itself to an unpredictable environment. The Popperian mind (after the philosopher Karl Popper) can represent possibilities and test them internally before responding (this is the metarepresentational System 2 of dual-process theory). The Gregorian mind (after the psychologist Richard Gregory) exploits the mental tools discovered by others to aid in the pretesting of responses. As discussed in the previous chapter, the computational power available to sustain the serial simulation of the Popperian mind is indexed by fluid intelligence in the CHC model of intelligence, whereas the power of the cultural tools (knowledge) used during serial simulation—the Gregorian mind, in Dennett's model—is in part indexed by differences in crystallized intelligence in the CHC model.

When we say that heuristics and biases tasks were designed for organisms with Popperian and Gregorian minds (humans), we do not mean that rationality cannot be assessed in nonhuman animals. To the contrary, the axiomatic approach to rationality assessment we described in chapter 1 allows the rationality of nonhuman animals to be assessed as well as that of humans, because it defines instrumental rationality as adherence to certain types of consistency and coherence relationships (see Kacelnik, 2006; Luce & Raiffa, 1957; Savage, 1954). In fact, many animals appear to have a reasonable degree of instrumental rationality; research has established that the behavior of many nonhuman animals does in fact follow pretty closely the axioms of rational choice (Alcock, 2005; Binmore, 2009; Dukas, 1998; Fantino & Stolarz-Fantino, 2005; Hurley & Nudds, 2006; Real, 1991; Schuck-Paim & Kacelnik, 2002; Schuck-Paim, Pompilio, & Kacelnik, 2004).

The adaptively shaped behavior of nonhuman animals can, in theory, deviate from the axioms of rational choice because it is possible for the optimization of fitness at the genetic level to dissociate from optimization at the level of the organism (Barkow, 1989; Dawkins, 1982; Houston, 1997; Marcus, 2008; Over, 2002; Skyrms, 1996; Stanovich, 2004; Waksberg, Smith, & Burd, 2009). Although such deviations do occur in the animal world (Bateson, Healy, & Hurly, 2002, 2003), it is nonetheless true that, as Satz and Ferejohn (1994) have noted, "pigeons do reasonably well in conforming to the axioms of rational-choice theory" (p. 77).

So although the assessment of nonhuman rationality is possible, the really interesting issues of rationality arise when we have an organism with Popperian and Gregorian minds riding on top of the lower-level minds. Such a situation raises the possibility that the Gregorian/Popperian minds might synthesize a better solution to some problem than the Skinnerian/Darwinian minds. In such a situation (the situation that spawns dual-process conceptualizations), assessing which of the minds wins out becomes of immense interest (and diagnostic of degrees of rationality). It is just this situation that heuristics and biases tasks put under the microscope. These tasks, interpreted within a dual-process framework (Kahneman, 2011), end up being diagnostic of the dominance of Type 1 versus Type 2 processing.

Two contextualizing points are necessary here in light of the purpose for which we are using these methods and ideas: to operationalize the concept of rational thinking so that we might assess individual differences. First, our model of rationality stipulates that issues of rationality are assessed at a personal level, not a subpersonal one (Bermudez, 2001; Frankish, 2009). A memory system in the human brain is not rational or irrational—it is merely efficient or inefficient (or of high or low capacity). Thus, *subprocesses* of the brain do not display rational or irrational properties per se. Rationality concerns the actions of an entity in its environment that serve that entity's goals. We of course could extrapolate the notion of environment to include the interior of the brain itself and then talk of a submodule that chose strategies rationally or not. But this move creates two problems. First, what are the goals of this subpersonal entity—what are its interests that its rationality is trying to serve? This is unclear in the case of a subpersonal entity (see Stanovich, 2004). Second, such a move regresses all the way down. We would need to talk of a neuron's firing as either rational or irrational—a semantic stretch that seems deeply wrong. As Oaksford and Chater (1998) put it, "The fact that a model is optimizing something does not mean that the model is a rational model. Optimality is not the same as rationality. … Stomachs may be well or poorly adapted to their function (digestion), but they have no beliefs, desires or knowledge, and hence the question of their rationality does not arise" (pp. 4, 5).

The second important contextualizing point to emphasize here is that it is wrong to strictly equate Type 2 processing with normatively correct responding and Type 1 processing with normatively incorrect responding (Evans & Stanovich, 2013). It is possible for a situation to trigger a Type 1 response that is normatively correct and a Type 2 response that is normatively incorrect (as when Type 2 processes are applying contaminated mindware; see the discussion later in this chapter). The claim that most dual-process theorists make is that the converse situation—Type 1 processing priming a normatively incorrect response and Type 2 processing overriding with the correct response—is statistically more likely. Additionally, as Evans and Stanovich (2013) have pointed out, some tasks (both in the laboratory and in real life) *require* Type 2 processing for their solution.

Our point here illustrates why we prefer the Type 1/Type 2 terminology rather than some other popular terms in the literature. For example, Epstein's (1994) terminology (experiential system and rational system) mistakenly implies that Type 2 processing always yields a response that is normatively rational (and perhaps pragmatically that the experiential system does not). Gibbard's (1990) labeling of Type 2 processing as emanating from a "normative control system" mistakenly implies the same thing (that Type 2 processing is always normative), as does Klein's (1998) labeling of Type 2 strategies as "rational choice strategies." As we said before, rationality is an organismic-level concept and should never be used to label a subpersonal process. Rather, Type 2 and Type 1 processing both contribute to an *overt response* that may be assessed in terms of degrees of rationality.

The Logic of Heuristics and Biases Tasks

So if we were only System 1 organisms, we would not need heuristics and biases tasks to assess our capabilities. Or, if we *were* Popperian organisms, but Type 1 processing always yielded the correct response, then we would not need heuristics and biases tasks for assessment either.[1] Heuristics and biases tasks are for Popperian organisms in hostile environments where Type 1 processing will sometimes produce important errors.

As discussed in chapter 2, for a person who defaults often to Type 1 processing, environments can be either benign or hostile. A benign environment is an environment that contains useful cues that, via practice or evolutionary history, have been well represented in Type 1 subsystems.

Additionally, for an environment to be classified as benign, it must not contain other individuals who will adjust their behavior to exploit those relying only on Type 1 processing. We would argue (Stanovich, 2004; Stanovich & West, 2000) that the modern world is somewhat hostile to Type 1 processing in critical ways (advertising, for example), thus making it important to assess rational thinking tendencies via the logic of heuristics and biases tasks.

It is also appropriate here to emphasize another way in which intelligence tests fail to tap important aspects of rational thinking. The novice reader might have thought at this point that it seems that intelligence tests clearly measure Type 2 reasoning—that is, conscious, serial simulation of imaginary worlds in order to solve problems. This is all true, but there is a critical difference. Intelligence tests contain salient warnings that Type 2 reasoning is necessary. It is clear to someone taking an intelligence test that fast, automatic, intuitive processing will not lead to superior performance. Most tests of rational thinking do not strongly cue the subject in this manner. Instead, many heuristics and biases tasks suggest a compelling intuitive response that happens to be wrong. In heuristics and biases tasks, unlike in intelligence tests, the subject must detect the inadequacy of the Type 1 response and then must use Type 2 processing to both suppress the Type 1 response and to simulate a better alternative.

With these clarifications in mind, let us revisit the three mental characteristics required for rational thinking that we discussed in chapter 2 (see also Stanovich & West, 2008b), and see how they are reflected in discussions of heuristics and biases tasks. First, the necessity of overriding Type 1 processing must be detected; the subject must be able to detect a conflict between his or her intuitive response to a problem and the response dictated by learned normative rules. Second, the mindware that allows the computation of more rational responses needs to be available and accessible during simulation activities. Third, algorithmic-level cognitive capacity is needed so that override and simulation activities can be sustained.

The first characteristic—conflict detection—is gaining increased recognition in the literature (De Neys 2014; De Neys & Glumicic, 2008; Stanovich & West, 2008b; Thompson, 2009; Thompson & Johnson, 2014; Thompson & Morsanyi, 2012). The second (the importance of prelearned mindware) is still underappreciated in the literature. Likewise, the third mental ability (sustained cognitive decoupling) is also sometimes ignored. For example, speculating on the reason for positive correlations between cognitive ability and normative responding on heuristics and biases tasks, Kahneman and Frederick (2002) argue that people with higher cognitive abilities "are more likely to possess the relevant logical rules and also to recognize the applicability of these rules in particular situations" (p. 68).

The phrase "possess the relevant logical rules and also to recognize the applicability of these rules in particular situations" suggests two conditions (mindware and conflict detection) that have to be fulfilled for a Type 1 response to be replaced by Type 2 processing. But the statement misses the third condition. Even after the necessity for override has been detected and the relevant mindware is available, the conflict has to be resolved. Resolving the conflict in favor of the normative response may require cognitive capacity, especially if cognitive decoupling must take place for a considerable period of time. Cognitive decoupling is involved in inhibiting the intuitive response and also in simulating alternative responses (as discussed in chapter 2). Recent work on inhibition and executive functioning has indicated that such cognitive decoupling is very capacity demanding and that it is strongly related to individual differences in fluid intelligence. This is the third mental characteristic required for rational thinking—sustained cognitive decoupling. The failure of this component in a particular instance will be termed sustained override failure.

The Interdependence of Mindware, Conflict Detection, and Override

It is important to understand that the presence of mindware, detection of the need for override, and sustained override capability are not just three separate categories of cognitive requirements; rather, they are intertwined in important ways. Here we will consider several such dependencies, each of increasing complexity. First, if the relevant mindware is not available, the person must, of necessity, emit the response primed by Type 1 processing systems. It is immaterial whether the person has the capacity to sustain override if the normatively appropriate response is simply not available. Conflict detection abilities will also not be assessed in such a situation, because the mindware necessary to generate a conflicting response is simply not present.

If the relevant mindware is in fact available, then the detection of conflict can be assessed. However, even if the relevant mindware is present, if the subject detects no reason to override the intuitive response, then sustained override capability will not come into play. In short, whether or not a given task (for a given person) assesses certain "downstream" capabilities (sustained override) depends on whether certain "upstream" capabilities (conflict detection, presence of mindware) are present for a given task.

Note in particular the trade-off-type relationship between the failure of override and the absence of mindware. Errors made by someone with well-learned mindware are more likely to be due to override failure than to mindware gaps. Conversely, override errors are less likely to be attributed to people with little, or poorly learned, mindware. Of course, the two categories trade off in a continuous manner with a fuzzy boundary between them. A well-learned rule not appropriately applied is a case of override failure. As the rule is less and less well instantiated, at some point it is so poorly compiled that it is not a candidate for retrieval in the override process and thus the override error becomes an error due to a mindware gap. In short, a process error has turned into a knowledge error.

The next couple of figures serve to illustrate the interdependence and complex relationships between mindware presence, conflict detection, and sustained override capability. Figure 3.1 is organized around a continuum reflecting how well the mindware in the relevant problem has been instantiated.[2]

At the far left of the continuum in figure 3.1, the mindware is totally absent. As the relevant mindware becomes practiced and stored in long-term memory, it becomes available for retrieval by Type 2 processing. In the middle of the continuum (mindware learned but not automatized), the mindware must be retrieved by expensive Type 2 processing to aid in creating what might be called a *computed response* to compete with what might be called the *intuitive* response that is naturally emitted by System 1.

On the far right of the continuum (mindware automatized), the relevant mindware has been so overly practiced that it has entered System 1 and is triggered automatically and autonomously (like other Type 1 responses). This mindware is so overly practiced that it can often automatically trump the intuitive response from System 1 without needing to invoke a taxing Type-2 override procedure. In short, the far right of the continuum is the area where no sustained override is needed. The intuitive Type 1 response is automatically trumped by the more recently learned mindware that now is the dominant response. Subjects will almost always solve the problem

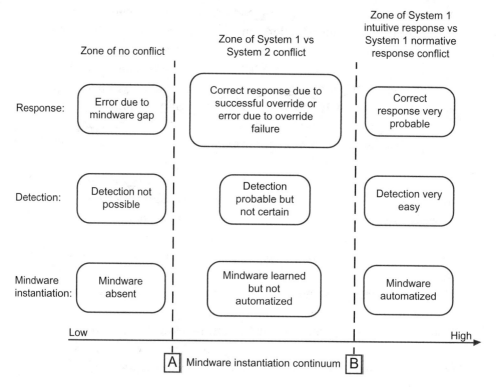

Figure 3.1
Processing states on the mindware continuum.

correctly while in this part of the continuum—when the normative mindware is so well instantiated. This situation contrasts sharply with that on the far left of the continuum. Here, the mindware is so little practiced that no conflict detection will occur and the subject will always make an error because of a mindware gap. Override of the Type 1 response is not possible, because the mindware is not instantiated well enough for the reflective mind to detect the necessity for override.

The middle section of the figure represents the zone of conflict between System 1 priming an intuitive response and System 2 suppressing it while simulating a normative response. Here, whether the subject responds correctly or not will depend on the success of sustained override. This zone of System 1/System 2 conflict is defined by the area demarcated by thresholds A and B on the mindware instantiation continuum. To the left of dotted line A, the mindware is not well enough instantiated to be retrieved from long-term memory during the simulation of a superior—that is, normative—response. To the right of dotted line A and to the left of dotted line B, the mindware is instantiated enough for retrieval but not so fully automatized that it has entered System 1. This is the zone of potential conflict between System 1 priming an intuitive response and System 2 computing an alternative normative response via decoupled simulation. To the right of dotted line B is the area discussed before, where an automatized normative response trumps the intuitive response.

It is important to note that figure 3.1 illustrates that the ease of conflict detection and the degree of mindware instantiation will be highly correlated, if not substantially the same thing (or at least difficult to separate). This will become an important theme in the final section of this chapter—*knowledge considerations and processing considerations are very difficult to separate in many of the heuristics and biases tasks that involve conflict.*

Figure 3.2 presents the logic of heuristics and biases tasks in even more detail. Here, the letters W through Z mark the criterion values on the mindware continuum that help define five different processing states. Criterion W marks the place on the mindware instantiation continuum where conflicts become possible. To the left of this criterion, the mindware is so little established that it should be considered absent and hence no conflict detection is possible. Conflict detection—that is, the subject's detection of a conflict between an intuitive response and learned normative rules—is *possible* to the right of criterion W. However, conflict is only *actually* detected by this particular subject on this particular task (see note 2 of this chapter) to the right of criterion X. This criterion creates an area (to the right of W and to the left of X) where the subject *could* make a computed response because the mindware is there for retrieval, but he or she detects no conflict and hence does not even attempt sustained override. This processing state represents what might be called a "detection error."

Conflict detection *is* probable to the right of criterion X in figure 3.2, making override a possibility in that part of the graphic. Criterion Y demarcates successful from unsuccessful sustained override. To the left of criterion Y (and to the right of X) is the processing state that De Neys has explored in numerous studies (De Neys, 2006a, 2006b, 2014; De Neys & Franssens, 2009; De Neys & Glumicic, 2008; De Neys, Moyens, & Vansteenwegen, 2010; see also Thompson & Johnson, 2014). His research group has demonstrated, with several heuristics and biases tasks, that various implicit

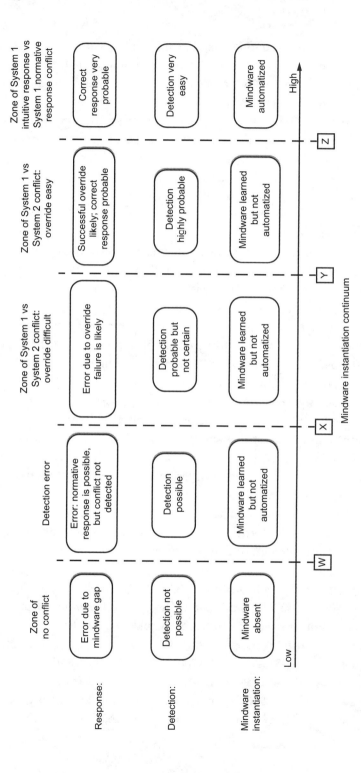

Figure 3.2
Processing states on the mindware continuum.

measures of performance (decision latencies, unannounced recall, brain activation, autonomic arousal) indicate that a subject sometimes detects conflict even in cases where he or she did not successfully carry out the override of Type 1 processing. In contrast, to the right of criterion Y and to the left of Z is the area of the mindware instantiation continuum where override is likely to be successful. Well-instantiated mindware makes detection of conflict easy and highly probable. Well-instantiated mindware also makes its sustained retrieval during the simulation process easy, leading to a high probability of successful override.

Finally, to the right of criterion Z is the processing state that we described before and illustrated in figure 3.1—where no processing override of the intuitive response is necessary because the normative mindware is instantiated in System 1 and is the dominant response. To summarize, in figure 3.2: the area to the left of W denotes an error due to missing mindware; the area between W and X an error due to detection failure; between X and Y an error due to override failure; between Y and Z a correct response achieved by sustained override; and to the right of Z a correct response due to automatic retrieval of the normative response from System 1. Moving from left to right in figure 3.2, the likelihood of deriving a correct response increases as the mindware becomes more automatized, conflict detection becomes easier, and override becomes easier as well. A novel contribution of figure 3.2 is that of demonstrating how the cognitive failures (and successes) differ as we move along the continuum from left to right.

Toward a Taxonomy of Thinking Errors

With what has been presented so far, we can begin to sketch a taxonomy of thinking errors. We present the beginning of such a taxonomy in table 3.1.

Table 3.1
Types of reasoning errors

	Miserly Processing			Mindware Problems	
Default to the autonomous mind	Failure of sustained override	Default to serial associative cognition with a focal bias		Mindware gaps	Contaminated mindware

Building on our previous discussion, we differentiate processing problems from content/knowledge problems. Keep in mind the linkage discussed previously, though—less well-instantiated mindware makes override harder and defaulting to Type-1 processing more likely. Processing problems we generically call the "tendency toward miserly processing" and content/knowledge problems we generically call "mindware problems."

That humans are cognitive misers has been a major theme throughout the past forty years of research in psychology and cognitive science (see Dawes, 1976; Kahneman, 2011; Simon, 1955, 1956; Taylor, 1981; Tversky & Kahneman, 1974; for the evolutionary reasons why, see Stanovich, 2004, 2009). When approaching any problem, our brains have available various computational mechanisms for dealing with the situation. These mechanisms embody a trade-off, however (Tomlin et al., 2015), between power and expense. Some mechanisms have great computational power—they can solve a large number of novel problems. However, these mechanisms take up a great deal of attention, tend to be slow, tend to interfere with other thoughts and actions we are carrying out, and they require great concentration that is often experienced as aversive (Kahneman, 1973; Kurzban, Duckworth, Kable, & Myers, 2013; Navon, 1989; Westbrook & Braver, 2015).

Humans are cognitive misers because their basic tendency is to default to processing mechanisms of low computational expense. These are often Type 1 processes, but not always. Sometimes more than one Type 2 processing mode is available (e.g., simulation or serial associative cognition), and the miserly choice here is again to choose the one of less computational expense (serial associative cognition).[3] Table 3.1 identifies three types of miserly processing, two of which we have discussed previously. The first is the tendency to default to the response options primed by the autonomous mind. It represents the shallowest kind of thinking error, because no Type 2 processing is done at all. It occurs when mindware is present for a normative response, but the necessity for override is not detected.

Note that in the dual-process literature, this type of processing default has sometimes been termed a "failure to override." That is, some previous theorists have collapsed two areas in figure 3.2 that we differentiate (the two areas demarcated by criterion X). However, in the taxonomy we are presenting here, for something to be considered a sustained override failure, Type 2 processing must lose out to Type 1 processing in a conflict of discrepant outputs. If Type 2 processing is not engaged at all, then it is not considered an override failure in our view.

In fact, the early heuristics and biases researchers were clearer on this point than many later dual-process theorists. The distinction between impressions and judgments in the early heuristics and biases work (for a discussion, see Kahneman, 2003; Kahneman & Frederick, 2002, 2005) made it clearer that nonnormative responses often result not from a Type 2/Type 1 struggle, but from intuitive impressions that are left uncorrected by the rules and strategies that Type 2 processing can recruit. Kahneman (2011) was especially clear on this point in his landmark book surveying the entire decades-long heuristics and biases tradition.

Some previous authors may have been taking a more phenotypic approach to their terminology. That is, they fused together situations where the subject simply defaults to Type 1 processing with those in which the subject attempts to override Type 1 processing but fails. In both of these cases, the subject makes an error, so many authors have lumped these situations together. However, we prefer to mark the different underlying reasons why the error was made, so we do separate these categories in our taxonomy. The second category from the left in table 3.1 represents, in our view, true override failure. It represents a less miserly tendency than simple default to the autonomous mind. Inhibitory Type 2 processes try to take the Type 1 processing of the autonomous mind offline in these cases (and keep it offline until better responses are synthesized via hypothetical reasoning), but they fail. In override failure, cognitive decoupling does take place, but it fails to suppress the Type 1 processing of the autonomous mind.

Failure of sustained override in this taxonomy encompasses what folk theory would call problems of willpower (see Ainslie, 2001, 2005, for a nuanced discussion of the folk concept of willpower in light of modern cognitive science). Psychologists and economists often term these situations as reflecting problems of self-control (Baumeister & Vohs, 2003, 2007; Hofmann, Friese, & Strack, 2009; Loewenstein, Read, & Baumeister, 2003). However, this category comprises more than just willpower and self-control issues. Sloman (1996) points out that at least for some subjects, the Linda conjunction problem (described in chapter 5; see Tversky & Kahneman, 1983) is the quintessential dual-process conflict. He quotes Stephen Gould's introspection that "I know the [conjunction] is least probable, yet a little homunculus in my head continues to jump up and down, shouting at me—'but she can't be a bank teller; read the description'" (Gould, 1991, p. 469). For sophisticated subjects such as Gould, resolving the Linda problem clearly involves a Type 1/Type 2 processing conflict, and in his case a conjunction error on the task would represent a true case of override failure. However, for the majority of subjects in this task, there may well be no conscious introspection going on—Type 2 processing is either not engaged or engaged so little that there is no awareness of a cognitive struggle. For such subjects, committing the conjunction error in the Linda problem represents a case of defaulting to the processing of the autonomous mind.

The third type of miserly processing represented in table 3.1 is to engage in serial associative cognition with a focal bias. This characteristic represents a tendency to overeconomize during Type 2 processing—specifically, to fail to engage in the full-blown simulation of alternative worlds or in fully disjunctive reasoning (Shafir, 1994; Toplak & Stanovich, 2002). Our discussion of the four-card selection task in chapter 2 provides an example. This category of error illustrates that not all miserly processing involves direct conflict detection.

The fourth category in table 3.1 is that of a mindware gap. It is represented in the far left areas in both figure 3.1 and figure 3.2. The subject responds with the intuitive response generated by System 1 and thus makes an error on the task. No issue of conflict detection or sustained override is relevant because the subject has not acquired the necessary mindware to an extent that would make conflict possible. Lacking knowledge in the areas of probabilistic reasoning, causal reasoning, logic, and scientific thinking could result in less rational thought or behavior because it is not available during simulation operations. This missing mindware has real-world implications. The study of pathological gambling behavior, for instance, has uncovered a lack of knowledge about probability and probabilistic events (Keren, 1994; Rogers, 1998; Toneatto, 1999; Toplak et al., 2007; Wagenaar, 1988). Likewise, missing mindware in the area of personal finance results in problems of money management and economic life planning (Lusardi & Mitchell, 2014).

One category in table 3.1 that we have not yet discussed is contaminated mindware. So far, we have discussed only situations where subjects failed to make a normative response because certain helpful mindware was missing (i.e., there were mindware gaps). But the term "contaminated mindware" serves to remind us that some acquired mindware can be the direct cause of actions that thwart our goals. This is why Evans and Stanovich (2013) reminded researchers that we cannot always equate Type 2 processing with rationality: sometimes Type 2 processing retrieves an alternative response that causes a *less* rational choice. In short, a subject may hold many specific clusters of misinformation that would make his or her behavior less rational. For example, the gambler's fallacy and many of the other misunderstandings of probability that have been studied in the heuristics and biases literature would fit here (Nickerson, 2004; Roney & Trick, 2009; Xu & Harvey, 2014). Of course, the line between missing mindware and contaminated mindware gets fuzzy in some cases, and that may be especially true in the domain of probabilistic reasoning.

Examples of the Three Types of Miserly Processing

It might help at this point to see an example of each type of miserly processing. The quintessential task for measuring the tendency to default to the autonomous mind because of failure to detect conflicting responses has become Frederick's (2005) cognitive reflection test (CRT), and the quintessential item from the test has become the bat-and-ball item: A bat and a ball cost $1.10 in total. The bat costs $1 more than the ball. How much does the ball cost?

When answering this problem, many people give the first response that comes to mind—10 cents—without thinking further and realizing that this cannot be right. The bat would then have to cost $1.10, and the total cost would then be $1.20 rather than the required $1.10. People often do not think deeply enough to realize their error, and they fail to check this math. These subjects certainly do not explicitly consider 5 cents as an alternative response, because the math is simple enough that its correctness would be immediately discerned. The subjects who miss this problem do not engage in an override process that fails—the right answer has not lost out in a cognitive struggle. There was no struggle in the first place.[4]

The bat-and-ball item contrasts strongly with a common paradigm used to assess belief bias in syllogistic reasoning—pitting logic against the believability of the conclusion to see whether logic or believability will win the struggle to determine the response. When subjects fail this task, it usually indicates the failure of sustained override. Consider, for example, the well-known "rose" syllogism (Markovits & Nantel, 1989; Stanovich & West,

1998c): All flowers have petals; all roses have petals; therefore, all roses are flowers. This syllogism is invalid. Subjects must suppress the tendency to endorse a valid response because of the "naturalness" of the conclusion—all roses *are* flowers. This response must be held in abeyance while reasoning procedures work through the partially overlapping set logic indicating that the conclusion does not necessarily follow and that the syllogism is thus invalid. The reasoning process may take several seconds of perhaps somewhat aversive concentration—seconds during which the urge to foreclose the conflict by acceding to the natural tendency to affirm "All roses are flowers" (by responding "valid") must be suppressed. The processing struggle described here represents a classic case of override failure. This is especially true given that the instructions to the task usually sensitize the participants to potential conflict (between argument validity and the truth of argument components).[5]

Finally, framing effects represent the classic example of serial associative cognition with a focal bias. In the so-called disease framing problem introduced by Tversky and Kahneman (1981), people respond differently depending on different wordings of an equivalent problem. This and other similar problems will be discussed in chapter 8. When presented with both versions of the problem together, most people agree that the problems are simply redescriptions of the same situation and that the alternative phrasing should not have made a difference.

In discussing the mechanisms that cause framing effects, Kahneman has stated that "the basic principle of framing is the passive acceptance of the formulation given" (2003, p. 703). The frame presented to the subject is taken as focal, and all subsequent thought derives from it rather than from alternative framings because the latter would require more thought. Take, for example, the fact that many people are willing to pay more for hamburger meat described as 94 percent fat free than they would have been willing to pay for the same meat described as containing 6 percent fat. When presented with the 94 percent fat free condition, a subject does not experience a conflict between a Type 1 and a Type 2 response. The subject instead fails to transform the situation into another via simulation—specifically, transforming the 94 percent fat free into 6 percent fat and asking himself whether this gives the same impression as the 94 percent fat-free figure. Such framing problems require the subject to generate the conflicting response rather than giving the subject conflicting information explicitly (just as in the famous engineer/lawyer problem of Kahneman & Tversky, 1973).

Anchoring effects provide another example of serial associative cognition with a focal bias. These will be discussed in chapter 8, but here we just note that the anchoring and adjustment heuristic reveals the operation of an inappropriate focal bias. The subject is putting too much weight on what is "given." Or, as Kahneman (2011) calls it in his book, they are victims of the WYSIATI tendency: What you see is all there is. As he notes, the WYSIATI tendency means that we are radically insensitive to the quality of the information that we automatically process via System 1; or, as he says, "our thoughts and our behavior are influenced, much more than we know or want, by the environment of the moment" (p. 128).

Knowledge and Process Are Intertwined in Rational Thinking Tasks

Having explicated table 3.1, we must emphasize that a particular task is not uniquely associated with any one error category. Given that we have differentiated the categories with particular tasks—for example, using belief bias in syllogistic reasoning to illustrate the failure of sustained override or the CRT to illustrate default to the autonomous mind—this is an even more important caveat for us to stress. In short, saying that a task in part implicates a category of miserliness does not mean that it does not implicate mindware difficulties as well (this has been one important point stressed throughout this chapter). Because this is such an important caveat—that is, because the intertwined nature of knowledge/mindware issues and processing issues in heuristics and biases tasks is such an *essential* part of their logic—we are going to try to illustrate the point with a variety of graphics.

Figures 3.1 and 3.2 have already displayed the dependence between mindware instantiation and the difficulty of System 1 override, as well as the dependence between mindware instantiation and the difficulty of conflict detection. For example, those figures illustrate that sustained override becomes easier as the relevant mindware for normative responding becomes more well instantiated—to the point where, in the extreme, fully automatized mindware renders override unnecessary.

Figure 3.3 represents another way of looking at the dependence between knowledge and process in heuristics and biases tasks. This figure, adapted from a discussion in Stanovich and West (2008b), is constructed in terms of

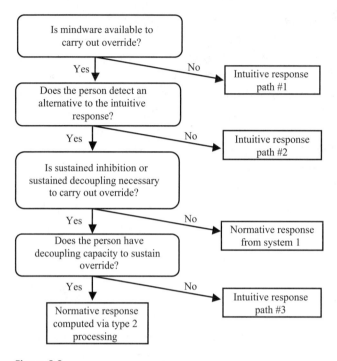

Figure 3.3
Dependence between knowledge and process in heuristics and biases tasks.

which paths lead to the incorrect, intuitive response, and which paths lead to the substitution of a normative response. Figure 3.3 shows that detection and override depend on knowledge and that override depends on detection. We can see this by proceeding downward, sequentially. The question addressed in the first stage of the framework at the top of figure 3.3 is whether, for a given task and subject, the mindware is available to carry out override (whether the subject has available relevant declarative knowledge to be substituted for the intuitive response). If the relevant mindware is not available, then the person must, of necessity, respond intuitively.

If the relevant mindware is in fact available, then the next question that becomes operative is whether or not the person detects the need to override the intuitive response. Even if the relevant mindware is present, if the subject does not detect any reason to override the intuitive response, then that response will be emitted (this is path 2 to the intuitive response as labeled in figure 3.3).

The next choice point in the figure concerns where the subject lies on the mindware learning continuum for that particular task. If the relevant mindware is present and if the subject has noted an alternative to the intuitive response, the question then becomes whether or not the task requires sustained inhibition (cognitive decoupling) in order for the subject to carry out the override of the intuitive response. If the mindware is so automated that it trumps the intuitive response, then no override is needed and a normative response is emitted from System 1 (see figures 3.1 and 3.2).

In contrast, if the normative mindware is not instantiated in System 1 and if the task requires sustained decoupling in order to carry out override, then we must ask whether the subject has the necessary cognitive capacity to solve the problem. If so, then the normative response will be given via Type 2 decoupling and simulation. If not, then the intuitive response will be given (path 3 to the intuitive response in figure 3.3)—despite the availability of the relevant mindware and the recognition of the need to use it.

Figure 3.3 captures the dependence between mindware, detection, and override. It makes clear that missing mindware renders questions about detection and override irrelevant. Mindware present at some minimal level at least provides the possibility of a detection error (or a correct response). The degree of mindware presence that enables (possible) detection then, indirectly, enables the possibility of sustained override (or override failure).

The Continua of Process Dependence and Knowledge Dependence

Saying generically that knowledge and process are intertwined in heuristics and biases tasks is just the first step in understanding their logic. This is because it is not the case that the dependence on knowledge and the dependence on process are the same for each and every task. Some heuristics and biases tasks are more process dependent than knowledge dependent. Others are more knowledge dependent than process dependent. Still others seem to stress both knowledge and process quite strongly.

In figure 3.4 we portray the two dimensions of process and knowledge dependence in which heuristics and biases tasks array themselves. The x-axis indexes how knowledge dependent the task is, ranging from high knowledge dependence on the far right to low knowledge dependence on the far left. The y-axis portrays the degree to which success on the task depends

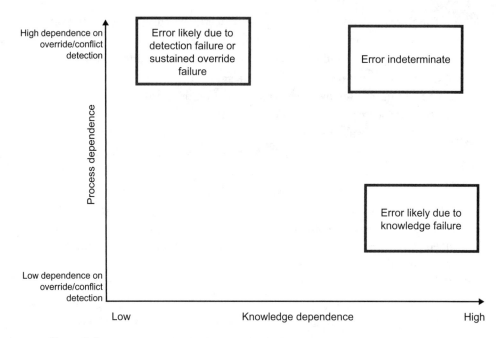

Figure 3.4
Space of process dependence and knowledge dependence in which heuristics and bases tasks array themselves.

on conflict detection and sustained override (i.e., process). Tasks in which conflict detection and override are highly salient—that is, key determinants of task success—are at the upper end of this continuum; whereas tasks that are not as dependent on conflict detection and override are at the bottom of this continuum. The entire figure thus shows the two-dimensional space of knowledge dependence crossed with the degree of process dependence.

Inside the two-dimensional space of figure 3.4, three rectangles serve to demarcate three areas of process/knowledge dependence that result in different types of errors. The area in the upper left represents tasks that depend heavily on conflict detection and override for success, but are less dependent on the presence of specific mindware for their solution. When an error occurs on such tasks, it is more likely to be a process error than a mindware error. Remember, however, that no rational thinking task is a pure process measure. Every rational thinking task will have a process component and a knowledge component, which will arrange themselves on a continuum based on how much they stress override/detection tendencies and how much they stress the presence of mindware.

The bottom right area of figure 3.4 represents tasks that depend heavily on specific knowledge and in which conflict detection and override are not as salient, usually because the subject does not have to overcome a conflicting response from System 1. Errors on these types of tasks are more likely due to missing mindware. The upper right of the figure represents the part of the two-dimensional space that contains tasks where any error made will be indeterminate. That is, these tasks are highly knowledge dependent, but they also require conflict detection and difficult sustained override processes (except for those who have thoroughly automated the relevant knowledge). They have high knowledge dependence and high process dependence—so that whether the subject fails the task because of knowledge or because of conflict/override cannot be determined simply on the basis of observing an error. The lower left area is blank, because it would seem like there would be no place for a task that was low in knowledge dependence and low in conflict/override dependency. However, some ancillary measures (thinking dispositions) do occupy this space, and we will discuss those subsequently.

As we shall see in subsequent illustrations, heuristics and biases tasks tend to cluster in the upper right quadrant of figure 3.4. That is, most of the tasks involve at least a moderate dependence on both knowledge and conflict override. This is what has made it so hard, in the analyses of many tasks in the literature, to pinpoint the exact source of the error response (Stanovich, West, & Toplak, 2011b; Toplak, West, & Stanovich, 2014b). It is also why we issued the warning earlier that no heuristics and biases task can be taken as a specific and unique indicator of a particular *type* of error. For example, we have discussed how performance on Frederick's (2005) CRT (which we exemplified by the example of the bat-and-ball problem) is often taken as an indicator of the tendency to default to the autonomous mind. However, while conflict detection is no doubt implicated substantially in performance on this task, it is not the case that it is independent of mindware. Many of the problems other than the bat-and-ball problem involve content areas, the familiarity of which might be implicated in performance. Even the bat-and-ball problem itself will be affected by the differential instantiation of numeracy. That some people are highly automated at doing subtractions and others are not will affect how easy the problem is. That some people are highly practiced at reaching for numerical checks to their answer ("10¢ + $1.10 = $1.20—oh, that can't be right") could affect accuracy on this problem. So although it is true that the CRT is not as knowledge saturated as some heuristics and biases tasks, it still has some degree of knowledge saturation. Figure 3.5 illustrates where we would position the CRT in the two-dimensional space.

Intelligence tests are different from rational thinking measures in this respect. First, they are somewhat different in that IQ tests certainly involve System 2, but they force the use of Type 2 processing because they contain strong hints and demand characteristics indicating that such processing is necessary. Intelligence tests do not require conflict detection and do not require special mindware (probabilistic thinking, scientific thinking, etc.). Measures of fluid intelligence—like Raven's Matrices—do not require specific content knowledge to replace an intuitive response. They require System 2 processing, but no sustained override. Many cognitive ability measures do tap crystallized knowledge, but that is general knowledge, rather than the domain-specific components of rational thought.

With the caveats and cautions of this chapter in mind, we will proceed in this volume using the terms "miserly processing" and "mindware problems," but with the realization that no task purely separates miserly processing from mindware. The converse is, however, a little easier. That is, it is easier to assess some of the knowledge structures necessary for rationality without heavily implicating detection/override issues. For example, when we introduce the framework for the CART in the next chapter, one critical component of mindware will be probabilistic numeracy—a central knowledge domain that is responsible for less than rational judgment and decision making. Following on much previous empirical work with measures of numeracy (Peters, 2012; Reyna & Brust-Renck, 2015; Weller et al., 2013), we have developed our own measure of this subcomponent of rational thinking. Figure 3.5 illustrates, however, that measures of numeracy reside in a different part of the conceptual space of rational thinking tasks than does the CRT, for example.

In the upper right-hand corner of figure 3.5 are positioned the two categories of probabilistic reasoning and scientific reasoning that encompass a large number of heuristics and biases tasks—for example, probabilistic reasoning problems such as the famous lawyer/engineer problem and Linda problem from the work of Kahneman and Tversky (1973; Tversky

Overcoming Miserly Processing

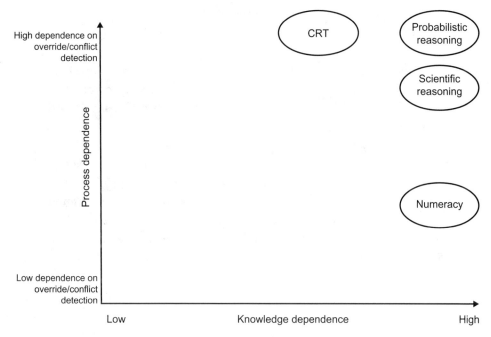

Figure 3.5
Different task types in different locations in the process/knowledge space.

& Kahneman, 1983), and scientific reasoning problems such as Wason's (1960, 1966) four-card selection task and 2–4–6 problem. Placing these in the upper right of figure 3.5 marks the fact that these tasks depend heavily on the presence of mindware as well as the ability to detect and override Type 1 responses. In contrast, the CRT heavily implicates conflict detection and override. However, the CRT does not rely on specific knowledge bases as much as most probabilistic reasoning and scientific reasoning tasks in the heuristics and biases literature. Most numeracy measures have the opposite profile of the CRT. They depend heavily on declarative knowledge of percentages and fractions, but they do not stress as much conflict detection and override as the CRT—because most items on numeracy measures do not contain an especially attractive foil or tempting intuitive response. The relationships portrayed in figure 3.5 reflect these facts, as the CRT is located to the left of probabilistic reasoning but numeracy is located below probabilistic reasoning. In the next chapter, we will situate the wide range of different tasks that compose the CART in the two-dimensional space defined in figure 3.5.

Summary

It is possible to define several ways in which thinking goes wrong on heuristics and biases tasks. In this chapter, we have defined two broad categories of thinking error: problems with miserly processing and problems with mindware. We have defined three types of miserly processing and two kinds of mindware problem (see table 3.1). Although these thinking errors are theoretically differentiable, it is important to realize that none of the five categories of error is defined by an error on a *particular* task. Normative responding on all the tasks that we will use is multiply determined.[6] Furthermore, it is important to recognize the dependencies between mindware, conflict detection, and override, which we have stressed throughout this chapter. The continuous conception we have described here will form the basis of the framework for the assessment of rational thinking that we will describe in chapter 4.

4 A Framework for the Comprehensive Assessment of Rational Thinking (CART)

In the previous chapter, we developed a taxonomy of rational thinking errors, summarized in table 3.1. We also spent considerable time explicating the logic of these different errors. In this chapter, we will transform this taxonomy of thinking errors into a positively stated framework for the assessment of rational thinking. Note, however, that there is historical precedent for beginning our discussion with an analysis of errors in chapter 3: this has been a common strategy in the heuristics and biases literature (see Kahneman, 2011).

In chapter 1, we discussed the axiomatic approach to measuring utility maximization—that if people's preferences follow the axioms of choice (transitivity, independence, etc.) and show descriptive and procedural invariance (Tversky & Kahneman, 1986), then they are behaving as if they are maximizing utility. This is what makes people's degrees of rationality measurable by the experimental methods of cognitive science. Although it is difficult to assess utility maximization directly, it is much easier to assess whether one of the axioms of rational choice is being *violated*. This is much like our judgments at a sporting event, where, for example, it might be difficult to discern whether the quarterback has put the ball perfectly "on the money," but it is not difficult at all to detect a bad throw.

Researchers in the heuristics and biases tradition have often been criticized for what some thought was an overemphasis on errors, but in many research domains this is a common strategy. In fact, in many domains of life it makes more sense to seek out errors than to pursue the optimal. This is because it is often difficult to specify what the very *best* response might be, but performance *errors* are much easier to spot. Postman (1988) has argued, for instance, that educators might adopt a stance more similar to that of physicians or attorneys. He points out that doctors would find it hard to define "perfect health," but, despite this, they are quite good at spotting disease. Likewise, lawyers are much better at spotting injustice and lack of citizenship than defining "perfect justice" or ideal citizenship. The literature on the psychology of rationality has followed this logic in that it has focused on identifying thinking errors, just as physicians focus on disease. Degrees of irrationality can be assessed in terms of the number and severity of such cognitive errors. Conversely, the *avoidance* of error becomes a measure of *rational* thought. That is precisely the logic of our rational thinking assessment battery.

The Framework for the Tasks and Subtests in the CART

Table 4.1 presents the overall framework for the CART, as well as some indication of the tasks used for assessment and the assessment domains. The framework relies heavily on the taxonomy of errors developed in chapter 3 and also the explication of the logic of heuristics and biases tasks contained in that chapter (particularly figures 3.4 and 3.5, which will be elaborated in this chapter). For example, it should not be surprising that the left column of table 4.1 serves to represent tasks saturated with processing requirements. Nor should it be surprising that the second column from the left represents tasks that are relatively saturated with knowledge from specific rational thinking domains. However, the key point from the last chapter was the intertwined nature of process and knowledge in heuristics and biases tasks. All heuristics and biases tasks involve both, to various extents. Two domains of rational thinking—probabilistic and statistical reasoning and scientific reasoning—have process and knowledge so intertwined that they span both columns in table 4.1 to emphasize this point. These domains of rational thinking will be discussed in chapters 5 and 6, respectively.

The domains exclusively in the first column of table 4.1 represent subtests of the CART that have heavy processing requirements. Several of these have been interpreted by researchers as indicators of miserly information processing: a reflection versus intuition measure (similar to the CRT), syllogisms that require overcoming belief bias, an avoidance of ratio bias task, and disjunctive reasoning. Continuing down the first column of table 4.1 are some other tasks that are best viewed as indirect measures of the avoidance of miserly processing. All are heavy in their processing requirements.

Table 4.1

Framework for classifying the types of rational thinking tasks and subtests on the CART

Tasks Saturated with Processing Requirements (Detection, Sustained Override, Hypothetical Thinking)	Rational Thinking Tasks Saturated with Knowledge	Avoidance of Contaminated Mindware	Thinking Dispositions That Foster Thorough and Prudent Thought, Unbiased Thought, and Knowledge Acquisition
Probabilistic and Statistical Reasoning subtest		Superstitious Thinking subtest	Actively Open-minded Thinking scale
Scientific Reasoning subtest		Antiscience Attitudes subtest	Deliberative Thinking scale
Avoidance of Miserly Information Processing subtests: • Reflection versus Intuition • Belief Bias Syllogisms • Ratio Bias • Disjunctive Reasoning	Probabilistic Numeracy subtest	Conspiracy Beliefs subtest	Future Orientation scale
Absence of Irrelevant Context Effects in Decision Making subtests: • Framing • Anchoring • Preference Anomalies	Financial Literacy and Economic Knowledge subtest	Dysfunctional Personal Beliefs subtest	Differentiation of Emotions scale
Avoidance of Myside Bias: • Argument Evaluation subtest	Sensitivity to Expected Value subtest		
Avoiding Overconfidence: • Knowledge Calibration subtest	Risk Knowledge subtest		
Rational Temporal Discounting subtest			

All of these tasks and their associated effects, although involving miserly processing in some way, are still quite complex tasks; more than miserly processing is going on when someone answers suboptimally on them. All of these tasks still have contentious theorizing surrounding them, and investigators have put forth many alternative models of task performance. Our only theoretical claim is quite minimal—it is that, whatever else is responsible for task performance, they all have miserly processing somewhat involved. All of these tasks will be discussed in detail in chapters 7 and 8.

Finally, it is important to stress one further thing. All of the remaining subtests in the first column of table 4.1 are important measures of rational thinking in their own right, whether or not we are correct about the involvement of miserly information processing. Our focus here is not on resolving the theoretical disputes surrounding every one of these effects. For example, the measurement of overconfidence would be part of our rational thinking assessment battery regardless of what the explanation for the effect turns out to be. Its status as a component of rational thinking is what is critical for our assessment battery, not our theoretical guesses as to the source of the effect.

With that caveat in mind, the first column of table 4.1 shows several important categories of our assessment battery: the absence of irrelevant context effects in decision making; the avoidance of myside bias; the avoidance of overconfidence in knowledge calibration; and rational temporal discounting of future rewards. All of these represent important aspects of rational thinking. They will all be discussed in detail in chapter 8.

The domains exclusively in the second column of table 4.1 are four components of the CART that depend heavily on knowledge bases. This is not to say that these components are completely independent of miserly processing issues, but they are considerably less dependent on processing considerations and much more dependent on the presence of certain specific types of declarative knowledge than are other subtests. These subtests of the CART tap probabilistic numeracy, financial literacy and economic knowledge, sensitivity to expected value, and risk knowledge. They will be discussed in chapter 9.

In the previous chapter, we discussed how suboptimal thinking is potentially caused by two types of mindware problems. Missing mindware or mindware gaps reflect the most common type—where Type 2 processing does not have access to adequately compiled declarative knowledge from which to synthesize a normative response to use in the override of Type 1 processing. However, we discussed in that chapter how not all mindware is helpful or useful in fostering rationality. Indeed, the presence of certain kinds of mindware is often precisely the problem. We coined the category label "contaminated mindware" for the presence of declarative knowledge bases that foster irrational rather than rational thinking.

There are probably dozens of different kinds of contaminated mindware if one looks very specifically at narrow domains of knowledge. It would obviously be impossible for a test of rational thinking to encompass all of these. Instead, we have focused on just a few of the broader categories of contaminated mindware that might have some domain generality in their effects. Of course, rational thinking, as indicated by CART performance, is defined as the *avoidance* or *rejection* of these domains of contaminated mindware. The third column in table 4.1 lists the four categories of contaminated mindware that we assess in the CART: superstitious thinking; antiscientific attitudes; conspiracy beliefs; and dysfunctional personal beliefs. These categories of contaminated mindware will all be discussed in chapter 10.

Finally, the fourth column of table 4.1 shows a set of supplementary measures that are included in the CART but are not part of the overall rational thinking scores on the test itself. These are some thinking dispositions that we measure by self-report questionnaires. The field of psychology has identified many different thinking dispositions, as we will discuss in chapter 11; however, we have chosen those specifically relevant to rational thinking. For example, we have focused on thinking dispositions that foster prudent thought, unbiased thought, and unbiased knowledge acquisition. The four thinking dispositions that we have focused on are actively open-minded thinking, deliberative thinking, future orientation, and the differentiation of emotions.

It is again important to reiterate a point that we made about thinking dispositions in chapter 2. These self-report measures are different from the other performance measures on the CART, which is why they are not part of the overall score on the test, but instead provide supplementary information. They are not part of the total score on the test because, among other things, the maximum score on a thinking disposition measure is not to be equated with maximal rationality. Optimal functioning on these measures is traced instead by an inverted U-shaped function. Maximizing these dispositions is not the criterion of rational thought itself. Thinking dispositions such as these are a means to rationality, not ends in themselves.

Subtests on the CART in Terms of Process and Knowledge

Figure 4.1 (a fleshed-out version of figure 3.5) provides a more continuous look at how the subtests of the CART arrange themselves in the process/knowledge space. This figure has the same axes as figures 3.4 and 3.5 of the previous chapter. The x-axis indexes how knowledge dependent the task is, ranging from high knowledge dependence on the far right to low knowledge dependence on the far left. The y-axis portrays how much the task depends on conflict detection and sustained override (i.e., process) for successful performance. The entire figure thus shows the two-dimensional space of process dependence crossed with knowledge dependence.

It should not be surprising, in light of the discussion in chapter 3, that some of the most central subtests on the CART cluster at the top and to the right of figure 4.1. That a substantial number of important subtests are in the upper right quadrant is because, as discussed in chapter 3, the logic of many types of heuristics and biases tasks intertwine processing and knowledge issues.

Because figure 4.1 is a fairly dense and complicated figure, we will take a brief tour around it starting at the upper right. Here, we find the probabilistic reasoning and scientific reasoning subtests that are at the very heart of the heuristics and biases literature. For that reason, these subtests are heavily weighted in the CART scoring. These subtests are very knowledge saturated, yet they often involve conflict detection and often require sustained override procedures in order to be answered correctly.

Down the right-hand side of the figure are all of the subtests listed in table 4.1 that are very knowledge saturated. These subtests rely less on processes of conflict detection and override than do the quintessential tasks of the heuristics and biases literature—probabilistic and scientific reasoning. So, for example, numeracy and financial literacy both depend heavily on declarative knowledge, but doing well on either of these subtests is nowhere near as dependent on conflict detection and override as are more classic heuristic and biases tasks involving probabilistic reasoning. Finally, in the far lower right of the figure are the subtests that assess contaminated

A Framework for the CART

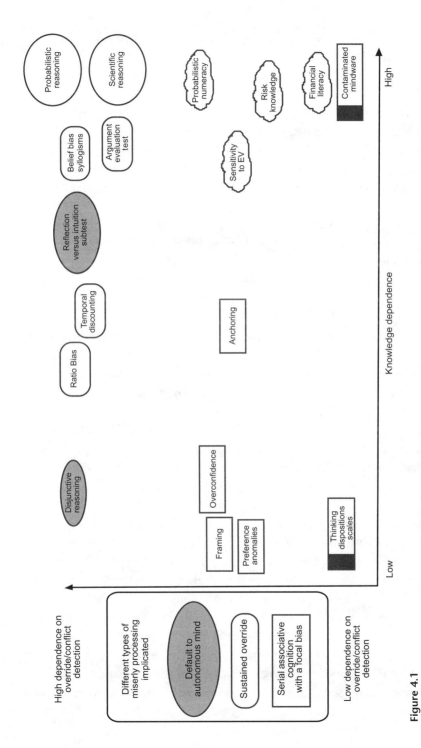

Figure 4.1
CART subtests arrayed in the process/knowledge space.

mindware. Performance on these subtests and questionnaires is largely a function of knowledge.

Moving now along the top part of the figure, we have represented a group of subtests that strongly tap miserly information processing. However, they are relatively spread out, because they differ in the degree to which they depend on acquired declarative knowledge. Cross-referencing table 4.1, we can see that four direct measures of miserly information processing are arrayed across the top of the table: disjunctive reasoning; ratio bias; the reflection versus intuition measure; and belief bias in syllogistic reasoning. The subtests are arrayed the way they are because it is conjectured that, for example, the types of tasks that are used in syllogistic reasoning paradigms are more knowledge dependent than are those used in disjunctive reasoning paradigms. However, it must be stressed that the placement of subtests in figure 4.1, and those like it in this book, are very much initial conjectures about the properties of the tasks used in these subtests.

The four direct measures of the avoidance of miserly processing that were just mentioned differ somewhat in the type of miserly processing that they tap (see table 3.1 for a list of the types of miserly processing errors). This is reflected in figure 4.1, which represents disjunctive reasoning and the reflection versus intuition measure with an oval and the ratio bias task and belief bias syllogisms with a curved rectangle. The latter denotes subtests where a miserly thinking error is most likely due to the failure of sustained override (see Bonner & Newell, 2010; Corser & Jasper, 2014; De Neys, 2006b; De Neys & Franssens, 2009; Stanovich & West, 2008b). The ovals, in contrast, denote subtests where a miserly thinking error is most likely due to defaulting to the autonomous mind. Also at the top of figure 4.1 (see also table 4.1) are subtests measuring rational temporal discounting and the Argument Evaluation subtest (a measure of myside bias).

The bottom left of figure 4.1 represents the thinking disposition scales. As discussed previously, these are self-report measures that do not depend on knowledge or the override of a conflicting Type 1 response. As mentioned above, they are not included as part of our overall rational thinking composite score. They are supplemental measures, and will be discussed in chapter 11. We include them here for completeness.

Directly above the thinking dispositions in the figure are four subtests represented by sharp-edged rectangles. Three of these subtests (all but anchoring) do not show a large degree of knowledge dependence, but they do depend more on conflict detection and override than do the thinking disposition scales. These subtests use four paradigms that tap miserly processing due to defaulting to serial associative cognition rather than full-blown simulation and hypothetical reasoning (see the discussion in chapter 3 and table 3.1). Three of these paradigms assess whether the subject can avoid irrelevant context effects in decision making: framing tasks, anchoring tasks, and tests for the presence of preference anomalies. Another indirect effect of miserly processing is the presence of overconfidence in judgment and decision making. One theoretical account of these effects (Koriat, Lichtenstein, & Fischhoff, 1980) is that they result from a type of focal bias that follows from defaulting to serial associative cognition rather than a fully disjunctive exploration of the problem space.

Again, we strongly caution that this figure contains many theoretical conjectures that await empirical confirmation. Our purpose here is to contextualize the kinds of tasks we are using in the CART. Figure 4.1 does convey the wide array of subtests used and their varying characteristics in terms of process and knowledge. It also captures the fact that most of these rational thinking tasks involve knowledge at least to some extent and processing issues at least to some extent. Thus the figure helps to drive home the point that there will be few rational thinking subtests that are "process-pure" (i.e., unique indicators of a specific process and nothing else).

Even more importantly, figure 4.1 and the associated table 4.1 illustrate the wide span of rational thinking tasks that are included in the CART. Knowledge bases of critical importance to epistemic rationality are amply represented—domains of knowledge having to do with probabilistic reasoning, numeracy, financial literacy, and risk. Testing epistemic conjectures via scientific principles is extensively examined in our scientific reasoning measure, as will be discussed in chapter 6. Regarding instrumental rationality, our test battery examines the ability to adhere to many of the important strictures of axiomatic utility theory, such as descriptive invariance, procedural invariance, maximizing expected value, and avoiding irrelevant context effects.

Table 4.2 lists the number of points allocated to each subtest of the CART. The details of turning raw CART subtest scores into CART points are discussed in the chapter that describes the specific subtest.

Table 4.2
CART points allocated to each subtest

CART Subtest	CART Points
Probabilistic and Statistical Reasoning	18
Scientific Reasoning	20
Reflection versus Intuition	10
Belief Bias in Syllogistic Reasoning	8
Ratio Bias	5
Disjunctive Reasoning	5
Framing	6
Anchoring	3
Preference Anomalies	3
Argument Evaluation Test	5
Knowledge Calibration	6
Rational Temporal Discounting	7
Probabilistic Numeracy	9
Financial Literacy and Economic Knowledge	10
Sensitivity to Expected Value	5
Risk Knowledge	3
Rejection of Superstitious Thinking	5
Rejection of Antiscience Attitudes	5
Rejection of Conspiracy Beliefs	10
Avoidance of Dysfunctional Personal Beliefs	5
Total CART Points	148

Summary and Conclusions

The CART assesses both epistemic rationality and instrumental rationality. Aspects of epistemic rationality that are assessed by our instrument include the tendency to make incoherent probability assessments, the tendency toward overconfidence in knowledge judgments, the tendency to ignore base rates, the tendency not to seek falsification of hypotheses, the tendency to try to explain chance events, the tendency to evaluate evidence with a myside bias, and the tendency to ignore the alternative hypothesis.

Additionally, the CART assesses aspects of instrumental rationality and irrationality, such as the ability to display disjunctive reasoning in decision making, the tendency to show inconsistent preferences because of framing effects, the tendency to substitute affect for difficult evaluations, the tendency to overweight short-term rewards at the expense of long-term well-being, the tendency to have choices affected by vivid stimuli, and the tendency for decisions to be affected by irrelevant context.

Figure 4.1 represents an initial effort to parsimoniously classify heuristics and biases tasks. The particular task choices displayed in table 4.1 reflect our attempt to make the CART comprehensive yet logistically tractable. As will be discussed in chapter 14, we have deliberately omitted some tasks from table 4.1, usually because of logistical constraints. The specific choices we have made for assessing rational thinking will be discussed in part 2 of the book where we take up each of the components of table 4.1.

II The Components of Rational Thought Assessed by the CART

5 Probabilistic and Statistical Reasoning

Probabilistic and statistical reasoning is one of the most thoroughly investigated areas in the heuristics and biases literature. This is not surprising because, as we outlined in chapter 1, probability assessment is central to the achievement of both epistemic and instrumental rationality. The expected utility of an action involves multiplying the probability of an outcome by its utility and summing across possible outcomes. Thus, determining the best action involves estimating the probabilities of various outcomes. These estimations are not typically conscious calculations, of course—they are beliefs about states of the world and the confidence that a person has in them.

If our probabilistic judgments about the states of the world are wrong, decision making will not maximize one's utility—our actions will not result in our getting what we most want. It is in this sense that rationality of belief—epistemic rationality—is one of the foundations for rationality of action. Rationality of belief is assessed by looking at a variety of probabilistic reasoning skills and statistical thinking skills, which are assessed on the subtest of the CART that will be discussed in this chapter.

To calibrate one's probabilistic beliefs rationally, it is not necessary to be a calculating genius; but, as with axiomatic utility theory, a few qualitative principles must be followed—the rules of the so-called probability calculus. For our purposes, the rules of the probability calculus culminate in the most important probability rule for assessing this type of reasoning: Bayes' rule.

Bayesian Reasoning

For us to be epistemically rational, our probability estimates must follow the rules of objective probabilities. Most of these rules are quite intuitive. Here are a few of the most important:

Probabilities vary between 0 and 1. So $0 \leq P(A) \leq 1$, where $P(A)$ is the probability of event A.

If an event is certain to happen, then its probability is 1.0. So $P(A) = 1$ when A is certain.

If an event is certain *not* to happen, then its probability is 0. So $P(A) = 0$ when A is certain not to happen.

If event A and event B cannot *both* happen, they are said to be mutually exclusive. When event A and event B are mutually exclusive, then the probability of one *or* the other occurring is the probability of each added together:

$$P(A \text{ or } B) = P(A) + P(B)$$

Decision theorists have shown that if our probability judgments do not follow the simple rules of probability just outlined, there are certain untoward consequences (the "money pump" and the "Dutch Book"; see Maher, 1993, Osherson, 1995, Resnik, 1987, and Schick, 1986). If you feel that the probability that the New England Patriots will win the next Super Bowl is 0.25 and you feel that the probability that the Detroit Lions will win the next Super Bowl is 0.10, then you had better also think that the probability of the Patriots *or* the Lions winning is 0.35. If you violate this stricture you will not be epistemically rational, and any action that you take on the basis of these probabilities will be suboptimal—it will not maximize your expected utility.

Continuing with our rules of the probability calculus, an important concept is that of conditional probability. Conditional probabilities concern the probability of one event (A) given that another (B) has occurred, $P(A/B)$. When A and B are mutually exclusive, then $P(A/B) = 0$, because if B has occurred A cannot (and likewise, $P(B/A) = 0$). However, when A and B are not mutually exclusive, the formula for the conditional probability is:

$$P(A/B) = \frac{P(A \text{ and } B)}{P(B)}$$

Note that, in general, P(A/B) is not necessarily the same as P(B/A), because the formula for the latter has a different denominator:

$$P(B/A) = \frac{P(A \text{ and } B)}{P(A)}$$

We can, however, write one of the conditional probabilities in terms of the other. When we do, after a little simple algebra we come to one of the most famous theorems in decision theory, Bayes' theorem, sometimes called Bayes' rule (discovered by the Reverend Thomas Bayes of Tunbridge Wells, England, in the eighteenth century; see Stigler, 1983, 1986):

$$P(A/B) = \frac{P(A)*P(B/A)}{P(A)*P(B/A) + P(\sim A)*P(B/\sim A)}$$

The formula has only one term we have not seen before, ~A, which means not A. Thus, P(~A) is the probability of event A not occurring.

All of these rules of probability are important, but for judgment and decision making, Bayes' rule has special salience. Bayes' theorem is used as the formal standard for the important task of belief updating—how we should update our belief in a particular hypothesis based on new evidence that is relevant to the hypothesis. All we have to do to see this is to substitute the A and B in the formula with two fundamental concepts: the focal hypothesis under investigation (labeled H) and a set of collected data that are relevant to the hypothesis (labeled D). Then we have:

$$P(H/D) = \frac{P(H)*P(D/H)}{P(H)*P(D/H) + P(\sim H)*P(D/\sim H)}$$

In the formula, P(H) is the probability estimate that the focal hypothesis is true *prior* to collecting the data, and P(~H) is the probability estimate that the *alternative* hypothesis (i.e., ~H) is true prior to collecting the data. Additionally, a number of conditional probabilities come into play. For example, P(H/D) represents the probability that the focal hypothesis is true *subsequent* to, or given, the data pattern actually observed (this is sometimes termed the *posterior* probability). P(D/H) is the probability of observing that particular data pattern given that the focal hypothesis is true, and P(D/~H) is the probability of observing that particular data pattern given that the alternative hypothesis is true. It is important to realize that P(D/H) and P(D/~H) are *not* complements (they do not add up to 1.0). The data might be likely, given both the focal and alternative hypothesis, or unlikely, given both the focal and alternative hypotheses.

People often have trouble following the strictures of Bayes' rule, as we shall see in several examples in this chapter and the next. However, it is important to emphasize that when we say that people make mistakes in Bayesian reasoning, we are not saying that they are *calculating* incorrectly, or getting precise figures incorrect. Instead, what is meant is that people are making rather large qualitative mistakes in their assessment of probabilities. The scoring of most of the Bayesian reasoning items on the CART requires only that the respondent get in the right ballpark of the correct answer. There is no requirement that the subject actually know Bayes' rule. The only requirement is that the subject's probability estimates are roughly in accord with what the rule prescribes. In short, it is whether people have a *feel* for Bayesian thinking—whether they are sensitive to the right variables and respond in the right general direction to evidence presented. This will become clearer as we discuss specific problems. Formal Bayesian statistics involve calculation to be sure, but to escape the thinking errors surrounding probability, a person needs only to have learned the *conceptual* logic of how correct thinking about probabilities works.

We will now detail some of the important areas in which heuristics and biases researchers have uncovered errors in probabilistic reasoning. All of these areas, plus one or two others, will be tapped by items on the Probabilistic and Statistical Reasoning subtest of the CART. Again, referring to figure 4.1 and table 4.1, we would reiterate the point that items on this subtest strongly implicate knowledge, but they also strongly implicate conflict detection, override, and hypothetical thinking.

The fact that the items on this subtest tap process as well as knowledge is specifically intended (as it was in the original heuristics and biases literature) and is not a flaw. It is a designed feature, not a drawback. In this domain of rational thinking, we are interested in individual differences in *sensitivity* to probabilistic reasoning principles. People can have knowledge of these principles without having the propensity to use them. That is, they can have the knowledge but not the propensity to see situations in terms of probabilities. A typical heuristics and biases task, like the items on our subtest, will pit a statistical way of viewing a problem against a nonstatistical way of viewing a problem in order to see which kind of thinking dominates in the situation. So, for example, we would not design an item where the subject chooses between a nine out of ten chance of winning versus a three out of ten chance of winning, with no other context provided. Such a problem would be one of pure probabilistic numeracy (more like the Probabilistic Numeracy subtest that we will describe in chapter 9). Instead, on most of our subtest items, statistical information is given, but also a *nonstatistical* way of thinking about the problem. People who may get the pure mathematics of statistical reasoning correct might not see certain problems themselves as probabilistic. It is just this variance in sensitivity to seeing a problem as probabilistic that we want to assess.

We will now proceed to detail some of the common errors in probabilistic reasoning that are assessed in the CART.

Problems with Probabilities: Base Rate Neglect

Assigning the right probability values to events is a critical aspect of rational thought. Interestingly, research has shown that people are sometimes quite good at dealing *implicitly* with probabilistic information (when it needs only to be tracked by the autonomous mind). However, when probabilities must be reasoned about *explicitly*, people often have considerable difficulty. The difficulties that people have with probabilistic information are illustrated in two examples of problems that have received intense scrutiny. The first, the so-called cabs problem (Bar-Hillel, 1980; Koehler, 1996; Lyon & Slovic, 1976; Macchi, 1995; Tversky & Kahneman, 1982), has been the subject of over three decades of research:

A cab was involved in a hit-and-run accident at night. Two cab companies, the Green and the Blue, operate in the city in which the accident occurred. You are given the following facts: 85 percent of the cabs in the city are Green and 15 percent are Blue. A witness reported that the cab in the accident was Blue. The court tested the reliability of the witness under the same circumstances that existed on the night of the accident and concluded that the witness correctly identified each of the two colors 80 percent of the time. What is the probability (expressed as a percentage ranging from 0 percent to 100 percent) that the cab involved in the accident was Blue?

Subjects are simply asked to give a "guesstimate" on the problem—they are told that they do not have to calculate the answer precisely. The point is not to get the precise answer, so much as to see whether people are in the right ballpark. The answers of many people are not.

In the cab problem, Bayes' theorem provides the optimal way of combining the two pieces of information that have been given:

1. Overall, 15 percent of the cabs are Blue.
2. A witness (whose identification accuracy is 80 percent) identified the cab in question as Blue.

Most people do not naturally combine the two pieces of information optimally. In fact, many people are surprised to learn that the probability that the cab is Blue is 0.41, and that, despite the witness's identification, it is still more likely that the cab involved in the accident was Green (0.59) rather than Blue (0.41). The reason is that the prior probability that the cab is Green (85 percent) is higher than the credibility of the witness's identification of Blue (80 percent). Without using the formula, we can see how the probability of 0.41 is arrived at. In 100 accidents of this type, 15 of the cabs will be Blue, and the witness would identify 80 percent of them (12) as Blue. Furthermore, out of 100 accidents of this type, 85 of the cabs will be Green, and the witness will identify 20 percent of them (17) as Blue. Thus, 29 cabs (12 + 17) will be identified as Blue, but only 12 of them will actually be Blue. The proportion of cabs identified as Blue that actually are Blue is 12 out of 29, or 41 percent.

In terms of Bayes' rule, here is how the calculation goes:

$$P(H/D) = \frac{P(H)P(D/H)}{P(H)P(D/H) + P(\sim H)P(D/\sim H)}$$

$$P(H/D) = \frac{(.15)(.80)}{(.15)(.80) + (.85)(.20)} = .41$$

Less than half of the subjects given this problem produce answers between 0.20 and 0.70 (Stanovich & West, 1998c, 1999). Most answer around 0.80—in short, they answer with the figure indicating the witness's accuracy without discounting this figure (as they should) because the prior probability (0.15) is quite low. That is, most people greatly overestimate the probability that the cab is Blue. They overweight the witness's identification and underweight the base rate, or prior probability, that the cab is Blue. This is an example of a tendency to overweight concrete and vivid single-case information when it must be combined with more abstract probabilistic information.

Consider another such problem, one that shares the logic of the cabs problem but is more relevant to everyday life. It concerns the estimation of medical risk, and it too has been the focus of considerable research (Casscells, Schoenberger, & Graboys, 1978; Cosmides & Tooby, 1996; Sloman, Over, Slovak, & Stibel, 2003; Stanovich & West, 1999):

Imagine that the XYZ virus causes a serious disease that occurs in one in every 1,000 people. Imagine also that there is a test to diagnose the disease that always indicates correctly that a person who has the XYZ virus actually has it. Finally, imagine that the test has a false-positive rate of 5 percent. This means that the test wrongly indicates that the XYZ virus is present in 5 percent of the cases where the person does not have the virus. Imagine that we choose a person randomly and administer the test, and that it yields a positive result (indicating that the person is XYZ-positive). What is the probability (expressed as a percentage ranging from 0 percent to 100 percent) that the individual actually has the XYZ virus, assuming that we know nothing else about the individual's personal or medical history?

The most common answer is 95 percent. The correct answer is approximately 2 percent! People vastly overestimate the probability that a positive result truly indicates the XYZ virus because of the same tendency to overweight the case information and underweight the base rate information that we saw in the cabs problem. Although the correct answer to this problem can again be calculated by Bayes' rule, some simple numerical reasoning can help to illustrate the profound effect that base rates have on probabilities. We were given the information that, of 1,000 people, just one will actually be XYZ-positive. If the other 999 (who do not have the disease) are tested, the test will indicate incorrectly that approximately 50 of them have the virus (0.05 multiplied by 999) because of the 5 percent false-positive rate. Thus, of the 51 patients testing positive, only one (approximately 2 percent) will actually be XYZ-positive. In short, the base rate of XYZ is so low that the vast majority of people do not have the virus. This fact, combined with a substantial false-positive rate, ensures that, in absolute numbers, the majority of positive tests will be of people who do not have the virus.

In terms of Bayes' rule, here is how the calculation goes:

$$P(H/D) = \frac{P(H)P(D/H)}{P(H)P(D/H) + P(\sim H)P(D/\sim H)}$$

$P(H/D) = (.001)(1.0)/[(.001)(1.0) + (.999)(.05)] = .001/.001 + .04995 = .0196$

In both of these problems many people display a tendency to overweight individual-case evidence and to underweight statistical information. The case evidence (the witness's identification, the laboratory test result) seems "tangible" and "concrete" to most people—it is more vivid. In contrast, the probabilistic evidence seems, well—probabilistic! This reasoning is of course fallacious because case evidence itself is always probabilistic. A witness can make correct identifications with only a certain degree of accuracy, just as a clinical test misidentifies the presence of a disease with a certain probability. The situation is one in which two probabilities must be combined if one is to arrive at a correct decision: the probable diagnosticity of the case evidence and the prior probability.

The problems presented thus far are often termed "noncausal base rates"—those involving base rates with no obvious causal relationship to the criterion behavior (Ajzen, 1977; Barbey & Sloman, 2007; Bar-Hillel, 1980, 1990; Koehler, 1996; Tversky & Kahneman, 1982). The *causal* variant of the cabs problem substitutes for the first fact (that 85 percent of cabs in the city are Green) the fact that "although the two companies are roughly equal in size, 85% of cab accidents in the city involve Green cabs and 15% involve Blue cabs" (Tversky & Kahneman, 1982, p. 157). In this version, the base rates seem more relevant to the probability of the Green cab being involved in the accident. People are more prone to use causal base rates when estimating probabilities than noncausal ones.

Another type of causal base rate problem is structured so that the participant has to make an inductive inference in a simulation of a real-life decision. The information relevant to the decision is conflicting and of two different types. One type of evidence is statistical: either probabilistic or aggregate base rate information that favors one of the bipolar decisions. The other type of evidence is a concrete case or personal experience that points in the opposite direction. The classic "Volvo versus Saab" item (Fong, Krantz, & Nisbett, 1986) provides an example. In this problem, a couple are deciding to buy one of two otherwise equal cars. Consumer surveys, statistics on repair records, and polls of experts favor the Volvo over

Probabilistic and Statistical Reasoning

the Saab. However, a friend reports experiencing a severe mechanical problem with his Volvo. The participant is asked to provide advice to the couple. Preference for the Volvo indicates a tendency to rely on the large-sample information in spite of salient personal testimony. A preference for the Saab indicates reliance on the personal testimony over the opinion of experts and the large-sample information. The Probabilistic and Statistical Reasoning subtest of the CART contains both causal and noncausal base rate items.

In all versions of these base rate problems, the situation is one in which two probabilities, the probable diagnosticity of the case evidence and the prior probability, must be combined if one is to arrive at a correct decision. The right way is to use Bayes' rule, or more specifically, the *insight* from Bayes' rule—that the diagnosticity of the evidence must be amalgamated with the base rate. We do not wish to imply in this discussion of Bayesian reasoning that people do, or should, always explicitly calculate using the Bayesian formula in their minds. It is enough that people learn to "think Bayesian" in a qualitative sense—that they have what might be called "Bayesian instincts." It is enough, for example, simply to realize the importance of the base rate. That would allow a person to see the critical insight embedded in the XYZ virus problem—that when a test with a substantial false alarm rate is applied to a disease with a very small base rate, then the majority of individuals with a positive test will not have the disease.

In short, the issue is whether people's natural judgments of probabilities follow—to an order of approximation—the dictates of the theorem. It is understood that people making probabilistic judgments are making spontaneous "guesstimates." This is the sense in which we employ the phrase "following Bayes' rule." The probability judgments of people might be described as consistent with Bayes' rule without them having *any* knowledge of the formula or being aware of any conscious calculation. All such items on this CART subtest require only that the respondent's answer be in the right ballpark—not that it be precisely correct.

Problems with Probabilities: The Conjunction Fallacy

Another problem that is famous in the literature of cognitive psychology is the so-called Linda problem. The Linda problem was first investigated by Tversky and Kahneman (1983). The literature on it is voluminous (e.g., Dulany & Hilton, 1991; Girotto, 2004; Mellers, Hertwig, & Kahneman,

2001; Politzer & Macchi, 2000; Tentori & Crupi, 2012; Tentori, Crupi, & Russo, 2013):

Linda is thirty-one years old, single, outspoken, and very bright. She majored in philosophy. As a student, she was deeply concerned with issues of discrimination and social justice, and also participated in antinuclear demonstrations. Please rank the following statements by their probability, using 1 for the most probable and 8 for the least probable.

a. Linda is a teacher in an elementary school. ____
b. Linda works in a bookstore and takes Yoga classes. ____
c. Linda is active in the feminist movement. ____
d. Linda is a psychiatric social worker. ____
e. Linda is a member of the League of Women Voters. ____
f. Linda is a bank teller. ____
g. Linda is an insurance salesperson. ____
h. Linda is a bank teller and is active in the feminist movement. ____

Most people make what is called a "conjunction error" on this problem by rating h as more probable than c or f. Alternative h (Linda is a bank teller and is active in the feminist movement) is the conjunction of alternatives c and f. Thus, the probability of h cannot be higher than that of either c (Linda is active in the feminist movement) or f (Linda is a bank teller). All feminist bank tellers are also bank tellers, so h cannot be more probable than f—yet over 80 percent of the subjects in studies rate alternative h as more probable than f, thus displaying a conjunction error.

When subjects answer incorrectly on this problem, it is often because they have engaged in attribute substitution (see Kahneman & Frederick, 2002, 2005). Attribute substitution occurs when a person needs to assess attribute A, but finds that assessing attribute B (which is correlated with A) is cognitively easier and so uses B instead. In simpler terms, attribute substitution amounts to substituting an easier question for a harder one. In this case, rather than think carefully and see the Linda problem as a probabilistic scenario, subjects instead answer on the basis of a simpler similarity assessment. A "feminist bank teller" seems to overlap more with the description of Linda than does the alternative "bank teller."

Of course, logic dictates that the subset-superset relationship should trump assessments of similarity when judgments of probability are at issue. That is, the category "feminist bank teller" is a subset of the category "bank teller." Nevertheless, even though most subjects see this logic when it is pointed out to them, for many, probabilistic thinking is not instantiated well enough to trump the easier strategy of similarity assessment.

This is the sense in which, as discussed above, we need to emphasize that heuristics and biases tasks measure people's *sensitivity* to probabilities—not merely numerical reasoning with probabilities when they are *told* that the problem involves probabilities. This point is often ignored by critics of the heuristics and biases literature who fail to appreciate the conceptual history behind how these tasks were originally designed. It is common to hear such critics claim something like: "But you can't fault the subject for their response, because the subject didn't actually *see* the problem as involving probability. The subject just saw it as a similarity task and matched feminist bank teller to Linda's description." Such critics fail to understand that the subject not seeing the problem as involving probabilities is precisely the thinking defect that the original heuristics and biases researchers sought to reveal with this problem! From our standpoint of designing a subtest for the CART, we are likewise trying to measure individual differences in the likelihood with which different people bring to bear probabilistic notions to problems they encounter. It is precisely the difference between a subject who sees the Linda problem as involving probabilities and the subject who does not that we wish to measure and quantify.

Problems with Probabilities: The Gambler's Fallacy

Consider the following two problems:

Problem A: Imagine that we are tossing a fair coin (a coin that has a 50–50 chance of coming up heads or tails) and it has just come up heads five times in a row. For the sixth toss, do you think that:

____ It is more likely that tails will come up than heads.
____ It is more likely that heads will come up than tails.
____ Heads and tails are equally probable on the sixth toss.

Problem B: When playing slot machines, people win something 1 out of every 10 times. Julie, however, has just won on her first three plays. What are her chances of winning the next time she plays? ____ out of ____

These two problems probe whether a person is prone to the so-called gambler's fallacy (or to its companion, belief in the "hot hand")—the tendency for people to see links between events in the past and events in the future when the two are really independent (Ayton & Fischer, 2004; Barron & Leider, 2010; Burns & Corpus, 2004; Croson & Sundali, 2005; Roney & Sansone, 2015; Xu & Harvey, 2014). Two outcomes are independent when the occurrence of one does not affect the probability of the other. Most games of chance that use proper equipment (no "crooked" roulette wheels, loaded dice, or fake coins, etc.) have this property. For example, the number that comes up on a roulette wheel is independent of the outcome that preceded it. Half the numbers on a roulette wheel are red and half are black (for purposes of simplification, we will ignore the green zero and double zero), so the odds are even (0.50) that any given spin will come up red. Yet after five or six consecutive reds, many bettors switch to black, thinking that it is now more likely to come up. This is the gambler's fallacy: acting as if previous outcomes affect the probability of the next outcome when the events are independent. In this case, the bettors are wrong in their belief. The roulette wheel has no memory of what happened previously. Even if fifteen reds in a row come up, the probability of red's coming up on the next spin is still 0.50. In problem A above, some people think that it is more likely that either heads or tails will come up after five heads. The correct answer is that heads and tails are equally probable on the sixth toss. Likewise, on problem B, the correct answer is 1 out of 10, but some subjects deviate from this stricture.

The gambler's fallacy (or belief in the "hot hand") is not restricted to the inexperienced. Research has shown that even habitual gamblers still display belief in the gambler's fallacy (Petry, 2005; Xu & Harvey, 2014; Wagenaar, 1988). In fact, research has shown that individuals with gambling problems are more likely to believe in the gambler's fallacy than are control subjects (Toplak et al., 2007).

It is important to realize that the gambler's fallacy is not restricted to games of chance. It operates in any domain in which chance plays a substantial role—that is, in almost everything. The genetic makeup of babies is an example. Psychologists, physicians, and marriage counselors often see couples who, after having two female children, are planning a third child because "We want a boy, and it's bound to be a boy this time." This, of course, is the gambler's fallacy. The probability of having a boy

Probabilistic and Statistical Reasoning

(approximately 50 percent) is exactly the same after having two girls as it was in the beginning (approximately 50 percent). The two previous girls make it no more likely that the third baby will be a boy.

Problems with Probabilities: Failure to Use Sample Size Information

Consider these two problems, developed by Tversky and Kahneman (1974):

1. A certain town is served by two hospitals. In the larger hospital, about 45 babies are born each day, and in the smaller hospital, about 15 babies are born each day. As you know, about 50 percent of all babies are boys. However, the exact percentage varies from day to day. Sometimes it is higher than 50 percent, sometimes lower. For a period of one year, each hospital recorded the days on which more than 60 percent of the babies born were boys. Which hospital do you think recorded more such days?

a. The larger hospital
b. The smaller hospital
c. About the same

In this problem, the majority of people answer "about the same." People not choosing this alternative pick the larger and the smaller hospital with about equal frequency. The correct answer is the smaller hospital, but approximately 75 percent of subjects given this problem answer incorrectly. These incorrect answers result from an inability to recognize the importance of sample size in the problem. Other things being equal, a larger sample size always more accurately estimates a population value. Thus, on any given day, the larger hospital, with its larger sample size, will tend to have a proportion of births closer to 50 percent. Conversely, a small sample size is always more likely to deviate from the population value. Thus, the smaller hospital will have more days on which the proportion of births displays a large discrepancy from the population value (60 percent boys, 40 percent boys, 80 percent boys, etc.).

Consider another problem illustrating the importance of sample size:

Imagine an urn filled with balls, two-thirds of which are of one color and one-third of which are of another. One individual has drawn 5 balls from the urn and found that 4 are red and 1 is white. Another individual has drawn 20 balls and found that 12 are red and 8 are white. Which of the two individuals should feel more confident that the urn contains two-thirds red balls and one-third white balls, rather than vice versa? What odds should each individual give?

In this problem, most people feel that the sample of 5 balls provides more convincing evidence that the urn is predominantly red. Actually, the probabilities are in the opposite direction. The odds are 8 to 1 that the urn is predominantly red for the 5-ball sample, but they are 16 to 1 that the urn is predominantly red for the 20-ball sample. Even though the proportion of red balls is higher in the 5-ball sample (80 percent versus 60 percent), this is more than compensated for by the fact that the other sample is four times as large and, thus, is more likely to be an accurate estimate of the proportions in the urn. The judgment of most subjects, however, is dominated by the higher proportion of red in the 5-ball sample and does not take adequate account of the greater reliability of the 20-ball sample.

The sample-size problem in our Probabilistic and Statistical Reasoning subtest is like those in the heuristics and biases literature in that the respondent has to recognize that an issue of sampling is involved. As in the hospital problem discussed above, the subject must see the problem as involving samples and also recognize the implications of differential sample sizes. This again is how heuristics and biases tasks differ from pure statistical knowledge measures. We could simply ask the subject if larger or smaller sample sizes were better, as we might in a university statistics course. This would be a much easier question than a traditional heuristics and biases item, because there would be no issue of recognition involved—the respondent would not have to detect the *relevance* of sample size. The subject would be focused on its relevance by being asked directly about it. This of course makes for a much easier problem. As statistics instructors ourselves, we have commonly seen that a student may answer a direct question about sample size correctly in a multiple choice format, but then, when given something like the hospital problem, does not perceive the relevance of sample size and answers incorrectly.

Problems with Probabilities: Probability Matching

The probabilistic contingency experiment has many versions in psychology (Fantino & Esfandiari, 2002; Gal & Baron, 1996; Gao & Corter, 2015;

Koehler & James, 2014; Newell et al., 2013; Shanks, Tunney, & McCarthy, 2002; Tversky & Edwards, 1966). In one version, the participant sits in front of two lights (one red and one blue) and is asked to predict which of the lights will be flashed on each trial and is also told that there will be several dozen such trials. Participants are often paid money for correct predictions. The experimenter has actually programmed the lights to flash randomly, with the provision that the red light will flash 70 percent of the time and the blue light 30 percent of the time. Participants do quickly recognize that the red light flashes more often, and they predict that the red light will flash on more trials than the blue light. Most often, they switch back and forth, predicting the red light roughly 70 percent of the time and the blue light roughly 30 percent of the time. This strategy of probability matching is suboptimal because it ensures that, in this example, the participant will correctly predict only 58 percent of the time ($0.7 \times 0.7 + 0.3 \times 0.3$) compared to the 70 percent hit rate that could be achieved by predicting the more likely color (red) on each trial.

An alternative procedure is to ask subjects about their global strategy, instead of requiring trial-by-trial responses. In these paradigms, verbal problems are presented with the frequencies of hypothetical outcomes either directly given or easily inferable from the outset. Gal and Baron (1996), for example, asked participants what global strategy they would use when betting on the most likely outcome for each roll of a die in a hypothetical game. In this game, a die with four red faces and two green faces was to be rolled several times. Approximately two-thirds of college students failed to use the maximizing strategy of predicting the most probable color for each roll of the die (Gal & Baron, 1996).

Similarly, West and Stanovich (2003) used two different probabilistic choice tasks and found that the utility-maximizing strategy of choosing the most probable alternative was not the majority response. In a story-problem version of a probabilistic choice task where participants chose from among five different strategies, the maximizing response and the probability-matching response were each selected by a similar number of students (roughly 35 percent of the sample selected each). In a more continuous trial-by-trial task, the utility-maximizing response was chosen by only one-half as many subjects as the probability-matching response.

West and Stanovich (2003) also found that, in both versions of the task, participants preferring the utility-maximizing response were significantly higher in cognitive ability than participants showing a probability-matching tendency. Likewise, Koehler and James (2010) found that maximizers scored higher on Frederick's (2005) cognitive reflection measure than did matchers. Rakow, Newell, and Zougkou (2010) found that higher working-memory capacity (strongly related to cognitive ability) was associated with the maximizing strategy, rather than matching.

These outcomes converge interestingly with the findings of Koehler and James (2009), that the nonnormative "probability matchers rate an alternative strategy (maximizing) as superior when it is described to them" (p. 123). Similarly, Koehler and James (2010) found that in a direct comparison of the two strategies, maximizing was endorsed by a majority of subjects after it was described to them. Also, probability matching is widely observed in the animal world (Gallistel, 1990), suggesting it originates in Type 1 processing. Thus, although there have been disputes about whether the maximizing strategy should really be considered normative in this task, the bulk of the evidence now available supports the standard practice of considering matching responses to be nonnormative and maximizing strategies to be normative (Koehler & James, 2014; West & Stanovich, 2003).

Previous Work with Tasks Measuring Probabilistic and Statistical Reasoning

From the very beginning of our work on individual differences in rational thought, we have included various probabilistic reasoning tasks in the performance batteries used in our lab (see table 1.1). Sometimes we have used several items to measure a single domain such as sensitivity to causal base rates. In other studies, we have only used a single item to measure a particular domain and then have amalgamated a variety of the domains into a composite measure of probabilistic reasoning performance.

For example, in one of our first studies (Stanovich & West, 1998c) we used seven items to form a sensitivity to causal base rates measure. One finding from this study that replicated several times was that this measure correlated with several different components of scientific thinking, which we will discuss in chapter 6 (Stanovich & West, 1998c, 1998d). This measure shows relationships with some of the measures of miserly processing, to be discussed in chapter 7 (Kokis et al., 2002; Stanovich & West, 1998c). It also displays a significant correlation with cognitive ability in the range of 0.30–.45 (Kokis et al., 2002; Stanovich & West, 1998c). Interestingly though, we have consistently found that some of the thinking dispositions, to be discussed in chapter 11, can predict variance in causal base rate usage after the variance due to cognitive ability has been partialled out.

In addition to causal base rates, we have studied many of the tasks discussed earlier in this chapter. Several composite variables of performance on heuristics and biases tasks that we have used (e.g., Toplak et al., 2011; West et al., 2008) include items tapping sample size, the gambler's fallacy, conjunction effects, probability matching, base rate usage, and regression to the mean. We have studied various aspects of probabilistic reasoning among children as well (Kokis et al., 2002; Toplak et al., 2014b), and also have occasionally studied special populations such as pathological gamblers (Toplak et al., 2007). These previous studies, plus a half dozen or more pilot investigations, formed the basis for our choice of items on the Probabilistic and Statistical Reasoning subtest of the CART.

The Probabilistic and Statistical Reasoning Subtest of the CART

The eighteen items on the Probabilistic and Statistical Reasoning subtest of the CART (see the appendix for sample items) were distributed as follows: four items assess the ability to avoid probability-matching tendencies and instead choose a maximizing strategy; five items assess the ability to avoid the gambler's fallacy; four items assess the ability to properly assign probabilities to conjunctions; three items assess sensitivity to base rates; one item assesses sensitivity to sample size considerations; and one item assesses the ability to see regression to the mean as an explanation of performance changes.

To assess the tendency toward probability matching versus maximizing, we used two formats common in this research tradition—a trial-by-trial selection procedure and an overall strategy choice procedure. In assessing base rate sensitivity, we used two causal base rate problems that had base rates implied but not numerically specified. We also used one noncausal base rate problem where the base rate was numerically specified. Our conjunction problems used both frequency formats (see Sloman et al., 2003) and probability estimation formats.

Each item on this subtest is scored as correct or incorrect, resulting in total raw scores that vary from 0 to 18. The subtest was run in an unpublished study in our laboratory, RT58,[1] involving 486 university students as subjects (approximately 70 percent were paid $30 for their participation and the rest were subject pool volunteers). The mean raw score was 10.6 (SD = 3.3). The reliability of our Probabilistic and Statistical Reasoning subtest was 0.73 (Cronbach's alpha). Items varied widely in difficulty, from 18.7 percent correct (on the noncausal base rate problem) to 91.2 percent correct (on one of the gambler's fallacy items).

In the same study, we examined the correlation of the Probabilistic and Statistical Reasoning subtest with other measures. These correlations are displayed in the first column of the correlation matrix presented in table 5.1.

The Probabilistic and Statistical Reasoning subtest displayed a correlation of 0.41 with SAT total scores and an even higher correlation of 0.51 with the eleven-item Reflection versus Intuition subtest of the CART described in chapter 7.

The Probabilistic and Statistical Reasoning subtest displayed significant correlations with three of the four measures of contaminated mindware discussed in chapter 10: −0.33 with the Superstitious Thinking subtest; −0.40 with the Antiscience Attitudes; and −.30 with the Conspiracy Beliefs subtest ($p < 0.001$ in all cases). Only the Dysfunctional Personal Beliefs subtest (omitted from the table) failed to correlate significantly with the Probabilistic and Statistical Reasoning subtest (−0.03).

Table 5.1
Correlations between the probabilistic and statistical reasoning subtest and other variables in RT58

	1	2	3	4	5	6	7
1. Probabilistic Reasoning							
2. SAT Total	.41						
3. Reflection vs. Intuition	.51	.49					
4. Superstitious Thinking	−.33	−.24	−.28				
5. Antiscience Attitudes	−.40	−.37	−.38	.42			
6. Conspiracy Beliefs	−.30	−.26	−.31	.45	.37		
7. Actively Open-Minded Thinking	.39	.27	.33	−.41	−.63	−.32	
8. Deliberative Thinking	.33	.28	.34	−.12	−.41	−.12	.53

Note: All of the correlations are statistically significant. Ns vary from 458 to 485.

The Probabilistic and Statistical Reasoning subtest displayed significant correlations with two of the four supplemental thinking dispositions assessed on the CART (see chapter 11): 0.39 with the Actively Open-Minded Thinking scale; and 0.33 with Deliberative Thinking subtest ($p < 0.001$ in both cases). The Future Orientation scale (0.06) and the Differentiation of Emotions scale (0.03) both failed to correlate significantly with the Probabilistic and Statistical Reasoning subtest.

Because previous investigations have found that males tend to score better in probabilistic reasoning and probabilistic numeracy than females (Gal & Baron, 1996; Weller et al., 2013; West & Stanovich, 2003), we analyzed scores on our Probabilistic and Statistical Reasoning subtest for these differences. Consistent with previous findings in the literature, we found that the mean score of the 181 males (11.7, SD = 3.5) was significantly higher than the mean score of the 305 females (10.0, SD = 3.0), $t(484) = 5.55$, $p < 0.001$, Cohen's $d = 0.521$. However, the males in our sample were also higher in cognitive ability. Of those reporting SAT total scores, we found that the mean score of the 172 males (1,185, SD = 139) was significantly higher than the mean score of the 295 females (1,127, SD = 136), $t(465) = 4.47$, $p < 0.001$. The correlation of –0.24 between sex and performance on the Probabilistic and Statistical Reasoning subtest was reduced to –0.18 when SAT total score was controlled but still remained significant at the 0.001 level.

As indicated in the appendix, the number of CART points on this subtest corresponds to the raw score on the subtest. We will present more data on the associations displayed by this subtest in chapters 12 and 13 when we report studies of a short-form version and full-form version of the CART.

6 Scientific Reasoning

Although they are central to rational thinking, the probabilistic and statistical judgment tasks discussed in the previous chapter still miss some areas of the heuristics and biases literature that are of immense importance. Many of these areas have to do with the evaluation of evidence, hypothesis formation, and theory testing. Our Scientific Reasoning subtest encompasses these domains.

Despite tapping different skills, the Scientific Reasoning subtest discussed in this chapter does, however, share certain characteristics with the Probabilistic and Statistical Reasoning subtest in one important way. Both strongly implicate knowledge, but they also strongly implicate conflict detection, override, and hypothetical thinking. As with probabilistic reasoning, the fact that the items on this subtest tap process as well as knowledge is *intended* and is not a flaw. People can have knowledge of the principles of scientific inquiry (control groups, falsification) without the propensity to use them, or without the propensity to see situations in terms of scientific thinking. A typical heuristics and biases task attempts to tap the propensity as well as the knowledge. So, for example, we would not include an item in our subtest where the subject simply *defines* the term "control group," as in a college course final exam. Instead, we might pit control group thinking against confounded group thinking in a task that does not make this distinction salient. People who may give an accurate definition might well tend not to see certain problems as *requiring* control group thinking. It is just this variance in sensitivity to seeing a problem in *need* of scientific thinking that we want to assess.

Our Scientific Reasoning subtest does assess Bayesian reasoning, but in a slightly different manner than the probabilistic reasoning tasks discussed in chapter 5. Whereas those tended to focus on the proper amalgamation of base rate information, scientific reasoning tasks tend to focus on the likelihood ratio of Bayes' rule. To understand what the likelihood ratio is, we have to introduce an alternate form of Bayes' theorem.

We have seen in the previous chapter how the failure to weight diagnostic evidence with the base rate probability is an error of Bayesian reasoning. However, sometimes it is not the base rate that is the problem but the processing of the data that should lead to belief updating. To illustrate this thinking error, we will utilize a different form of Bayes' rule—one arrived at by simple mathematical transformation. The formula in chapter 5 was written in terms of the posterior probability of the focal hypothesis (H) given a new datum (D). It is of course possible to write the formula in terms of the posterior probability of the *nonfocal* hypothesis (~H) given a new datum (D):

$$P(\sim H/D) = \frac{P(\sim H)*P(D/\sim H)}{P(\sim H)*P(D/\sim H) + P(H)*P(D/H)}$$

By dividing the two formulas we can arrive at the most theoretically transparent form of Bayes' formula (see Fischhoff & Beyth-Marom, 1983)—one that is written in so-called odds form:

$$\frac{P(H/D)}{P(\sim H/D)} = \frac{P(H)}{P(\sim H)} * \frac{P(D/H)}{P(D/\sim H)}$$

In this ratio, or odds form, from left to right, the three ratio terms represent the posterior odds favoring the focal hypothesis (H) after receipt of the new data (D); the prior odds favoring the focal hypothesis; and the so-called likelihood ratio (LR), composed of the probability of the data given the focal hypothesis divided by the probability of the data given the alternative hypothesis. Specifically:

posterior odds = $P(H/D)/P(\sim H/D)$
prior odds = $P(H)/P(\sim H)$
likelihood ratio = $P(D/H)/P(D/\sim H)$

The formula tells us that the odds favoring the focal hypothesis (H) after receipt of the data are arrived at by multiplying together the other two terms—the likelihood ratio and the prior odds favoring the focal hypothesis:

posterior odds favoring the focal hypothesis = prior odds × LR

Scientific Reasoning

We will now proceed to discuss several aspects of scientific thinking that are not assessed by the probabilistic reasoning measures described in chapter 5. However, because we have just introduced the odds form of Bayes' rule, we will begin with a reasoning error that is best understood in the context of this form of the rule. The error is that, when evaluating the diagnosticity of evidence (that is, the likelihood ratio: P(D/H)/P(D/~H)), people often fail to appreciate the relevance of the denominator term, P(D/~H). They fail to see the necessity of evaluating the probability of obtaining the data observed if the focal hypothesis were *false*.

Problems in Scientific Reasoning: Ignoring P(D/~H)

A large research literature has demonstrated that the tendency to ignore the probability of the evidence given that the nonfocal hypothesis is true—P(D/~H)—is a ubiquitous psychological tendency. For example, Doherty and Mynatt (1990) used a simple paradigm in which subjects were asked to imagine that they were a doctor examining a patient with a red rash. They were shown four pieces of evidence and were asked to choose which pieces of information they would need in order to determine whether the patient had the disease "Tigirosa."[1] The four pieces of information were:

The percentage of people with Tigirosa.
The percentage of people without Tigirosa.
The percentage of people with Tigirosa who have a red rash.
The percentage of people without Tigirosa who have a red rash.

These pieces of information corresponded to the four terms in the Bayesian formula: P(H), P(~H), P(D/H), and P(D/~H). Because P(H) and P(~H) are complements, only three pieces of information are necessary to calculate the posterior probability. However, P(D/~H)—the percentage of people who have a red rash among those *without* Tigirosa—clearly must be selected because it is a critical component of the likelihood ratio in Bayes' formula. Nevertheless, 48.8 percent of the individuals who participated in the Doherty and Mynatt (1990) study failed to select the P(D/~H) card (45.7 percent in the Stanovich & West, 1998d, study). Thus, to many subjects presented with this problem, the people with a red rash but without Tigirosa do not seem relevant—they seem (mistakenly) to be a nonevent.

The importance of P(D/~H) often seems counterintuitive. People have to be taught that it is important or else their default is to ignore it. Consider the following problem:

Imagine yourself meeting David Maxwell. Your task is to assess the *probability that he is a university professor* based on some information that you will be given. This will be done in two steps. At each step you will get some information that you may or may not find useful in making your assessment. After each piece of information, you will be asked to assess the probability that David Maxwell is a university professor.

Step 1: You are told that David Maxwell attended a party in which 25 male university professors and 75 male business executives took part, 100 people all together. Question: What do you think the probability is that David Maxwell is a university professor? ___ percent

Step 2: You are told that David Maxwell is a member of the Bears Club. 70 percent of the male university professors at the above-mentioned party were members of the Bears Club, and 90 percent of the male business executives at the party were members of the Bears Club. Question: What do you think the probability is that David Maxwell is a university professor? ___ percent

This problem is used in studies to assess whether people can deal correctly (or at all) with the P(D/~H) information (Beyth-Marom & Fischhoff, 1983; Stanovich & West, 1998d). The first step is simple. The probability of the focal hypothesis is 0.25, because 25 of the 100 are university professors. It is step two that is the tricky one. It might seem that, because 70 percent of the male university professors were members of the Bears Club and that this percentage is greater than 50 percent, the probability that David Maxwell is a university professor should go up—that it should now be judged to be higher than the base rate of 25 percent. But that would be making the error of ignoring P(D/~H). In fact, it is *more* likely that a business executive is a member of the Bears Club. Being a member of the Bears Club is more diagnostic of being a *business executive* than a university professor, so it actually *lowers* the probability of the latter. The likelihood ratio here is less than 1 (0.70/.90), so the odds against David Maxwell being a university professor, after the information about the Bears Club is received, get worse—from 1 to 3 against (0.25/.75) to:

posterior odds = likelihood ratio × prior odds
posterior odds = (0.70/.90) × (0.25/.75)
posterior odds = 0.175/.675 = 1 to 3.86

In terms of the probability version of Bayes' rule, the proper Bayesian adjustment is from 0.25 in step 1 to 0.206 in step 2 [(.70 × 0.25)/(.70 × 0.25 + 0.90 × 0.75)]. In a study that our research group ran using this problem (Stanovich & West, 1998d), we found that only 42 percent of the sample moved their probability assessments in the right direction and lowered it from 0.25 after receiving the evidence. Many subjects raised their probabilities after receiving the evidence, indicating that their focus was on the relatively high value of P(D/H)—.70—and that they failed to contextualize this conditional probability with the even higher value of P(D/~H).

Here is a parallel problem that is tricky, but in a slightly different way, and it again tests whether people understand the implications of the likelihood ratio, and more specifically, the importance of P(D/~H):

Imagine yourself meeting Mark Smith. Your task is to assess the *probability that he is a university professor* based on some information that you will be given.

Step 1: You are told that Mark Smith attended a party in which 80 male university professors and 20 male business executives took part, 100 people all together. Question: What do you think the probability is that Mark Smith is a university professor? ____

Step 2: You are told that Mark Smith is a member of the Bears Club. 40 percent of the male university professors at the above mentioned party were members of the Bears Club, and 5 percent of the male business executives at the party were members of the Bears Club. Question: What do you think the probability is that Mark Smith is a university professor? ____

In this problem, reliance on the base rate at step 1 would result in an estimate of 0.80. Step 2 is structured so that although the likelihood ratio is considerably greater than 1 (0.40/.05), P(D/H) is less than 0.50. This might suggest to someone ignoring P(D/~H)—which is in fact lower than P(D/H)—that these data should *decrease* the probability that David is a university professor. In fact, the proper Bayesian adjustment is from 0.80 in step one to 0.97 in step two ([.40 × 0.80]/[.40 × 0.80 + 0.05 × 0.20]). Any adjustment upward from step 1 to step 2 would suggest that the subject had been attentive to P(D/~H). Moving in the right direction is all that is necessary to show that one is a Bayesian thinker. However, in the study that our research group ran using this problem (Stanovich & West, 1998d), we found that only 30 percent of the sample moved their probability assessments in the right direction, that is, raised it from 0.80 after receiving the evidence. Note that these problems provide examples of the point made in the last chapter—that Bayesian reasoning is signaled simply by the qualitative movement in the right *direction*; there is no requirement that the subject actually use Bayes' rule or that he or she make a precisely accurate calculation. When scoring problems of this type on the CART, we require only that the subject moves in the right direction for the response to be scored as correct.

The failure to attend to the alternative hypothesis—to the denominator of the likelihood ratio when receiving evidence—is not a trivial reasoning error. Paying attention to the probability of the observation under the alternative hypothesis is a critical component of clinical judgment in medicine and many other applied sciences. It is the reason we use control groups. It is essential to know what would have happened if the variable of interest had not been changed. Both clinical and scientific inference is fatally compromised if we have information about only the treated group.

Problems in covariation detection paradigms (included in our Scientific Reasoning subtest) also reflect the tendency for people to ignore essential comparative (control group) information—the equivalent of ignoring P(D/~H). For example, in a much researched covariation detection paradigm (Levin, Wasserman, & Kao, 1993; Shanks, 1995; Wasserman, Dorner, & Kao, 1990), subjects are shown a 2 × 2 matrix summarizing the data from an experiment examining the relation between a treatment and patient response:

	Condition improved	No improvement
Treatment given	200	75
No treatment	50	15

The numbers in the matrix represent the number of people in each cell. Specifically, 200 people were given the treatment and their condition

Scientific Reasoning

improved, 75 people were given the treatment and no improvement occurred, 50 people were not given the treatment and their condition improved, and 15 people were not given the treatment and no improvement occurred. In covariation detection experiments, subjects are asked to indicate whether the treatment was effective. Many think that the treatment in this example is effective. They focus on the large number of cases (200) in which improvement followed the treatment. Secondarily, they focus on the fact that more people who received treatment showed improvement (200) than showed no improvement (75). Because this probability (200/275 = 0.727) seems high, subjects are enticed into thinking that the treatment works. Such an approach ignores the probability of improvement given that treatment was *not* given. Since this probability is even higher (50/65 = 0.769), the particular treatment tested in this experiment can be judged to be completely *ineffective*. The tendency to ignore the outcomes in the no-treatment cells and focus on the large number in the treatment/improvement cell seduces many people into viewing the treatment as effective when in fact it is not.

Hypothesis Testing and Falsifiability

Just as people have difficulty learning to assess data in light of an alternative hypothesis, people have a hard time thinking about evidence and tests that could falsify their focal hypotheses. Instead, people tend to seek to confirm theories rather than falsify them. Performance on Wason's (1966) four-card selection task is often interpreted as indicating this tendency. The task was described in chapter 2 and it has been investigated in dozens, if not hundreds, of studies (Evans, 1972, 1996, 2007a; Evans, Newstead, & Byrne, 1993; Johnson-Laird, 1999, 2006; Klauer, Stahl, & Erdfelder, 2007; Oaksford & Chater, 1994, 2007; Stanovich & West, 1998a). Recall that the abstract version of the problem is often presented as follows. Imagine four rectangles, each representing a card lying on a table. Each one of the cards has a letter on one side and a number on the other side. Here is a rule: "If a card has a vowel on its letter side, then it has an even number on its number side." Two of the cards are letter-side up, and two of the cards are number-side up. The subject's task is to decide which card or cards must be turned over in order to find out whether the rule is true or false. The four cards confronting the subject have the stimuli K, A, 8, and 5 showing.

The correct answer is A and 5 (the only two cards that could show the rule to be false), but the majority of subjects answer, incorrectly, A and 8 (showing a so-called matching bias). The next most common response selection is the A card only. The 8 and 5 cards are the difficult choices. Many people get these two cards wrong. They mistakenly think that the 8 card must be chosen. More importantly, the 5 card, which most people do not choose, is absolutely essential but is often not chosen. The 5 card might have a vowel on the back and, if it did, the rule would be shown to be false because all vowels would not have even numbers on the back. In short, to show that the rule is not false, the 5 card must be turned.

Although many alternative theories exist (see Evans, 2007a; Johnson-Laird, 1999, 2006; Klauer, Stahl, & Erdfelder, 2007; Oaksford & Chater, 1994, 2007; Stanovich, 1999), one of the oldest theories that certainly plays at least some role in the poor performance on the selection task is that people focus on confirming the rule. This is why they turn the 8 card (in hopes of confirming the rule by observing a vowel on the other side) and turning the A card (in search of the confirming even number). What they do not set about doing is selecting the card that would falsify the rule—a thought pattern that would immediately suggest the relevance of the 5 card (which might contain a disconfirming vowel on the back). Although people might perform poorly for many reasons, regardless of which of these descriptive theories explains the error, there is no question that a concern for falsifiability would rectify the error.

Understanding How Science Converges on an Explanation

The principle of converging evidence describes how research results are synthesized in science: No one experiment is definitive, but each helps us to rule out at least some alternative explanations and, thus, aids in the process of homing in on the truth. The use of a variety of methods makes scientists more confident that their conclusions rest on a solid empirical foundation.

Research is highly convergent when a series of experiments consistently supports a given theory while collectively eliminating the most important competing theory. Although no single experiment can rule out all alternative explanations, taken collectively, a series of partially diagnostic experiments can lead, if the data patterns line up in a certain way, to a strong conclusion. For example, suppose that five different theoretical accounts (call them A, B, C, D, and E) of a given set of phenomena exist at one time and are investigated in a series of experiments. Suppose that one experiment represents a strong test of theories A, B, and C, and that the data largely refute theories A and B and support C. Imagine also that another experiment is a particularly strong test of theories C, D, and E, and that the data largely refute theories D and E and support C. In such a situation, we would have strong converging evidence for theory C. Not only do we have data supportive of theory C, but we have data that contradict its major competitors. No one experiment tests all the theories, but taken together, the entire set of experiments allows a strong inference.

By contrast, if *both* experiments represented strong tests of B, C, and E, and the data of both experiments strongly supported C and refuted B and E, the overall support for theory C would be less strong than in our previous example. The reason is that, although we have data supporting theory C, we have no strong evidence ruling out two viable alternative theories (A and D). Thus, research is highly convergent when a series of experiments consistently supports a given theory while collectively eliminating the most important competing explanations.

Previous Work with Tasks Measuring Scientific Reasoning

From the very beginning of our work on individual differences in rational thought, we have included various scientific reasoning tasks in the performance batteries used in our lab, including some obvious ones not discussed here. For example, it will be unsurprising that items tapping the ability to avoid inferring causation from correlational evidence are part of the Scientific Reasoning subtest of the CART. Sometimes we have used a variety of items to measure a single domain such as covariation detection, for example. In other studies, we have only used a single item to measure a particular domain and then have amalgamated the domains into a composite measure of performance on a variety of scientific reasoning items.

For example, in some of our first rational thinking studies (Stanovich & West, 1998a, 1998c, 1998d) we examined covariation detection skill, efficient hypothesis testing (isolating variables), sensitivity to P(D/~H), and falsification tendencies in the four-card selection task. We found that all of these skills covaried significantly. These measures also showed relationships with some of the measures of miserly processing (to be discussed in chapter

7) (Sá et al., 1999; Stanovich & West, 1998c, 1998d). They also displayed significant correlations with cognitive ability (Stanovich & West, 1998c, 1998d, 1999, 2008b).

We have used several composite variables of performance on these tasks (often mixing them with probabilistic reasoning items) in our previous studies (e.g., Stanovich & West, 1998c; Toplak et al., 2011; West et al., 2008). We have studied various aspects of scientific reasoning in special populations such as pathological gamblers (Toplak et al., 2007). These previous studies, plus a half dozen or more pilot investigations, formed the basis for our choice of items on the Scientific Reasoning subtest of the CART.

The Scientific Reasoning Subtest of the CART

The Scientific Reasoning subtest (see the appendix for sample items) receives a maximum of 20 points on the CART. Seventeen of those 20 points were distributed across seventeen items as follows: four items tapping falsification tendencies in the four-card selection task (two deontic and two nondeontic); two items tapping knowledge of the logic of converging evidence; three items tapping the tendency to avoid drawing causal inferences from correlation evidence; five items tapping the tendency to accurately assess the likelihood ratio by processing $P(D/\sim H)$; and three problems that tap the tendency to use control-group reasoning. The remaining three points on the subtest were derived from a twenty-five-item measure of covariation detection ability (like that used by Stanovich & West, 1998d).

We have discussed most of these areas of scientific reasoning earlier in this chapter. One of the three control-group reasoning problems was inspired by the work of Lehman, Lempert, and Nisbett (1988) and two were inspired by the work of Tschirgi (1980). In the latter study, subjects were given vignettes in which a story character observed an outcome and had to test a hypothesis about the importance of variables to an outcome. The subject is asked to choose one of three ways to test this hypothesis the next time the situation occurs. The three alternatives correspond to three alternative hypothesis testing strategies: changing all the variables (CA), holding one thing at a time constant (HOLDONE), and vary one thing at a time (VARYONE). The CA strategy is clearly nonnormative. However, in this simplified situation, both the HOLDONE and VARYONE strategies are equally normative (see Baron, 1985, pp. 142–144). Therefore, we scored both the HOLDONE and VARYONE responses as correct.

The five items tapping the tendency to accurately assess the likelihood ratio by processing P(D/~H) used three paradigms: a choice paradigm like that of Doherty and Mynatt (1990), discussed previously; probability estimation items like the David Maxwell problem, discussed previously; and two questions were derived using a 2 × 2 matrix summarizing the data from an experiment examining the relation between a treatment and patient response, also discussed previously. In the latter, we used numbers in the matrix that had the logic of the above example. If subjects were ignoring P(D/~H), then the relatively large number (200) in the A cell would suggest an efficacious treatment. But if subjects were paying attention to P(D/~H)—that is, 50/65—then they would be able to see that the treatment is actually inefficacious.

Three points on the Scientific Reasoning subtest of the CART were derived from a twenty-five-item measure of covariation detection ability. Judging event interrelationships is a critical component of much thinking in the everyday world and has been the subject of much investigation (see Allan, 1993; Alloy & Tabachnik, 1984; Cheng & Novick, 1992). We used a paradigm where people are given covariation information that is presented in a 2 × 2 contingency table. In one of the classic studies using this paradigm, Wasserman, Dorner, and Kao (1990) devised a situation where the subject had to determine whether a particular drug improved the condition of psoriasis. The contingency information concerned the number of rats who had or had not been given the drug and the number who had improved or not improved. We adapted this paradigm by adding to each problem the context of a real-life issue.[2] Subjects were then asked to evaluate the degree of association between two variables in the data of twenty-five research studies. For example, in one problem the subjects were presented with data on whether couples who live together before marriage tend to have successful marriages. Subjects were told to imagine that a researcher had sampled 225 couples and found the following:

	Successful marriages	Divorced
Lived together	100	50
Did not live together	25	50

The subject then judged the extent of the relationship between living together before marriage and successful marriages in these data on a scale ranging from −10 (strong negative association) to +10 (strong positive association) and centered on 0 (no association). The remaining twenty-four problems dealt with a variety of hypotheses (e.g., that secondary smoke is associated with lung problems in children; that eating spicy foods is associated with stomach problems; that early birth order is associated with high achievement; that getting chilled is associated with catching a cold; that watching violent television is associated with violent behavior).

Previous experiments have indicated that subjects weight the cell information in the order: cell A > cell B > cell C > cell D (from upper left to lower right, reading across)—with cell D reliably receiving the least weight and/or attention (see Kao & Wasserman, 1993; Levin et al., 1993; Wasserman et al., 1990). The tendency to ignore cell D is nonnormative, as indeed is any tendency to differentially weight the four cells. The normatively appropriate strategy (see Allan, 1980; Kao & Wasserman, 1993; Shanks, 1995) is to use the conditional probability rule—subtracting from the probability of the target hypothesis when the indicator is present, the probability of the target hypothesis when the indicator is absent. Numerically, the rule amounts to calculating the Δp statistic of conditional probability: $[A/(A+B) - C/(C+D)]$ (see Allan, 1980). For example, the Δp value for the problem presented above is +.333, indicating a moderately strong positive association. The Δp values used in the 25 problems ranged from −0.600 (strong negative association) to 0.600 (strong positive association), and were a set of those used by Wasserman et al. (1990), but the absolute numbers in the cells were different. The main measure of each subject's performance on this task was the correlation of their individual 25 contingency judgments with the 25 Δp values.

In an unpublished study in our laboratory (RT58, the same study discussed in chapter 5) involving 486 university students as subjects, the mean of the subject's individual correlations between Δp and their individual contingency judgments was 0.540 (SD = 0.289) and the median correlation was 0.598. It is important to note that this mean of the individual correlations is substantially lower than the aggregate correlation between Δp and the mean item evaluation (0.93 collapsed across all 486 subjects).

As indicated in the appendix, the covariation detection task was weighted with a maximum of 3 points. Correlations of contingency judgments with

$\Delta p \geq 0.75$ received 3 points; correlations ≥ 0.45 and < 0.75 received 2 points; correlations ≥ 0.10 and < 0.45 received 1 point; and correlations < 0.10 received no points.

The 3 points allotted to covariation detection means that the Scientific Reasoning subtest has a maximum of 20 points. Like the Probabilistic and Statistical Reasoning subtest, the raw score on the Scientific Reasoning subtest and its CART points are the same. In the unpublished study mentioned previously, the mean raw score on this subtest was 10.7 (SD = 3.4). The reliability of our Scientific Reasoning subtest was 0.66 (Cronbach's alpha). Items varied widely in difficulty, from 22.4 percent correct (on the nondeontic abstract four-card selection task problem) to 92.6 percent correct (on one of the three control-group reasoning problems).

In the same study, we examined the correlation of the Scientific Reasoning subtest with other measures. Those correlations are displayed in table 6.1. The Scientific Reasoning subtest displayed a substantial 0.61 correlation with the Probabilistic and Statistical Reasoning subtest discussed in chapter 5. It displayed a correlation of 0.44 with SAT total scores and an even higher correlation of 0.55 with the eleven-item Reflection versus Intuition subtest of the CART described in chapter 7 ($p < 0.001$ in all cases).

The Scientific Reasoning subtest displayed significant correlations ($p < 0.001$ in all cases) with three of the four measures of contaminated mindware discussed in chapter 10: –0.35 with the Superstitious Thinking subtest; –0.43 with the Antiscience Attitudes; and –0.29 with the Conspiracy Beliefs

Table 6.1

Correlations between the scientific reasoning subtest and other variables in RT58

	Scientific Reasoning
Probabilistic Reasoning	.61
SAT Total	.44
Reflection vs. Intuition	.55
Superstitious Thinking	–.35
Antiscience Attitudes	–.43
Conspiracy Beliefs	–.29
Actively Open-Minded Thinking	.42
Deliberative Thinking	.37

Note: All of the correlations are statistically significant at the .001 level. Ns vary from 467 to 486.

subtest ($p < 0.001$ in all cases). The Scientific Reasoning subtest displayed significant correlations with two of the four supplemental thinking dispositions assessed on the CART (see chapter 11): 0.42 with the Actively Open-Minded Thinking scale; and 0.37 with Deliberative Thinking subtest.

Because there were sex differences on the Probabilistic and Statistical Reasoning subtest (see chapter 5), we also analyzed scores on the Scientific Reasoning subtest for these differences. We found that the mean score of the 181 males (11.2, SD = 3.6) was significantly higher than the mean score of the 305 females (10.3, SD = 3.2), $t(484) = 2.82$, $p < 0.001$. However, the magnitude of the difference was lower on the Scientific Reasoning subtest than on the Probabilistic and Statistical Reasoning subtest (Cohen's d = 0.264 versus 0.521). Similar effects sizes of 0.25, 0.32, and 0.38 will be reported in studies discussed in chapters 12 and 13.

As noted in chapter 5, the males in our sample were also higher in cognitive ability. In that chapter, we described how the sex difference on the Probabilistic and Statistical Reasoning subtest remains after the difference in SAT total scores was controlled. This was not the case with the Scientific Reasoning subtest. Here, the significant correlation of –0.13 ($p < 0.01$) between sex and performance on the Scientific Reasoning subtest was reduced to a nonsignificant –0.05 when SAT total score was controlled.

We will present more data on the associations displayed by this subtest in chapters 12 and 13, when we report studies of a short-form version and full-form version of the CART.

7 Avoidance of Miserly Information Processing: Direct Tests

In the previous two chapters, we have discussed components of the CART that are classic heuristics and biases tasks in that they heavily implicate detection of conflicting responses, override of intuitive responses, and the presence of mindware. That is, as can be seen in figure 4.1 (where these tasks occupy the upper right-hand corner), superior performance on the Probabilistic and Statistical Reasoning subtest as well as on the Scientific Reasoning subtest requires rational processing strategies as well as specific types of declarative knowledge. This is why these two domains span the two left columns of table 4.1. In this chapter and the next, we will introduce the components of the CART that comprise the far left column of table 4.1. That column represents tasks that are not as balanced in their processing and knowledge components as are probabilistic and scientific reasoning. Performance on these tasks depends much more heavily on conflict detection, override abilities, and hypothetical thinking than on the presence of specific knowledge bases.

In this chapter, we will take up four tasks that provide some of the more direct measures of the avoidance of miserly processing that have been introduced into the literature. The tasks are a reflection versus intuition measure; a syllogistic reasoning task with belief bias; a task assessing the avoidance of ratio bias; and a disjunctive reasoning measure. As is clear from figure 4.1, these tasks all have heavy processing requirements despite varying quite a bit in their dependence on knowledge.

The Reflection versus Intuition Subtest

We designed the Reflection versus Intuition subtest for the CART based on the original cognitive reflection measure introduced by Frederick (2005).

His Cognitive Reflection Test (CRT) was originally designed to measure the tendency to override a prepotent response alternative that is incorrect and to engage in further reflection that leads to the correct response. We discussed this three-item measure, in particular the bat-and-ball problem, in chapter 3. The bat-and-ball item was the quintessential item from the CRT and was first discussed by Kahneman and Frederick (2002) in a paper that reframed the heuristics and biases literature in terms of the concept of attribute substitution. This problem and the two others in the subtest (the widget problem and the lilypad problem) seem at first glance to be similar to the well-known insight problems in the problem solving literature, but they in fact display a critical difference. Classic insight problems (see Gilhooly & Fioratou, 2009; Gilhooly & Murphy, 2005) do not usually trigger an attractive alternative response. Instead, the participant sits lost in thought trying to reframe the problem correctly, as in, for example, the classic nine dot problem. Frederick's three cognitive reflection items became of great interest to researchers working in the heuristics and biases tradition because they required that a strong intuitive response be overridden.

In short, cognitive reflection problems fit the logic of heuristics and biases tasks, as articulated in chapter 3, seemingly in a fairly pure form. That is, such problems seemingly keep the influence of stored knowledge rather low (although in figure 4.1 we classify our version as intermediate in this respect). Because it seems a relatively potent measure of conflict detection, it has not been surprising that Frederick's measure has correlated with many rational thinking tasks, especially those deriving from the heuristics and biases tradition. Frederick (2005) observed that with as few as three items, his CRT could predict performance on measures of temporal discounting, the tendency to choose high expected-value gambles, and framing effects. Other studies have found cognitive reflection measures to be significantly associated with expected value choices; probabilistic reasoning; the endorsement of profit maximizing strategies; the avoidance of the illusion of explanatory depth; nonsuperstitious thinking; performance calibration; the avoidance of vividness effects; and general numeracy (Baldi et al., 2013; Cokely et al., 2012; Fernbach et al., 2013; Koehler & James, 2010; Liberali et al., 2012; Mata, Ferreira, & Sherman, 2013; Moritz et al., 2013; Pennycook et al., 2012; Shenhav, Rand, & Greene, 2012).

Our own studies of Frederick's three-item measure have been among the most comprehensive in the literature. Toplak, West, and Stanovich (2011) formed a composite variable of fifteen separate rational thinking tasks from many different domains in the heuristics and biases literature. Our composite measure was heavily weighted with probabilistic reasoning items, but contained some scientific reasoning items (e.g., covariation detection) as well as items tapping the avoidance of miserly processing (e.g., framing, ratio bias). The three-item CRT displayed a correlation of 0.42 with this rational thinking composite. It was a better predictor of rational thinking than either measures of intelligence or measures of executive functioning. Several of the regression analyses conducted indicated that the three-item CRT could predict rational thinking performance independent not only of intelligence but also of executive functioning and thinking dispositions. In fact, in all of the analyses, the CRT by itself accounted for more unique variance than the block of cognitive ability measures (intelligence and executive function)—an astoundingly predictive performance for a three-item measure! This is why we included such a cognitive reflection measure as part of our operationalization of rational thinking in the CART.

In a follow-up study, Toplak, West, and Stanovich (2014a) found a correlation between the three-item CRT and a somewhat different rational thinking composite score (one weighted more toward miserly thinking than toward probabilistic reasoning) of 0.52. This again was a higher correlation than that achieved by an intelligence measure (the Wechsler Abbreviated Scales of Intelligence, WASI).

Cognitive reflection measures do display moderate correlations with measures of intelligence. Frederick (2005) reported a correlation of 0.44 between three-item CRT performance and SAT total scores, as well as a 0.43 correlation between CRT scores and performance on the Wonderlic IQ test (see also Obrecht, Chapman, & Gelman, 2009).

The studies listed in table 7.1 are all studies in which we were experimenting with different items and versions of a cognitive reflection measure with a view toward creating a subtest for the CART. The items varied from scale to scale across the studies. Occasionally, we also examined the composite of the original three-item scale to give us an anchor during the development of new items. The cognitive ability measures were varied but included the Wechsler Abbreviated Scales of Intelligence, executive function measures, Raven's Matrices, self-reported SAT total scores, verified SAT total scores, and the Wonderlic IQ test. As can be seen from the table, the correlations ranged from 0.31 to 0.61 (18 of 23 correlations were between 0.35 and 0.55). The samples in table 7.1 were all university students.

Table 7.1
Correlations between versions of the cognitive reflection measures used in our lab and cognitive ability

Study	Cognitive Ability Measure	Cognitive Reflection Measure	Sample Size	Correlation
Toplak, West, & Stanovich (2014a)	Wechsler Abbreviated Scales of Intelligence (WASI)	7-item scale	160	.50
Toplak, West, & Stanovich (2014a)	WASI	original 3-item scale	160	.48
Toplak, West, & Stanovich (2014a)	WASI	new 4-item scale	160	.41
West, Meserve, & Stanovich (2012)	reported SAT total score	original 3-item scale	482	.45
Toplak, West, & Stanovich (2011)	WASI	original 3-item scale	346	.32
Toplak, West, & Stanovich (2011)	WASI + Executive function composite	original 3-item scale	346	.40
RT57	reported SAT total score	10-item scale	371	.45
RT56	reported SAT total score	8-item scale	238	.48
RT56	Raven's Matrices	8-item scale	162	.43
RT55	reported SAT total score	7-item scale	146	.43
RT55	verified SAT total score	7-item scale	101	.61
RT55	n-back task	7-item scale	142	.35
RT54	reported SAT total score	7-item scale	407	.43
RT54	verified SAT total score	7-item scale	345	.54
RT54	n-back task	7-item scale	396	.31
RT53	reported SAT total score	7-item scale	264	.37
RT53	Wonderlic IQ Test	7-item scale	264	.32
RT51	reported SAT total score	7-item scale	439	.37

Table 7.1 (continued)

Study	Cognitive Ability Measure	Cognitive Reflection Measure	Sample Size	Correlation
RT50	reported SAT total score	7-item scale	371	.41
RT50	Raven's Matrices	7-item scale	371	.37
RT48	reported SAT total score	7-item scale	181	.56
RT48	Raven's Matrices	7-item scale	294	.49
RT47	reported SAT total score	original 3-item scale	482	.45

Note: RT labels are the labels of unpublished studies from our lab.

The magnitude of these correlations (keeping in mind that they are attenuated given the restricted range) is perhaps surprising; it certainly seems on their face that most cognitive reflection items do not put extreme stress on sustained decoupling. It may be, though, that intelligence is related to conflict detection in this task. Evans (2007b) has labeled two of the possible loci of intelligence associations as the quality and quantity hypotheses. First, individuals higher in cognitive ability may be more likely to compute the correct response given that they have engaged in Type 2 processing (what Evans calls the "quality hypothesis" regarding cognitive ability). The second (what Evans calls the "quantity hypothesis") is that individuals higher in cognitive ability are more likely to see the need for Type 2 processing. The quality hypothesis might not be much implicated in CRT performance, but the quantity hypothesis certainly is—thus boosting the correlation with intelligence.

What Evans's (2007b) account leaves out, however, is the third locus: missing mindware. The presence of mindware might well be related to cognitive ability, creating an association with intelligence. In chapter 3, we speculated that some researchers may have underestimated the role of mindware in cognitive reflection. Cognitive reflection measures are often treated as pure measures of miserly processing. We have not made this assumption, however, as indicated in figure 4.1, where we posit that these measures are at least moderately dependent on mindware. For example, in chapter 3 we argued that even the simple bat-and-ball problem will be affected by the differential instantiation of numeracy skills. That some people find math calculations to be second nature while others do not will affect how easy the problem is. Cognitive reflection measures are probably robust predictors of performance on heuristics and biases tasks because they tap many of the different processing components discussed in chapter 3. Many investigators have suggested that the task is a complex indicator.

Investigations of the underlying processing components of cognitive reflection are still ongoing (Stupple, Gale, & Richmond, 2013), but it is now clear that whatever such tasks tap, it appears to be central to many aspects of thinking rationally (Frederick, 2005; Toplak et al., 2011, 2014a). Nevertheless, there are problems on the horizon for Frederick's three-item CRT. The items are becoming extremely well known—especially the famous bat-and-ball item. The latter is used in countless classroom demonstrations, and it has appeared in many magazines and famous books—most notably Daniel Kahneman's rightly lauded and extensively reviewed *Thinking, Fast and Slow* (2011). From the standpoint of reliability, three items is obviously too few. Finally, in some populations, the overall score on the three-item version might be floored. Frederick (2005) reported the mean performance on the three items across a variety of academic institutions and found that, for example, students at Michigan State University and Bowling Green State University on average got fewer than one item out of three correct. The mean for the University of Toledo was just 0.57. Clearly, using the three-item version in high schools and community colleges will be problematic in terms of floor effects.

For the purposes of the CART, the three-item CRT is clearly inadequate (see Primi et al., 2015). However, we have been working on an alternative version for some time, and it has become what we term the Reflection versus Intuition subtest of the CART. The history of our eleven-item Reflection versus Intuition subtest begins with the Toplak et al. (2014a) study, where we showed that it was possible to construct a seven-item cognitive reflection measure, one that includes the original three items reported by Frederick (2005) and four others without the extensive research track record of the original problems. Of the four new items in Toplak et al. (2014a), one is retained in the CART, and revised analogues of the other three were also used in the CART. In that study, we examined the ability of a seven-item version (CRT7) to predict performance on seven rational thinking tasks from the heuristics and biases literature and whether the four new items (CRT4) add to the variance explained by the original three (CRT3).

To situate the seven-item version within the overall space of individual differences, we also assessed cognitive ability (intelligence) and a variety of thinking dispositions.

Most indications were that the CRT4 items could be combined with the CRT3 items. Toplak et al. (2014a) found that the median correlation among the seven items of CRT7 was 0.27 and Cronbach's alpha was a substantial 0.72. The CRT7 was a more potent predictor of a rational thinking composite score (made up of several heuristics and biases tasks) than was either cognitive ability or thinking dispositions. After the latter variables were entered into the regression equation, CRT7 still explained substantial unique variance (11.4 percent).

The CRT7 did seem to be an improvement on the CRT3. There may be a variety of reasons why this is so. First, the longer measure should be more reliable. Also, as discussed earlier, the CRT3 is known to be a hard test and scores on it are very low even among elite student populations (Frederick, 2005). The CRT4, on the other hand, is at least somewhat easier than the CRT3. The mean probability of answering an item correctly on the CRT3 was 0.17, whereas the mean probability of answering an item correctly on the CRT4 was 0.24. The CRT3 and CRT4 had a substantial 0.58 correlation.

The CRT7 score had moderately sizable correlations (0.57 and 0.42, respectively) with two other tasks (belief bias syllogisms and ratio bias) that are also direct measures of miserly processing (see table 4.1). Table 7.2 presents the correlations between versions of cognitive reflection measures and both belief bias and ratio bias (to be discussed later in this chapter) in several of our published and unpublished studies. Also included in the table are correlations with a version of a disjunctive reasoning subtest that is included in the CART (also to be discussed later in this chapter).

The Reflection versus Intuition subtest of the CART contains eleven items (see the appendix for sample items). We included in this subtest analogues of the problems used in CRT4 of Toplak et al. (2014a). The remaining seven items were new, but some were based on the same strategies used by Frederick (2005) to construct the original CRT. Others were devised by us or inspired by the problem solving literature (but with the caution to try to avoid insight problems that do not prime an alternative response).

We investigated this eleven-item version in an unpublished study with the lab code RT58 in which 477 subjects (70 percent of them paid for their participation) completed the Reflection versus Intuition subtest. The mean

Table 7.2
Correlations between versions of cognitive reflection tasks and other measures of rational thinking with heavy detection, override, and hypothetical thinking requirements

Study	Rational Thinking Measure	Cognitive Reflection Measure	Sample Size	Correlation
Correlations with Belief Bias Syllogisms				
Toplak, West, & Stanovich (2014a)	Belief Bias in Syllogistic Reasoning	7-item scale	160	.57
Toplak, West, & Stanovich (2014a)	Belief Bias in Syllogistic Reasoning	original 3-item scale	160	.55
Toplak, West, & Stanovich (2014a)	Belief Bias in Syllogistic Reasoning	new 4-item scale	160	.48
Toplak, West, & Stanovich (2011)	Belief Bias in Syllogistic Reasoning	original 3-item scale	346	.36
RT55	Belief Bias in Syllogistic Reasoning	7-item scale	144	.36
RT54	Belief Bias in Syllogistic Reasoning	7-item scale	407	.40
RT51	Belief Bias in Syllogistic Reasoning	7-item scale	439	.34
RT50	Belief Bias in Syllogistic Reasoning	7-item scale	371	.37
RT48	Belief Bias in Syllogistic Reasoning	7-item scale	294	.42
Correlations with Ratio Bias				
Toplak, West, & Stanovich (2014a)	Ratio Bias (5 items)	7-item scale	160	.42
Toplak, West, & Stanovich (2014a)	Ratio Bias (5 items)	original 3-item scale	160	.37
Toplak, West, & Stanovich (2014a)	Ratio Bias (5 items)	new 4-item scale	160	.38
RT50	Ratio Bias (5 items)	7-item scale	371	.22
RT48	Ratio Bias (5 items)	7-item scale	294	.33

Table 7.2 (continued)

Study	Rational Thinking Measure	Cognitive Reflection Measure	Sample Size	Correlation
Correlations with Disjunctive Reasoning				
RT50	Disjunctive Reasoning (5 items)	7-item scale	371	.31
RT48	Disjunctive Reasoning (5 items)	7-item scale	294	.30

Note: RT labels are the labels of unpublished studies from our lab.

raw score was 3.1 (SD = 2.5). The reliability of our Reflection versus Intuition subtest was 0.73 (Cronbach's alpha). That these items were functioning as measures of reflection versus intuition and not as insight problems is indicated by the fact that the intuitive response was the dominant incorrect response in ten of the eleven items. For those ten, the intuitive response as a percentage of the incorrect responses was, in increasing magnitude: 62.8 percent, 72.6 percent, 73.4 percent, 73.7 percent, 75.3 percent, 87.3 percent, 89.9 percent, 92.0 percent, 92.1 percent, and 95.5 percent.

In the same study, the Reflection versus Intuition subtest displayed a correlation of 0.49 with SAT total scores and even higher correlations of 0.51 and 0.55 with the Probabilistic and Statistical Reasoning subtest and Scientific Reasoning subtest, respectively. We will present more data on the associations displayed by this subtest in chapters 12 and 13 when we report studies of a short-form version and full-form version of the CART. As indicated in the Appendix, the CART score on this subtest is equal to the number of items answered correctly, with a maximum score of 10.

The Belief Bias in Syllogistic Reasoning Subtest

In syllogistic reasoning tasks, subjects are presented with premises and then asked if the conclusion follows logically from them. To transform the task from a pure reasoning task (of the type that might be on an intelligence test) into a measure of rational thinking, researchers add a belief bias component. Specifically, they use conclusions that contradict world knowledge when the syllogism is valid and conclusions that are consistent with world knowledge when the syllogism is invalid.

Consider the following syllogism:

Premise 1: All living things need water.
Premise 2: Roses need water.
Therefore, roses are living things.

If you are like about 70 percent of the university students who have been given this problem, you will think that the conclusion is valid. If you did think that it was valid, you would be wrong.[1] Premise 1 states that all living things need water, not that all things that need water are living things. So, just because roses need water, it does not follow from Premise 1 that they are living things. Consider the following syllogism with exactly the same structure:

Premise 1: All insects need oxygen.
Premise 2: Mice need oxygen.
Therefore, mice are insects.

Here it seems pretty clear that the conclusion does not follow from the premises. The same thing that made the rose problem so hard made the mice problem easy. Logical validity is not about the believability of the conclusion—it is about whether the conclusion necessarily follows from the premises. In both of these problems, prior knowledge about the nature of the world (that roses are living things and that mice are not insects) is implicated in a type of judgment that is supposed to be independent of content: judgments of logical validity. In the rose problem, prior knowledge was interfering, and in the mice problem prior knowledge was facilitative. The rose syllogism exemplifies the logic of heuristics and biases tasks discussed in chapter 3: it creates a conflict between a natural, intuitive response and a more considered, normative response.

Syllogisms where validity and prior knowledge are in conflict assess an important thinking skill—the ability to maintain focus on reasoning through a problem without being distracted by our natural tendency to use the easiest cue (our natural tendency to be cognitive misers). Such problems probe our tendencies to rely on attribute substitution (Kahneman & Frederick, 2002) when the instructions tell us to avoid it. In these problems, the easiest cue to use is simply to evaluate whether the conclusion is true in the world. Validity is the harder concept to process, but it must be focused on while the easier cue of conclusion believability is ignored and/or suppressed. The task thus puts quite a stress on sustained decoupling, accounting for the finding (to be discussed below) that such tasks have substantial associations with measures of cognitive ability.

It is important to realize that the rose-type syllogism is not the type of syllogism that would appear on an intelligence test. It is the type of item more likely to appear on a critical thinking test, where the focus is on assessing thinking tendencies and cognitive styles. The relative openness of the item in terms of where to focus (on the truth of the conclusion or the validity of the argument) is a desired feature in a rational thinking test (where it's a feature, not a bug). The relative reliance on reasoning versus context when it is miserly to default to the latter is precisely what we want to assess on a rational thinking test. However, this relative openness would be unwanted on an intelligence test where the focus is on (ostensibly) the raw power to reason when there is no ambiguity about what constitutes optimal performance. On an intelligence test (or any aptitude measure or cognitive capacity measure) the syllogism would be stripped of content into "All As are Bs" form. Alternatively, unfamiliar content would be used, such as this example with the same form as the "rose" syllogism:

Premise 1: All animals of the hudon class are ferocious.
Premise 2: Wampets are ferocious.
Therefore, wampets are animals of the hudon class.

Items like this strip away the "multiple minds in conflict" aspect of the problem that is the distinguishing feature of the rose syllogism. Problems that do not involve such conflict tap only the power of the algorithmic mind and fail to tap important aspects of the reflective mind. For example, our research has shown that performance on rose-type syllogisms can be predicted by thinking dispositions that are part of the reflective mind such as cognitive flexibility, open-mindedness, context independence, and need for cognition (Kokis et al., 2002; Sá et al., 1999; Macpherson & Stanovich, 2007; Stanovich & West, 1998c, 2008a). Finally, although the rose syllogism may seem like a toy problem, it indexes a cognitive skill of increasing importance in modern society: the ability to reason from the information given and at least temporarily put aside what we thought before we received new information (Stanovich, 2004).

When we began investigating belief bias in syllogistic reasoning as an aspect of rational thinking from the perspective of individual differences (Stanovich & West, 1998c), there was already a substantial literature on belief bias effects (Evans, Barston, & Pollard, 1983; Evans et al., 1993; Markovits & Bouffard-Bouchard, 1992; Markovits & Nantel, 1989; Newstead, Pollard, Evans, & Allen, 1992; Oakhill, Johnson-Laird, & Garnham, 1989; Revlin, Leirer, Yopp, & Yopp, 1980). Subjects were known to be more likely to endorse a conclusion that was not logically valid if it conformed to world knowledge and/or was believable than if the conclusion contradicted world knowledge. The phenomenon was already known to be robust, but little attention had been paid to individual differences in the effect.

In our original investigation (Stanovich &West, 1998c), we adapted eight problems from the work of Markovits and Nantel (1989). Four of the problems had conclusions that did not follow logically but had believable conclusions. That is, they were like the rose syllogism discussed above. The other four problems were valid but had unbelievable conclusions (e.g., All mammals walk; Whales are mammals; Conclusion: Whales walk). Subjects were instructed to decide whether the conclusion follows logically from the premises (or not), assuming that the premises are all true.

We have used a variety of indices of the ability to overcome the effects of belief bias in syllogistic reasoning. As mentioned, in this initial investigation we simply used all items that had a conflict between the believability of the conclusion and the validity of the syllogism. In later investigations, we sometimes used both consistent and inconsistent syllogisms—that is, items where believability and validity primed the same response, and items where believability and validity were in conflict. In some investigations where we tested both types of syllogism, we used a difference score as the index of belief bias. Finally, in our most recent investigations (and in the CART), we have simply used the total score on an equal number of consistent and inconsistent items as the index of the ability to reason logically with content-laden material. We will explain the theoretical reasons for the choice of this scoring index shortly. However, the correlation of this task with other rational thinking tasks and with other cognitive ability and thinking dispositions measures depends little upon the scoring index that is used.

From the very beginning of our investigations, syllogistic reasoning with belief bias content has correlated with numerous other heuristics and biases tasks that we have studied. Importantly, though, and as we would expect from the nature of the task as well as Spearman's positive manifold, performance on belief bias syllogistic reasoning problems has also consistently shown moderate correlations with cognitive ability. As syllogistic reasoning is a quintessential reasoning task (at least with abstract content), we would be surprised if it did not correlate with cognitive ability.

Table 7.3 presents a series of correlations between performance on syllogisms with belief bias content and various measures of cognitive ability. The specific index of syllogistic reasoning with belief bias varied from study to study. However, as can be seen from the table, despite this variation, the correlations were consistently moderate in magnitude (the vast majority were between 0.30 and 0.50).

Throughout our work on belief bias in syllogistic reasoning we have consistently found correlations between reasoning performance and various thinking dispositions. Although these correlations are not nearly as large as those found with cognitive ability, they are often statistically significant. Sometimes we have found that various thinking dispositions can explain variance in performance after the variance accounted for by cognitive ability has been partialled out (Kokis et al., 2002; Sá et al., 1999; Stanovich & West, 1998c; West et al., 2008; Toplak et al., 2014b). The top part of table 7.4 presents some of the zero-order correlations with thinking dispositions that we have found.

From the very beginning of our investigations, we have found that performance on belief bias syllogisms tends to correlate with a variety of heuristics and biases tasks. However, as is clear from table 4.1, in our framework for the CART, we conceive of this syllogistic reasoning task as a task with heavy processing requirements relevant to the tendency to avoid miserly information processing. We saw in table 7.2 that it displays substantial correlations with the cognitive reflection measures, another key indicator of the ability to avoid miserly information processing. The other two tasks that are strong indicators of miserly processing are the ratio bias and disjunctive reasoning tasks, both to be discussed below. Table 7.4 also presents the correlations between these two measures of miserly processing and performance on belief bias syllogisms. These correlations were consistently positive and statistically significant.

Throughout the years, we have made small revisions in the original stimuli used by Stanovich and West (1998c). However, for the current CART, we have made more extensive revisions in how we measure belief bias with syllogisms. These improvements are reflected in the final set of stimuli that we chose for the Belief Bias Syllogistic Reasoning subtest of the CART.

Table 7.3
Correlations between versions of the belief bias syllogisms and measures of cognitive ability

Study	Cognitive Ability Measure	Syllogistic Reasoning with Belief Bias Measure	Sample Size	Correlation
Stanovich & West (1998c) Experiment 1	Composite Measure of Raven's Matrices, SAT total scores, and Nelson-Denny Comprehension	8 Inconsistent items	192	.50
Stanovich & West (1998c) Experiment 2	SAT total scores	8 Inconsistent items	529	.41
Stanovich & West (1998c) Experiment 4	Composite Measure of Raven's Matrices, SAT total scores, and Nelson-Denny Comprehension	8 Inconsistent items	211	.33
Sá, West, & Stanovich (1999)	Cognitive ability 1 (CA1): WAIS-R (block design and vocabulary) Cognitive ability 2 (CA2): Raven's Matrices and checklist vocabulary	Difference index: C–I 8 Inconsistent items 8 Consistent items 8 Neutral items	124	.45 (CA1) .50 (CA2)
Stanovich & West (2008a)	SAT total scores	Inconsistent correct 12 Inconsistent items 12 Consistent items 8 Neutral items	420	.46
Macpherson & Stanovich (2007)	WASI (matrix and vocabulary)	Inconsistent correct 8 Inconsistent items 8 Consistent items	195	.41

Table 7.3 (continued)

Study	Cognitive Ability Measure	Syllogistic Reasoning with Belief Bias Measure	Sample Size	Correlation
Kokis, Macpherson, Toplak, West, & Stanovich (2002)	WISC-III (block design and vocabulary)	Inconsistent correct (child adapted, logical alien version) 4 Inconsistent 4 Consistent (ceiling)	108 children (gifted classes, grades 5, 6, 8)	.43
Stanovich & West (2008b)	reported SAT total scores	Difference index: C–I 8 Inconsistent items 8 Consistent items 8 Neutral items	436	.28 (C–I) .45 (Inconsistent items) .25 (Consistent items) .39 (Neutral items)
Toplak, West, & Stanovich (2011)	WASI (matrix and vocabulary) CA2 (WASI and WM composite)	5 Inconsistent items	346	.35 (WASI) .36 (CA2)
Toplak, West, & Stanovich (2014a)	WASI (matrix and vocabulary)	8 Inconsistent items	160	.44
West, Toplak, & Stanovich (2008)	SAT total scores	12 Inconsistent items	793	.44
Toplak, West, & Stanovich (2014b)	WASI (matrix and vocabulary)	8 Inconsistent items	204 children (grades 2–9)	.42
RT50	reported SAT total scores	16 Items (Balanced Consistent and Inconsistent)	371	.31
RT50	Raven's Matrices	16 Items (Balanced Consistent and Inconsistent)	371	.26

Table 7.3 (continued)

Study	Cognitive Ability Measure	Syllogistic Reasoning with Belief Bias Measure	Sample Size	Correlation
RT49	reported SAT total scores	16 Items (Balanced Consistent and Inconsistent)	170	.42
RT49	Raven's Matrices	16 Items (Balanced Consistent and Inconsistent)	170	.53
RT48	reported SAT total scores	16 Items (Balanced Consistent and Inconsistent)	181	.44
RT48	Raven's Matrices	16 Items (Balanced Consistent and Inconsistent)	294	.35

Note: RT labels are the labels of unpublished studies from our lab. All correlations scored in positive direction.

Table 7.4
Correlations between versions of the belief bias syllogisms, thinking dispositions, and other measures of rational thinking with heavy detection, override, and hypothetical thinking requirements

Study	Measure	Syllogistic Reasoning with Belief Bias Measure	Sample Size	Correlation
Correlations with Thinking Dispositions				
Stanovich & West (1998c) Experiment 1	Thinking Dispositions Composite	8 Inconsistent items	192	.28
Stanovich & West (1998c) Experiment 2	Thinking Dispositions Composite	8 Inconsistent items	545	.33
Stanovich & West (1998c) Experiment 4	Thinking Dispositions Composite	8 Inconsistent items	211	.12

Table 7.4 (continued)

Study	Measure	Syllogistic Reasoning with Belief Bias Measure	Sample Size	Correlation
Sá, West, & Stanovich (1999)	Actively Open-Minded Thinking composite (AOT, 68-item version)	Difference index: C–I 8 Inconsistent items 8 Consistent items 8 Neutral items	124	.32
Stanovich & West (2008a)	AOT (41 items) Need for Cognition Scale (NCog, 18 items)	Inconsistent correct 12 Inconsistent items 12 Consistent items 8 Neutral items	420	.23 (AOT) .21 (NCog)
Macpherson & Stanovich (2007)	AOT (41 items) NCog (18 items)	Inconsistent correct 8 Inconsistent items 8 Consistent items	195	.21 (AOT) .19 (NCog)
Kokis, Macpherson, Toplak, West, & Stanovich (2002)	Child versions: AOT (30-item) NCog (9-item)	Inconsistent correct (child adapted, logical alien version) 4 Inconsistent	108 children (gifted classes, grades 5, 6, 8)	.40 (AOT) –.08 (NCog)
Toplak, West, & Stanovich (2011)	AOT (41 items) Consideration of Future Consequences (12-items)	5 Inconsistent items	346	.12 (AOT) .15 (CFC)
Toplak, West, & Stanovich (2014a)	AOT (41 items) NCog (18 items) Consideration of Future Consequences (12-items)	8 Inconsistent items	160	.26 (AOT) .24 (NCog) .26 (CFC)
West, Toplak, & Stanovich (2008)	AOT (41 items) NCog (18 items)	12 Inconsistent items	793	.21 (AOT) .24 (NCog)

Table 7.4 (continued)

Study	Measure	Syllogistic Reasoning with Belief Bias Measure	Sample Size	Correlation
Toplak, West, & Stanovich (2014b)	Child versions: AOT (30-item); NCog (9-item)	8 Inconsistent items	204 children (grades 2–9)	.31 (AOT) –.04 (NCog)
Correlations with Ratio Bias				
Toplak, West, & Stanovich (2014a)	Ratio Bias (5-items)	8 Inconsistent items	160	.33
Toplak, West, & Stanovich (2014b)	Ratio Bias (6-trials)	8 Inconsistent items	204 children (grades 2–9)	.27
RT50	Ratio Bias (5-trials)	16 Items (Balanced Consistent and Inconsistent)	371	.13
RT49	Ratio Bias (5-trials)	16 Items (Balanced Consistent and Inconsistent)	170	.44
RT48	Ratio Bias (5-trials)	16 Items (Balanced Consistent and Inconsistent)	294	.23
Correlations with Disjunctive Reasoning				
RT50	Disjunctive Reasoning (5-trials)	16 Items (Balanced Consistent and Inconsistent)	371	.19
RT48	Disjunctive Reasoning (5-trials)	16 Items (Balanced Consistent and Inconsistent)	294	.25

Note: All correlations scored in positive direction.

First, we have made some alterations in the instructions to emphasize the importance of ignoring the believability of the conclusion. We have also used items that have pronounceable nonwords for the middle term as a means of controlling for *premise* believability. This is a procedure recommended by Evans et al. (2001), Heit and Rotello (2014), and Newstead et al. (1992). We believe there is a modest advantage to these slightly new procedures even though the correlations obtained with these syllogisms are indistinguishable from those used in our older published and unpublished studies.[2]

We made a major change in the scoring of the syllogistic reasoning items for this subtest of the CART. In many of the studies listed in tables 7.3 and 7.4, we used only inconsistent syllogisms—syllogisms where the believability of the conclusion and the validity of the syllogism were in conflict. In such studies, we simply used the total number of correct responses to these inconsistent items. In other studies, however, we used equal numbers of inconsistent and consistent items. Often, however, even in these studies, we simply used the number of correct responses on inconsistent items as the key measure. Once in a while, as in the Sá et al. (1999) study in table 7.3, we used a difference score: the number of correct responses to the consistent items minus the number of correct responses to the inconsistent items (shown as C–I in the table).

We now believe that neither of these indices is optimal. The reason is that in the two-alternative forced-choice problems that we use, two different miserly processing biases are at work. The first is the one that is discussed extensively in the reasoning literature: the tendency to respond based on the believability of the conclusion. Such a bias would lead the subject to answer each of the consistent items correctly and each of the inconsistent items incorrectly. The latter are constructed so that valid items have unbelievable conclusions (leading such a subject to respond "invalid" incorrectly) and so that invalid items have believable conclusions (leading such a subject to respond "valid" incorrectly).

However, another bias plays a role in the task. In the survey literature, it would probably be called "response acquiescence." In our paradigm, it is the tendency to evaluate all the syllogisms as valid (see Evans et al., 1999). This is a marked tendency that shows up in all of our studies. For example, an optimal subject—one answering all of the items correctly—would say "valid" eight times and "invalid" eight times on a sixteen-item task with equal number of consistent and inconsistent items. Likewise, a subject with an extreme believability bias would answer "valid" eight times and "invalid" eight times. The trend in our data, however, is for the mean number of "valid" responses to be substantially above eight. This indicates that there is a miserly validity bias in our data, as well as a miserly believability bias. That there is a validity response bias operating in this task is also perhaps suggested by the fact that intelligence correlates with correct responding to items having "invalid" as the correct response (DA and AC items: denying the antecedent and affirming the consequent) but does not correlate with correct responding to MT items (modus tollens) where "valid" is the correct response (Attridge & Inglis, 2014; Newstead et al. 2004).

With these two biases operating simultaneously in our data, a problem with our previous scoring systems becomes apparent. The logic of this problem can be seen in table 7.5. The top row of the table indicates that a subject with the most extreme validity bias—that is, a subject who responded "valid" on every trial—would get four consistent items correct on a perfectly balanced sixteen-item test. The same subject would get four inconsistent items correct. The top row of the table indicates that such a subject would have a belief bias difference score of 0 and have a total score of 8 items correct. A subject with the other miserly processing bias—a believability bias—would score drastically differently. Such a subject would get 8 consistent items correct but no inconsistent items correct, thus having the same total score of 8 but an extremely different belief bias difference score of 8 as well. Finally, the last row of the table indicates the performance

Table 7.5

The effects of the two different miserly biases on syllogistic reasoning scores

	Consistent Items Correct	Inconsistent Items Correct	Belief Bias Difference Score	Total Items Correct
Validity Bias (subject says "valid" on every trial)	4	4	0	8
Believability Bias (subject goes with believability on every trial)	8	0	8	8
Optimal Responding (correct on every trial)	8	8	0	16

of the optimal subject, one answering correctly every time. Such a subject ends up with a belief bias difference score of 0 and sixteen items correct.

The problem with the previous two scoring systems that we have used is apparent from the rows of table 7.5. Using only the inconsistent items as an index of belief bias in syllogistic reasoning, we see that the optimal subject receives a score of 8 and that a subject with an extreme believability bias receives a score of 0, which seems right. However, what does not seem right is that a subject with a bias that seems at least equally miserly—a subject with a validity response bias—receives a score of 4 on this index. Four of the inconsistent items are valid and four are invalid, and thus if a subject with an extreme validity bias answers "valid" each time, he or she gets four of the eight inconsistent items correct, a substantially higher score than the subject with an extreme believability bias (who receives a score of 0).[3]

Using the belief bias difference score seems even worse. The optimally responding subject receives a belief bias difference score of 0, and the subject with extreme believability bias receives a belief bias difference score of 8. Again, this seems right. But notice that a subject with an extreme validity bias receives a belief bias difference score of 0—the same score received by a subject responding optimally. This, of course, is a drastic flaw in this particular index of performance. This is why we have come to rely on the total items correct as the primary indicator of performance in this task. As can be seen in the far right column of table 7.5, this index gives the optimally responding subject a score of 16 and subjects with each of the miserly biases receive a score of 8. Thus the total score does not favor one miserly bias over the other. This would seem to be what we are looking for in a subtest that is in part indexing the avoidance of miserly processing. Thus, the scoring of this task uses the total score, combining consistent and inconsistent items.

We investigated our sixteen-item Belief Bias in Syllogistic Reasoning subtest (see the appendix for sample items) in an unpublished study (Turk2) of 108 subjects on the Amazon Mechanical Turk (there were more subjects in this study, but they received different versions of this task). The mean raw score was 11.8 (SD = 2.9). The reliability of the subtest was 0.73 (Cronbach's alpha). In the same study, we examined the correlation of the Belief Bias in Syllogistic Reasoning subtest with other measures that were completed by the same subjects. It displayed a correlation of 0.57 with the Reflection versus Intuition subtest. The Belief Bias in Syllogistic Reasoning subtest displayed correlations of 0.31 and 0.48 with two subtests to be described later in this chapter: the Ratio Bias subtest and the Disjunctive Reasoning subtest. We will present more data on the associations displayed by this subtest in chapters 12 and 13 when we report studies of short-form and full-form versions of the CART. As indicated in the appendix, the maximum CART score on this subtest is 8 points; the appendix indicates how the raw score (which ranges from 0 to 16) is transformed into CART points.

The Ratio Bias Subtest

As previously discussed, rationality often requires that that we override responses based on Type 1 processing. But override is a capacity-demanding operation, and any tendencies toward miserly processing in a situation where override is required will likely result in a failure to substitute a superior response for the intuitive one. One of several phenomena in the heuristics and biases literature that illustrates the failure of sustained override is the phenomenon of ratio bias. Epstein and colleagues (Denes-Raj & Epstein, 1994; Kirkpatrick & Epstein, 1992; Pacini & Epstein, 1999) demonstrated that it can result in a startling failure of rational judgment.[4] Adults in several of these experiments were presented with two bowls that each contained clearly identified numbers of jelly beans. In the first were 9 white jelly beans and one red jelly bean. In the second were 92 white jelly beans and 8 red jelly beans. A random draw was to be made from one of the two bowls and if the red jelly bean was picked, the participant would receive a dollar. The participant could choose which bowl to draw from. Although the two bowls clearly represent a 10 percent and an 8 percent chance of winning a dollar, a number of subjects chose the 100-bean bowl (8 percent chance), thus reducing their chance of winning.

The majority did pick the 10 percent bowl, but a healthy minority (from 30 to 40 percent of the participants) picked the 8 percent bowl. Although most of these participants in the minority were at least somewhat aware (see Bonner & Newell, 2010) that the large bowl was statistically a worse bet, that bowl also contained more enticing winning beans—the 8 red ones instead of just one. In short, the tendency to respond to the absolute number of winners, for these participants, trumped the formal rule (pick the one with the best percentage of reds) that they knew was the better choice. The ratio bias phenomenon is often viewed as a form of denominator neglect (see Reyna & Brainerd, 2008), and sometimes goes by that name in the literature.

Many subjects were aware of the poorer probability but failed to resist picking the large bowl, as indicated by comments from some of them such as the following: "I picked the one with more red jelly beans because it looked like there were more ways to get a winner, even though I knew there were also more whites, and that the percents were against me" (Denes-Raj & Epstein, 1994, p. 823). Ratio bias thus seems to be a good exemplar of the failure of sustained decoupling. The simpler Type 1 tendency to respond to the absolute number of winners sometimes trumps the more analytic Type 2 process of calculating a ratio.

Bonner and Newell (2010) presented results that were consistent with this interpretation of the task. Using response latencies to make inferences about the processing requirements of the task, they concluded that even subjects who made the nonnormative large-tray choice detected the conflict between the numerical probabilities and the visually salient higher number of winners in the larger tray. The problem was not the failure to register the relevance of the probabilities, but instead the inability to override an intuitive response tendency that was stronger than the impact of the probabilities themselves (see also Corser & Jasper, 2014).

A typical item used in a ratio bias task is as follows:

Assume that you are presented with two trays of black and white marbles (pictured in figure 7.1). The small tray contains 10 marbles. The large tray contains 100 marbles. The marbles inside each tray will be randomly mixed up, and you must draw out a single marble from one of the trays without looking. If you draw a black marble you win $5.

In a real situation, which tray would you prefer to select a marble from?

a. Strongly prefer the small tray
b. Moderately prefer the small tray
c. Slightly prefer the small tray
d. Slightly prefer the large tray
e. Moderately prefer the large tray
f. Strongly prefer the large tray

Although several different experimental variables and individual difference variables have been linked to the avoidance of ratio bias (Bonner &

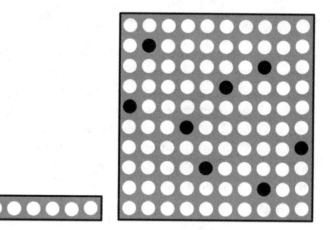

Figure 7.1
One black and 9 white marbles; 8 black and 92 white marbles.

Newell, 2008, 2010; Dale et al., 2007; Liberali et al., 2012; Mikels, et al., 2013; Okan et al., 2012; Yamagishi, 1997), we will focus here, as we did with the previous two tasks in this chapter, on cognitive ability and some of the other measures of rational thinking in the CART. Table 7.6 displays the correlations between a short measure of the avoidance of ratio bias that we have used in several studies and various indices of cognitive ability. The correlations will be positive, because the ratio bias task is always scored in the direction of avoidance.

These correlations are somewhat lower than we have seen in earlier tables, averaging around 0.25. Correlations between cognitive ability and cognitive reflection measures were closer to the 0.40 to 0.45 range. Similarly, belief bias in syllogistic reasoning also displayed correlations with cognitive ability of roughly 0.40 to 0.50. However, table 7.2 indicates that the avoidance of ratio bias had more substantial correlations with cognitive reflection measures (in the range of 0.30 to 0.40). Table 7.4 indicates that the avoidance of ratio bias had a variable correlation with belief bias in syllogistic reasoning (in the range of 0.20 to 0.40). At the bottom of table 7.6 it is indicated that the avoidance of ratio bias had small but significant correlations with a measure of disjunctive reasoning, to be discussed later in this chapter. Contributing to the modesty of some of these correlations, however, is that, as table 7.6 illustrates, we have often used short forms of the ratio bias task that involve no more than five items.

Table 7.6

Correlations between avoidance of ratio bias and measures of cognitive ability and other heuristics and biases tasks

Study	Cognitive Ability Measure	Avoidance of Ratio Bias Measure	Sample Size	Correlation
Toplak, West, & Stanovich (2014b)	WASI (matrix and vocabulary)	6 items, discrete scale	204 children (grades 2–9)	.36
Toplak, West, & Stanovich (2014b)	composite: three executive function tasks	6 items, discrete scale	204 children (grades 2–9)	.30
Toplak, West, & Stanovich (2014a)	WASI (matrix and vocabulary)	5 items, continuous scale	160	.37
Kokis, Macpherson, Toplak, West, & Stanovich (2002)	WISC-III (block design and vocabulary)	5 items, discrete scale	108 children (gifted classes, grades 5, 6, 8)	.28
Stanovich & West (2008b) Experiment 8	reported SAT total score	1 item, discrete scale	819	.17
West, Toplak, & Stanovich (2008)	reported SAT total score	1 item, discrete scale	793	.19
RT50	reported SAT total scores	5 items, continuous scale	371	.24
RT50	Raven's Matrices	5 items, continuous scale	371	.16
RT49	reported SAT total scores	5 items, continuous scale	170	.24
RT49	Raven's Matrices	5 items, continuous scale	170	.24
RT50	Disjunctive Reasoning (5 items)	5 items, continuous scale	371	.22
West, Toplak, & Stanovich (2008)	Syllogisms (12 Inconsistent items)	1 item, discrete scale	793	.15
RT48	Disjunctive Reasoning (5 items)	5 items, continuous scale	294	.16

Note: RT labels are the labels of unpublished studies from our lab. In the last three lines of the table the correlate is another heuristics and biases task.

The Ratio Bias subtest that we have constructed for the CART has been expanded to twelve scored items (see the appendix for sample items). To those twelve items we have added three filler items. The version of the ratio bias task that we use in the CART has several advantages over the versions used in the research displayed in table 7.6. First, as mentioned, we have expanded the number of critical items from five to twelve. Second, we have included three filler items in which it is rational to choose the large tray rather than the small tray. This was done to discourage the development of the response set (and corresponding mindless responding) that the smaller tray is always the correct one. Finally, we have altered the response scale from that used in previous research. Previous research often scored this task simply 0 or 1 depending on whether the subject chose the right tray. However, as indicated in the example given above, we now use a continuous response scale that may be more sensitive to individual differences, especially in a task that involves just a few trials.

All twelve of the critical items involved probability differences that favored the smaller tray. Some previous research has employed items where the small and large trays were equal in probability and the subject was given a third response option indicating that the chances of winning are equal (Klaczynski, 2001b; Liberali et al., 2012). We did not adopt this approach, but instead used critical trials where the probability was slightly in favor of the small alternative and the subject was forced to indicate at least a mild preference for one of the trays.

A recent study of 222 subjects on the Amazon Mechanical Turk (Turk2) indicated that the twelve-item (the three filler items are not scored) Ratio Bias subtest of the CART using the continuous scale has a high reliability (Cronbach's alpha of 0.89). Scoring the twelve items 0/1, rather than using the continuous scale, results in only a minor reduction in reliability—to a Cronbach's alpha of 0.87. Interestingly, scoring the items 0/1 rather than on a continuous scale did not reduce the correlations in this particular study. The correlation with the Reflection versus Intuition subtest of the CART was 0.34 for the continuously scored version and 0.38 for the discretely scored version. The correlation with the Belief Bias Syllogisms subtest of the CART was 0.32 for the continuously scored version and 0.34 for the discretely scored version. However, in other studies we have run, the continuously scored version sometimes resulted in higher correlations. For instance, in the study testing the short-form version of the CART (to be described in chapter 12), the correlation with the Reflection versus Intuition subtest of the CART was 0.32 for the continuously scored version and 0.26 for the discretely scored version. Similarly, the correlation with the Probabilistic and Statistical Reasoning subtest was 0.32 for the continuously scored version and 0.29 for the discretely scored version. Analogously, the correlation with the Scientific Reasoning subtest was 0.26 for the continuously scored version and 0.21 for the discretely scored version.

We will present more data on the associations displayed by this subtest in chapters 12 and 13 when we report studies of short-form and full-form versions of the CART. At the end of the appendix it is indicated that the maximum CART score on this subtest is 5 points, and the appendix indicates how the raw score is transformed into CART points.

The Disjunctive Reasoning Subtest

We define fully reflective disjunctive reasoning as the tendency to consider all possible states of the world when deciding among options or when choosing a problem solution in a reasoning task (see Johnson-Laird, 2006; Shafir, 1994; Toplak & Stanovich, 2002). Most decision-making situations can be thought of as disjunctions of possible states of the world. Thus, choosing optimally entails combining the probabilities of the states with the utilities of the outcomes under each of the decision options (Jeffrey, 1983; Savage, 1954). Many problem-solving situations can likewise be optimally evaluated by constructing all of the mental models that are consistent with the premises as presented (Johnson-Laird, 1983, 1999, 2006).

Despite the seeming obviousness of disjunctive reasoning as a general thinking strategy, Shafir (1994) demonstrated how, if one looks across the wide domain of reasoning tasks used in cognitive science, it is easy to find tasks in which people perform suboptimally because they reason nondisjunctively. These problems include prisoner's dilemmas (Rapoport & Chammah, 1965); framing problems; Newcomb's quasi-magical thinking problem (Nozick, 1969; Shafir & Tversky, 1992); probabilistic reasoning; Wason's (1966) four-card selection task; "knights and knaves" puzzles (Rips, 1994; Smullyan, 1978); and so-called double disjunction problems (Johnson-Laird, Byrne, & Schaeken, 1992).

Many of the cognitive phenomena discussed by Shafir (as well as those effects and biases studied by Toplak & Stanovich, 2002) are somewhat indirect effects of failures of disjunctive reasoning. For the CART, we wanted a task that was a more direct measure of disjunctive reasoning. Such a task would strongly tap processes of the reflective mind in that it would assess typical modes of processing rather than optimal modes. Only by using a task that taps typical modes (whether it is typical for the subject to default to the autonomous mind and not engage in Type 2 processing) can miserly tendencies be assessed.

The task we chose was based on two problems studied by Toplak and Stanovich (2002) that they took from the work of Levesque (1986, 1989). The one that we have used as an example in several of our publications is as follows:

Jack is looking at Ann but Ann is looking at George. Jack is married but George is not. Is a married person looking at an unmarried person?

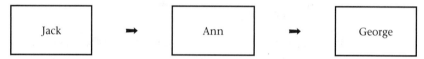

A. Yes
B. No
C. Cannot be determined

Answer A, B, or C before you look ahead. Many people who respond to this problem answer incorrectly. A large number of people usually answer C (cannot be determined) when in fact the correct answer is A (yes). The answer is easily revealed once we engage in fully disjunctive reasoning. To solve the problem, it is necessary to consider both possibilities for Ann's marital status (married and unmarried) to determine whether a conclusion can be drawn. If Ann is married, then the answer is "Yes" because she is looking at George, who is unmarried. If Ann is not married, then the answer is still "Yes" because Jack, who is married, is looking at Ann. Considering all the possibilities (the fully disjunctive reasoning strategy) reveals that a married person is looking at an unmarried person whether Ann is married or not. People respond "cannot be determined" because the problem does not *reveal* whether Ann is married or not. One has to avoid the automatic reaction: "Oh, since we don't know whether Ann is married or not we cannot determine anything."

Most people can carry out fully disjunctive reasoning when they are explicitly told that it is necessary. If *told* to reason through all of the alternatives, almost all adult subjects would have done so more efficiently.

However, without that instruction, many default to computationally simple cognition. Thus, more than sustained decoupling, these problems stress the ability to recognize the possibility of options that can be derived from deeper processing. Overall, in terms of task characteristics, the Disjunctive Reasoning subtest is located in the upper left of the 2 × 2 space of figure 4.1. That is, it is a subtest with high detection requirements, but low knowledge requirements.

In previous unpublished research we have used a disjunctive reasoning scale of five items structurally similar to the example just discussed. Such a scale has been observed to have low but significant correlations with intelligence. In one study (RT50) of 371 university students, this disjunctive reasoning scale displayed correlations of 0.19 with SAT total scores and 0.24 with scores on Raven's Matrices. In a second study (RT48) of 294 university students, the same five-item disjunctive reasoning scale displayed correlations of 0.25 with the SAT total score and 0.19 with scores on the Raven. A perusal of tables 7.4 and 7.6 indicates that disjunctive reasoning had roughly similar 0.20 correlations with performance on belief bias syllogisms and performance on the ratio bias task. However, table 7.2 indicates that disjunctive reasoning had a somewhat higher 0.30 correlation with a cognitive reflection measure.

Our current version of the Disjunctive Reasoning subtest of the CART is a six-item measure (see the appendix for sample items). These items are not presented in a block, as is the case with many other subtests of the CART. Instead, the items are dispersed—either dispersed within a long subtest (as in RT60 described in chapter 13) or each item is separated by a subtest (as in RT59 described in chapter 12). For administration purposes, it is critical that these six items be presented separately. Also, the "Ann is married or unmarried" problem that was discussed above is the last of the six items to appear. In previous studies, we have found that when this item is presented first it does not contribute to scale reliability or construct validity. Often, when this item is presented first, performance on it is extremely low (often only 10–15 percent answer correctly). Also, when presented as the initial item, this item often does not correlate with other items on the test and in fact reduces reliability. However, when presented as the sixth item, after the other five, performance on the item increases to about 40 percent correct or more (still the hardest item along with the blocks problem), and it converges with other items on the subtest.

In a recent study of 222 subjects (Turk2), all Amazon Mechanical Turk workers, the six-item scale displayed a reliability of 0.81 (Cronbach's alpha). Furthermore, in the same study, the Disjunctive Reasoning subtest displayed a correlation of 0.52 with the Reflection versus Intuition subtest of the CART, a correlation of 0.44 with the Belief Bias Syllogisms subtest of the CART, and a correlation of 0.28 with the Ratio Bias subtest of the CART. The appendix indicates how the 0–6 raw score on the subtest is transformed into points on the CART (a maximum of 5 points for this subtest).

8 Avoidance of Miserly Information Processing: Indirect Effects

In chapter 7 we discussed four subtests that provide some of the more direct measures of the avoidance of miserly processing on the CART. However, the CART also contains several other subtests that assess the ability to avoid suboptimal thought patterns that arise *indirectly* from miserly detection, miserly override, or miserly hypothetical thinking tendencies. The next cell down in the first column of table 4.1 lists three tasks that assess an important component of axiomatic utility theory: the ability to avoid being affected by irrelevant context when making decisions. The three tasks that measure the ability to avoid this tendency are a framing task, an anchoring task, and a measure of susceptibility to preference anomalies.

The avoidance of myside bias in argument evaluation is a fundamental component of performance in most discussions of rational thinking and critical thinking. We used our original version of an argument evaluation task (see Stanovich & West, 1997, 1998c) to measure this component of rational thinking. Myside bias might well be a multiply determined processing bias, implicating, for example, motivated reasoning in addition to miserly processing, but most models of this bias implicate miserly processing at least to some extent. The same is true for the next subscale on the CART, which will be discussed in this chapter—our measure of the ability to avoid overconfidence in knowledge calibration. The final important component of rational thinking tested by the CART that will be discussed in this chapter is the tendency toward the rational temporal discounting of reward. This task has heavy processing requirements (see figure 4.1) in that it involves sustained override in order to pass up an immediate reward. We will discuss each of these measures in this chapter.

Context Effects in Decision Making: Avoidance of Framing Effects

The standard view of so-called rational man in economics and decision theory traditionally assumes that people have stable, underlying preferences for each of the options presented in a decision situation (Thaler, 2015). That is, it is assumed that a person's preferences for the options available for choice are complete, well ordered, and well behaved in terms of the axioms of choice. All of the axioms of choice (independence of irrelevant alternatives, transitivity, independence, and reduction of compound lotteries, etc.), in one way or another, ensure that decisions are not influenced by irrelevant context (Stanovich, 2013). Well-behaved internal preferences imply that a person is a utility maximizer—he or she acts to get what he or she most wants.

The main problem with this conception is that three decades of work by Kahneman and Tversky (2000; see Kahneman, 2011) and a host of other cognitive and decision scientists (Dawes, 1998; Gilovich, Griffin, & Kahneman, 2002; Lichtenstein & Slovic, 1971, 2006; Shafir & LeBoeuf, 2002; Shafir & Tversky, 1995; Slovic, 1995) have brought this view of rational economic man into question. This work has shown that people's choices—sometimes choices about very important things—*can* be altered by irrelevant changes in how the alternatives are presented to them. This problem is illustrated when people violate one of the simplest strictures of normative rationality, the principle of *descriptive invariance*: "that the preference order between prospects should not depend on the manner in which they are described" (Kahneman & Tversky, 1984, p. 343). As Arrow (1984) describes it, "A fundamental element of rationality, so elementary that we hardly notice it, is, in logicians' language, its extensionality. The chosen element depends on the opportunity set from which the choice is to be made, independently of how that set is described" (p. 268).

Empirically, though, people display framing effects that violate the principle that has been described by Arrow (1984) and Kahneman and Tversky (1984). Tversky and Kahneman (1981) have provided the most famous demonstration. The two problems are usually separated in time:

Problem 1. Imagine that the United States is preparing for the outbreak of an unusual disease, which is expected to kill 600 people. Two alternative programs to combat the disease have been proposed. Assume that the exact scientific estimates of the consequences of the programs are as follows: If Program A is adopted, 200 people will be saved. If Program B is adopted, there is a one-third probability that 600 people will be saved and a two-thirds probability that no people will be saved. Which of the two programs would you favor, Program A or Program B?

Problem 2. Imagine that the United States is preparing for the outbreak of an unusual disease, which is expected to kill 600 people. Two alternative programs to combat the disease have been proposed. Assume that the exact scientific estimates of the consequences of the programs are as follows: If Program C is adopted, 400 people will die. If Program D is adopted, there is a one-third probability that nobody will die and a two-thirds probability that 600 people will die. Which of the two programs would you favor, Program C or Program D?

Many subjects select alternatives A and D in these two problems despite the fact that the two problems are redescriptions of each other and that Program A maps to Program C rather than D. This response pattern violates descriptive invariance. The violation of descriptive invariance is a very fundamental one. If choices flip-flop based on problem characteristics *that the subjects themselves view as irrelevant* (Koehler & James, 2009; Shafir, 1993; Shafir & Tversky, 1995; Stanovich & West, 1999)—then people cannot possibly be maximizing expected utility. More practically, however, people subject to framing are not in control of their own preferences. If they accept every frame as it is given to them, then they are under the control of whatever party is providing the frame.

The literature on framing effects is enormous and beyond our scope here.[1] An excellent review (Levin et al., 1998) and a study (Levin et al., 2002) by Levin's group discuss the two categories of framing effect that we include in our subtest: risky-choice framing and attribute framing. Importantly for the CART, framing effects represent a classic example of miserly information processing. In discussing the mechanisms causing framing effects, Kahneman has stated that "the basic principle of framing is the passive acceptance of the formulation given" (2003, p. 703). In short, the frame presented to the subject is taken as focal, and all subsequent thought follows from it rather than from alternative framings, because the latter would require more thought. This means that framing effects derive from a particular kind of miserly processing—serial associative cognition with a focal bias—as outlined in chapters 2 and 4 (see figure 2.3, table 3.1, and figure 4.1). Recall that serial associative cognition with a focal bias is a form of Type 2 cognition that rather inflexibly locks into an associative mode that takes as its starting point the easiest model of the world to construct. Consistent with this interpretation is the finding that being forced to take more time or to provide a rationale for selections reduces framing effects (Miller & Fagley, 1991; Sieck & Yates, 1997; Takemura, 1992, 1993, 1994), although these effects sometimes interact with cognitive styles and thinking dispositions (Simon, Fagley, & Halleran, 2004). Regarding the latter, the need for cognition thinking disposition (see Cacioppo et al., 1996) has sometimes associated with the magnitude of framing effects, but the results here are complex (LeBoeuf & Shafir, 2003; Levin et al., 2002; Shiloh et al., 2002; Simon et al., 2004; Smith & Levin, 1996).

Critical for our purposes, reliable individual differences in framing effects have been observed in numerous studies (e.g., Levin et al., 2002; Mahoney et al., 2011; Schneider, 1992; Stanovich & West, 1998b), although it is often the case that framing effects are sometimes shown only by a minority of subjects (Frisch, 1993; Levin et al., 2002; Stanovich & West, 1998b). However, when aggregated, the effects are most often statistically significant. Importantly, though, what this shows is that a number of subjects are in fact able to avoid the effect of irrelevant contexts in decision making, whereas other subjects are quite susceptible to these nonnormative effects. Not surprisingly, between-subjects experiments show larger framing effects (Mahoney et al., 2011; Stanovich & West, 1998b). Nonetheless, even within-subjects experiments yield a substantial number of inconsistent responses—many subjects switch their preferences depending on the phrasing of the question. This is absolutely critical for our purposes, as we need an index of the avoidance of framing effects for each individual CART respondent. Thus, it is necessary that any individual differences measure of this response tendency be run within subjects.

The Framing subtest of the CART is composed of eleven pairs of items (see the appendix for sample items). They are adaptations of several different types of problems that have been studied in the literature (Highhouse & Paese, 1996; Kahneman & Tversky, 1984, 2000; Levin & Gaeth, 1988; Levin et al., 1985; Levin, Schnittjer, & Thee, 1988; Roszkowski & Snelbecker, 1990; Schneider, 1992; Schoorman et al., 1994; Tversky & Kahneman, 1981). Seven of the pairs are designed to assess risky choice framing and four of the pairs are designed to assess attribute framing. Of course, pairs of framing items are not presented contiguously in within-subjects framing assessments. As is the case with most laboratory demonstrations of within-subjects framing effects, the twenty-two items are presented in two blocks of eleven items—with many other subtests intervening between the two blocks. Each block of eleven items contains one item from each pair. When the positive frame of an attribute framing pair appears in one block, the negative frame appears in the other block. Likewise, when the gain frame of a risky-choice pair appears in one block, the loss frame appears in the other block. Positive frames and gain frames appear roughly equally in each of the two blocks.

In some previous research, framing effects have been scored discretely as 0 or 1 depending on whether the subject displayed a framing effect in the expected direction. The expected direction of the framing effects on the attribute framing problems was that subjects would be more positively disposed toward the positively worded item in the pair than the negatively worded item. The expected direction of the framing effects on the risky-choice framing problems was that the subject was expected to be more positively disposed toward the risky option in the loss framing than in the gain framing (see Kahneman & Tversky, 1979; Tversky & Kahneman, 1981).

As with the Ratio Bias subtest discussed in chapter 7, we use a continuous six-point response scale in the Framing subtest that may be more sensitive to individual differences (see Bruine de Bruin et al., 2007). Problems were scored by subtracting their negative frame ratings (loss in risky-choice and negative frame in attribute framing) from their corresponding positive frame ratings (gain in risky-choice and positive frame in attribute framing). Given the direction of the response scales presented in the appendix, negative difference scores thus indicated a framing effect in the expected direction and represented a violation of the principle of descriptive invariance. Difference scores of 0 indicated the absence of a framing effect. Given the direction of the response scales, positive difference scores indicated a framing effect in the unexpected direction. It is important to note that displaying a framing effect on this continuous difference-score metric does *not* necessarily mean that the subject has shown an outright preference reversal; it shows only that the frame has at least altered the subject's choice on the response scale.

In an unpublished study in our laboratory (RT53) involving 264 university students as subjects, significant framing effects in the expected direction were found for each of the eleven framing pairs, although they varied in size (M differences = –0.13 to –0.54; $t(263)$ = –1.86 to –8.88). For each of the eleven problem pairs, the modal score was 0 (indicating no framing effect), but for each of the problems more subjects were framed in the expected direction than in the unexpected direction.

For the purposes of the composite score, we multiplied by –1 the positive difference scores of subjects showing framing effects in the unexpected direction. This was so that subjects who showed reverse framing effects would not receive higher scores than those who displayed descriptive invariance (no framing effect). It also treats either type of framing effect as equally nonnormative. After multiplying the positive scores by –1, we formed a composite framing score by summing the eleven difference scores. The mean composite score was –8.55 (SD = 5.0) and Cronbach's alpha was 0.61. This reliability is understandably moderate given that there are only eleven scores contributing and that each of the scores represents a difference score. It is similar to the result obtained by Bruine de Bruin et al. (2007) using a thirteen-pair measure and a similar continuous scale. They obtained a Cronbach's alpha of 0.62 and a test-retest reliability of 0.58. In the appendix, the composite score is translated into the six-point weighting that this subtest is given on the CART.

Also in RT53, the framing composite score showed very modest correlations with measures of cognitive ability: 0.10 with SAT total scores; 0.15 with the Wonderlic Test; 0.15 with Shipley Vocabulary; and 0.23 with Shipley Abstract Reasoning (Shipley, Gruber, Martin, & Klein, 2009). With the large sample size, the latter three were still statistically significant. A composite cognitive ability measure combining all four of these indices displayed a significant 0.22 correlation with the Framing subtest composite score.[2]

Although correlations with cognitive ability were modest, the framing composite score did correlate with an earlier cognitive reflection measure (similar to the seven-item version used in Toplak et al., 2014a) at the level of 0.21 ($p < 0.005$). It also displayed statistically significant correlations with several thinking dispositions: 0.15 with need for cognition and 0.18 with the actively open-minded thinking scale used by Stanovich and West (2007).

Context Effects in Decision Making: Avoidance of Anchoring Effects

Many Type 1 processes and characteristics have the property that they automatically contextualize a problem-solving situation. This automatic tendency to amalgamate as much contextual information as possible was termed the "fundamental computational bias" by Stanovich (2003, 2004). Of course, the tendency to use all the contextual information available to supplement problem solving is more often a help than a hindrance. Certainly such a processing bias makes evolutionary sense (e.g., Buss, 2005, 2009; Buss & Schmitt, 2011; Pinker, 1997; Tooby & Cosmides, 1992). It is resident in the brain because it was adaptive in the so-called environment of evolutionary adaptedness (EEA). Nevertheless, despite making evolutionary sense, the modern world sometimes presents situations in which the type of contextualization rendered by the fundamental computational bias proves extremely problematic. These situations may be numerically minority situations, but they tend to be ones where a misjudgment might have disproportionately large consequences for a person's future utility maximization. It is just such hostile environments for Type 1 processing that many heuristics and biases tasks are trying to simulate (recall our discussion of hostile environments for Type 1 processing in chapters 2 and 3).

All situations where the contextual information is seemingly relevant but actually is not will be hostile processing situations for these types of Type 1 heuristics. One of several such effects in the heuristics and biases literature is the anchoring effect introduced by Tversky and Kahneman (1974). The avoidance of such anchoring effects is what we measure in the subtest of the CART that is the focus of this section.

In their famous experiment, Tversky and Kahneman (1974) had subjects watch a spinning wheel and when the pointer landed on a number (rigged to be the number 65), they were asked whether the percentage of African countries in the United Nations was higher or lower than this percentage. After answering this question, the subjects then had to give their best estimate of the percentage of African countries in the United Nations. Another group of subjects had it arranged so that their pointer landed on the number 10. Now it is clear that because a spinning wheel was used, the number involved in the first question is totally irrelevant to the task of answering the second question. Yet the number that came up on the spinning wheel affected the answer to the second question. The mean estimate of the first group (the group where the spinning wheel stopped at 65) turned out to be significantly larger (45) than the mean estimate (25) for the second group. Both groups were using the anchoring and adjustment heuristic—the high anchor group adjusting down and the low group adjusting up—but their adjustments were "sticky." They were not adjusting enough because they had failed to fully take into account that the anchor was determined in a totally random manner. This anchoring phenomenon reveals a miserly tendency to rely too much on an anchor regardless of its relevance.

As always in cognitive psychology, after the initial discovery of an important phenomenon, our understanding of the phenomenon quickly "complexifies." For example, sometimes anchoring appears to derive from insufficient adjustment from an anchor and other times it is due to the increased accessibility of anchor-consistent information (the former when the anchor is self-generated and the latter in the standard paradigm, see Epley & Gilovich, 2001, 2004, 2006). A more fine-grained view of how anchoring and adjustment works (and the theoretical complications surrounding various theoretical interpretations) is provided in many other publications and is beyond our scope here.[3]

The large research literature on anchoring has shown that many of the effects generalize to real-life situations (see Stewart, 2009). Also, it has been found that even when an anchor is not randomly determined, the cognitive miser tends to rely on it too much. For example, it has been found that even experienced real estate agents are overly affected by the listing price of a home when trying to assess its actual value (Northcraft & Neale, 1987). Likewise, in personal injury cases, the amount of compensation requested affects the judgment itself as well as the amount awarded to the plaintiff. Also, it has been shown that, statistically, prosecution requests for sentences affect the sentencing of judges as well as bail decisions. Judges appear to be cognitive misers too—they succumb to simple heuristics that promise to lighten the cognitive load (Englich, Mussweiler, & Strack, 2006).

We view anchoring effects as a member of a class of effects related to the mindless use of reference points.[4] Paying attention to irrelevant reference points (as was discussed in the context of framing) renders many choices and evaluations nonnormative. Also, similar to the case of framing effects, we view anchoring effects as instances of the class of miserly processing that we have termed serial associative cognition (see figure 2.3 and table 3.1).

The Anchoring subtest of the CART contains eight items (see appendix for sample items). We used a standard format and very typical questions for our subtest. Each item has two parts. The first is a question containing the quantitative anchor (either high or low), and the question always has dichotomous alternative answers (usually just yes or no). Following the two-choice question, the subject is asked for an estimation of the quantity in question. There are four low anchors and four high anchors in the eight items of the anchoring subtest of the CART. As is the case with most experiments subsequent to those of Tversky and Kahneman (1974), the instructions make no effort to convince the subjects that the anchors were drawn at random (unlike the example presented previously with the spinning wheel).

One difficulty with assessing anchoring in a situation like our CART subtest is that we need an avoidance of anchoring score for each individual who takes the subtest. Those familiar with the anchoring literature will immediately see the problem here. The study of anchoring effects in the heuristics and biases literature has overwhelmingly relied on between-subjects designs, where one group of subjects is presented the item with a high anchor and the other group of subjects is presented the item with the low anchor. The presence of an anchoring influence is inferred using between-subjects statistics. Such a design fails to provide an anchoring score for an individual subject—it fails to indicate whether a particular subject has been anchored more than another subject.

We see two ways to address this problem. One is to try to run an anchoring assessment within subjects, somewhat like what was done for the Framing subtest. With this method, subjects would get a second block involving the same estimations but now preceded by the other anchor—that is, the anchor they did not receive for that item on the first block. As with the study of framing effects within subjects, this type of design logic inserts a period of time and other tasks between the two blocks. Nonetheless, such a design quite substantially changes the psychological nature of the anchoring task. It now seems to become more of a task involving response consistency—that is, a task assessing the felt need for such consistency across the two responses and the information-processing capability (that is, memory) to remember the previous response.

The other method is to give the subject only one of the anchors (either high or low) for each of the items as in the standard between-subjects design, but to try to score the degree of anchoring of a given subject on each item (thereby allowing the summation of anchoring effects across items to form a composite score). We attempted to use this method by using the notion of a calibration group introduced by Jacowitz and Kahneman (1995). A calibration group is a group of subjects who make the estimation judgments but do not receive the prior anchoring question. The performance of the calibration group tells the investigator what normal estimations are on that item without the influence of an arbitrary anchor. With this piece of information, along with the knowledge of the actual correct value for the estimation, an arbitrary cut point can be determined on one side of which the subject is scored as avoiding anchoring (one point) and on the other side of which the subject is scored as not avoiding anchoring (scored as zero points).

The procedure can best be illustrated with specific examples from our Anchoring subtest.[5] The eight items were run in a laboratory study involving 139 university students (RT55). We obtained information from a separate calibration group who completed the estimation task without anchoring. This calibration sample was a group of 200 subjects who completed the task on the Amazon Mechanical Turk.

Consider, for example, the redwood tree item in our subtest. The actual height of the tallest redwood is 370 feet. With the high anchor used in this item (1,000 feet) the median value in the sample in our study was 975 feet. Only 16.6 percent of the sample gave a value below 400 feet. In the calibration sample that we studied on the Amazon Mechanical Turk, the median value given was 250 feet. Thus, the calibration sample shows evidence of a considerable anchoring effect for this item. We decided to use a liberal scoring cutoff for this item, scoring every estimate of 500 feet or below as one point (some avoidance of anchoring) and every estimate over 500 feet as zero (little avoidance of anchoring). Even with this liberal scoring criterion, only 26.6 percent of the sample scored a point on this item.

The total raw score on this subtest thus ranges from 0 to 8. The mean score was 3.29 (SD = 1.72) in RT55. The reliability of the total score was quite modest (Cronbach's alpha of 0.42). The raw total score on the Anchoring subtest displayed a correlation of 0.19 with verified SAT total scores, and a correlation of 0.22 with performance on sixteen belief bias syllogisms. It displayed its highest correlation (0.35) with a seven-item cognitive reflection measure similar to that used by Toplak et al. (2014a).

Because of the low reliability observed in this study, we altered the items and scoring in the study of the full-form test (RT60) that is reported in chapter 13. The scoring rules were slightly different in RT60, which resulted in a slightly higher reliability. The appendix presents two examples of the scoring cutoffs for RT60. Below the cutoff values in the appendix, the raw score is translated into the three-point weighting that this subtest is given on the CART.

One final caveat regarding this task concerns its rather obvious dependence on prior knowledge (see Smith & Windschitl, 2015). If one knows the exact (or close approximate) value of the quantity in question, then one will not be anchored by the priming question. Thus, as we have discussed extensively throughout this volume, the Anchoring subtest taps a complex combination of knowledge and the tendency to be unduly influenced by irrelevant information.

Context Effects in Decision Making: Avoidance of Preference Anomalies

Anchoring effects, as they are studied in cognitive psychology, represent an example of irrelevant context affecting our judgments. Likewise, framing effects represent a case of irrelevant context affecting our preferences. Framing, as studied in the literature we have reviewed, is only one type of preference anomaly that has been examined in the heuristics and biases literature; there are many more (see Kahneman, 2011; Kahneman & Tversky, 2000; Thaler, 2015). In the Preference Anomalies subtest of the CART, we assess the ability to avoid several other anomalies in decision making. This subtest is a bit of a potpourri, combining anomalies of various kinds. However, many of them do arise from similar psychological mechanisms. Several are consistent with the "constructed preference" model of how people make decisions, the dominant model of preference anomalies in cognitive science (Kahneman & Tversky, 2000; Krantz & Kunreuther, 2007; Lichtenstein & Slovic, 2006; Shafir & Tversky, 1995; Simonson, 2008; Slovic, 1995; Tversky & Kahneman, 1981, 1986; Tversky & Thaler, 1990).

Instead of assuming a set of preferences that exist prior to the choice situation in a stable, well-ordered form, the contemporary view is that preferences are constructed "online"—in direct response to the probe for a decision. Importantly, the preferences are constructed in part by using cues in the elicitation situation to help access decision-relevant information in memory (Schwarz, 1996). While it creates problems for viewing humans as purely rational creatures, the constructed preference view does map nicely onto contemporary dual-process views of cognition (Evans & Stanovich, 2013). To see this connection, it is important to realize that when presented with the rational choice principle (transitivity, descriptive invariance, dominance, etc.) that they have just violated in a choice situation, most subjects will actually endorse the axiom (Koehler & James, 2009; Shafir, 1993, 2003; Stanovich & West, 1999; Tversky, 1996). That subjects endorse the strictures of rationality when presented with them explicitly suggests that most subjects' analytic abilities are such that they can appreciate the intellectual force of the axioms of rational choice. Our dual-process interpretation of the constructed preference view centers around the idea of biased sampling procedures (Stanovich, 2004; see also Simonson, 2008).

For any given choice option, there is, throughout the brain, the sum total of information relevant to the option (call this the "network total"). The preference for the option is the abstraction that is the valenced total of all of the information in the connectionist networks relevant to it. This quantity is not the measure of a single entity, but is instead a theoretical abstraction somewhat like the center of gravity (Dennett, 1991). Preference reversals and other preference anomalies result from sampling procedures that bias the information recruitment process in systematic ways. Even though two choice options are formally equivalent and thus map into the same network total, it is not the network total that determines the response, but a sample of information from it. The subset of information that might be recruited for two equivalent versions of the problem might be different because their descriptions led to sampling from the network in two different ways.

The main assumption in the classical view that is incorrect is the assumption that System 2 has exhaustive and reliable access to *all* information relevant to a decision. In fact, the analytic system samples based on the retrieval cues it is primed with, and problems worded in different ways often contain different retrieval cues. Once evidence has to be sampled, differentially biased retrieval becomes an obvious possibility.[6]

With this as our theoretical background, to construct the Preference Anomalies subtest of the CART, we sampled a wide variety of effects from the literature (Kahneman & Tversky, 2000; Lichtenstein & Slovic, 2006). Most probably result from biased sampling of cues triggered by the specific wording of the question. We were interested in measuring the ability to avoid being biased by context in decision making. In fact, the specific model of these effects is of less pertinence than the effects themselves (regardless of their theoretical explanation). Hence, we did not seek to explore only a theoretically coherent class of effects. Indeed, some of these items might be considered examples of framing effects, under a looser definition of what a framing effect represents (Frisch, 1993; Stanovich & West, 1998b).

Our final version of the Preference Anomalies subtest of the CART has nine items (see the appendix for a sample item). The nine items on this subtest assess a variety of effects, including the certainty effect in decision making (Kahneman & Tversky, 1979); outcome bias (Baron & Hershey, 1988); the undue weighting of the word "free" effect (Ariely, 2008); omission bias (Baron & Ritov, 2004; Zikmund-Fisher et al., 2006); and fairness reversals (two items; see Kahneman, Knetsch, & Thaler, 1986). Finally, three problems in this subtest concerned so-called less-is-more effects that have been demonstrated in the preference anomalies literature—that, given the proper context, sometimes people can prefer less to more. For example, Slovic, Finucane, Peters, and MacGregor (2002) found that people rated a gamble with 7/36 chance to win $9 and 29/36 to lose 5¢ more favorably than a gamble with 7/36 chance to win $9 and 29/36 chance to win nothing. Presumably, the representation of the numerically small loss highlights the magnitude and desirability of the $9 to be won, leading subjects to rate that option more highly. Note that the subjects violated the dominance stricture (Dawes, 1998; Savage, 1954: Tversky & Kahneman, 1986), a fundamental rule of rational choice.

Another less-is-more phenomenon relies on the fact that the cognitive miser uses the easily processed cue of salience—a cue that can sometimes overly bias processing. A study by Yamagishi (1997) demonstrated that people rated a disease that killed 1,286 out of 10,000 people as more dangerous than one that killed 24.14 percent of the population. Again, the vividness of representing 1,286 actual people rather than an abstract percentage is presumably what triggers an affective response that leads to a clearly suboptimal judgment. We used two items of Yamagishi type and one of the Slovic type in the Preference Anomalies subtest of the CART.

The nine items were run in a laboratory study involving 322 university students (RT57). The total raw score on this subtest ranges from 0 to 10 because the omission bias item has a maximum of 2 points. The mean total score was 4.86 (SD = 1.76). The reliability of the total score was very low (Cronbach's alpha of 0.28). In the appendix, the raw score is translated into the three-point weighting that this subtest is given on the CART. In the same laboratory study, the raw total score on the Preference Anomalies subtest displayed a correlation of 0.11 with SAT total scores and a correlation of 0.14 with the Reflection Versus Intuition subtest of the CART.

Avoiding Myside Bias: The Argument Evaluation Subtest

Critical thinking is often thought to entail the ability to decouple prior beliefs and opinions from the evaluation of evidence and arguments.[7] The literature on Bayesian reasoning (e.g., de Finetti, 1989; Earman, 1992; Fischhoff & Beyth-Marom, 1983; Howson & Urbach, 1993) provides justification for the emphasis on unbiased evidence evaluation in the critical thinking literature. The key reasoning principle captured by Bayes' theorem (see chapter 6) is that the evaluation of the diagnosticity of the evidence (the likelihood ratio) should be conducted *independently* of the prior odds favoring the focal hypothesis.

In sum, prior beliefs are encompassed in one of two multiplicative terms that lead to the posterior probability, but the diagnosticity of the evidence should be assessed separately from the prior belief. Nevertheless, people often fall short of this rational ideal by displaying both belief bias and myside bias (Thompson & Evans, 2012). Belief bias occurs when people have difficulty evaluating conclusions that conflict with what they think they know about the world. The Belief Bias in Syllogistic Reasoning subtest of the CART discussed in chapter 7 is a direct measure of the ability to avoid such biased reasoning. Relatedly, people display myside bias when they evaluate evidence, generate evidence, and test hypotheses in a manner biased toward their own opinions (Stanovich, West, & Toplak, 2013). Myside bias has sometimes been viewed as a subclass of confirmation bias (see Hahn & Harris, 2014; McKenzie, 2004) and is related to the construct of actively open-minded thinking (Baron, 2008).

Myside bias has been amply demonstrated in numerous empirical studies (e.g., Greenhoot et al., 2004; Taber & Lodge, 2006; Toplak & Stanovich, 2003; Wolfe & Britt, 2008). For example, Klaczynski (1997; Klaczynski & Lavallee, 2005; Klaczynski & Robinson, 2000) presented subjects with flawed hypothetical experiments that led to either opinion-consistent or opinion-inconsistent conclusions and evaluated the quality of the reasoning used when the subjects critiqued the flaws in the experiments. Klaczynski and colleagues found a tendency to critique opinion-inconsistent experimental results more harshly than opinion-consistent ones.

In several studies using an argument evaluation paradigm (Stanovich & West, 2008a), we had subjects rate the quality of arguments about abortion (and another issue—lowering the drinking age—that yielded similar results). Arguments were one-sided (all pro-choice or all pro-life statements) and two-sided (an equal number of pro-choice and pro-life statements). The pro-choice and pro-life arguments were prejudged by experts to be approximately equivalent in quality and strength. Consistent with some previous research (Baron, 1995), we found that one-sided arguments were preferred to two-sided arguments (regardless of direction). Also, we observed a strong myside bias. That is, participants rated the arguments consistent with their own position as better than the arguments not consistent with their own position. In another study, we found that American subjects were more likely to approve the United States banning of a very dangerous German car than they were of the Germans banning of an equally dangerous American car (Stanovich & West, 2008b).

The examples discussed here do not begin to exhaust the demonstrations of myside bias in cognitive psychology (Baron, 1995, 2008), social psychology (Ditto & Lopez, 1992; Edwards & Smith, 1996; Lord, Ross, & Lepper, 1979; Munro, 2010), political science (Taber & Lodge, 2006), cognitive neuroscience (Westen et al., 2006), or the informal reasoning literature (Kuhn, 1991, 1993; Slusher & Anderson, 1996). It would seem that there are many myside paradigms to choose from when constructing a subtest for the CART. However, only quite recently has attention been paid to individual differences in such effects (Stanovich, West, & Toplak, 2013); and when we start to focus on individual differences, we in fact find several problematic features of the empirical literature.

The first problem is that the vast majority of experimental demonstrations of myside bias use between-subjects designs. Such designs do not yield a myside bias susceptibility score for each subject—something that is essential in an assessment device like the CART. Take, for example, a between-subjects design used by Stanovich and West (2007, 2008a) to study so-called natural myside bias. Natural myside bias is the tendency to evaluate propositions from within one's own perspective when given no instructions to avoid doing so and when there are no implicit cues (such as within-subjects conditions) to avoid doing so. Stanovich and West (2007) defined the participant's perspective as their previously existing status on four variables: their sex, whether they smoked, their alcohol consumption, and the strength of their religious beliefs. Participants then evaluated a proposition relevant to each of these demographic factors. For example, the proposition for the demographic variable sex was: "The gap in salary between men and women generally disappears when they are employed in the same position." Myside bias was defined between subjects as the mean difference in the evaluation of the proposition between groups with differing prior status on the variable. This is fine for questions regarding generic theories of myside bias, but it is unsatisfactory for cognitive assessment because it does not yield a score for a given subject that indicates his or her propensity to avoid myside bias.

Not all studies in the literature use solely between-subjects designs. Some have measured myside bias in within-subject situations that could, in principle, yield a myside susceptibility score for a particular individual. In these types of studies, however, another problem arises. Often, the logistics of the study dictate that not many different myside issues can be studied. Constructing a score on the basis of a single issue or a small set of issues is problematic because questions of generality arise. For example, we (Stanovich & West, 2008a) adapted a paradigm originally used by Baron (1995) in which he had participants evaluate the thinking of hypothetical students on the issue of abortion. Participants evaluated paragraphs that were transcriptions of the arguments made by the hypothetical students while thinking about the issue. Participants graded the paragraphs on an A to F scale. The results revealed a myside bias—participants gave higher grades to thinking that coincided with their own opinion on the issue.

In theory, one could use the difference between the myside and otherside conditions in this experiment as an index of myside bias for a given subject, because every subject received both conditions. In fact, in this very study we did just that and found minimal correlations with individual difference variables (both cognitive ability and thinking dispositions) that were most often not significant. However, the danger in using such an index based on a single issue was illustrated by another aspect of our results: the pro-life group displayed more myside bias than did the pro-choice group. Of course we checked the lack of overall correlation between the SAT total score and myside bias by computing it separately for the pro-life and pro-choice groups (it failed to correlate in both instances).

More importantly, a more refined analysis (see Stanovich & West, 2008a) revealed further how myside bias was related to the content and the strength of the prior opinion rather than to cognitive ability. In a subsequent experiment we examined the abortion issue again but also looked at another issue—the legal age for drinking—that is not as emotionally charged as the abortion issue. Certainly from the standpoint of a laboratory demonstration, extending our exploration of myside bias in the Stanovich and West (2008a) study from one issue to two was an important improvement. However, we are still a long way away from an index that we can hope to have any generalizability. This is even more true in light of evidence (Toplak & Stanovich, 2003) that there is considerable specificity in myside bias across issues. Nevertheless, the logistical constraints of the Baron (1995) paradigm mitigate against expanding the number of issues, because of the length of the paragraphs that the subjects must read. Also problematic is that balanced sets of arguments on either side of the issue have to be constructed for as many issues as are examined.

Because of the limitations observed in other methods, in searching for a paradigm for the subtest of the CART that measures the avoidance of myside bias, we were drawn to an argument evaluation task that we had invented some time ago (see Stanovich & West, 1997). It is an informal reasoning paradigm similar to some critical thinking measures (Watson & Glaser, 2006), but it has several advantages. For example, it allows an aggregate score to be calculated for each subject and involves almost two dozen different issues. We will now describe the unique features of this measure as well as some of its complexities and drawbacks.

Because the avoidance of belief bias and myside bias is central to most definitions of rational thinking, it is critical to use material that would be most likely to provoke potential biases. The problem is that previous assessment devices tend not to *measure* the prior belief, but instead try to distribute arguments of varying quality across issues that vary in their controversial nature, hoping that, for a given subject, items balance out (in terms of prior agreement). Any lack of balance will make the argument evaluation task unusually difficult for some subjects and unusually easy for others.

The logic of the task that we designed (Stanovich & West, 1997) was aimed at directly addressing this problem. With this task—the Argument

Evaluation Test (AET)—we introduced an analytic technique for deriving an index of a person's reliance on the quality of an argument independent of his or her own personal opinion about the issue in question. Our methodology involved assessing, on a separate instrument, the participant's prior opinions about a series of propositions. On an argument evaluation measure, administered at a later time, the participants evaluated the quality of arguments related to the same propositions. The arguments had an operationally determined objective quality that varied from item to item. Our analytic strategy was to regress each subject's evaluations of the arguments simultaneously on the objective measure of argument quality and on the strengths of the opinions he or she had about the propositions prior to reading the arguments. The standardized beta weight for argument quality then becomes an index of that subject's reliance on the quality of the arguments independent of his or her opinions on the issues in question.

The Stanovich and West (1997) methodology is different from the traditional logic used in critical thinking tests, and it is a more sensitive one for measuring individual differences. Rather than merely trying to balance opinions across items by utilizing a variety of issues and relying on chance to ensure that prior belief and strength of the argument are relatively balanced from respondent to respondent, in our task the prior opinion is *measured* and taken into account in the analysis. The technique allowed us to examine thought processes in areas of "hot" cognition where biases are most likely to operate (Babad & Katz, 1991; Kunda, 1990; Pyszczynski & Greenberg, 1987). With one exception, the propositions used in the Stanovich and West (1997) experiment all concerned real social and political issues on which people hold varying, and sometimes strong, beliefs (e.g., gun control, taxes, university governance, crime, automobile speed limits). These items were slightly modified for use in the Argument Evaluation subtest of the CART. For example, in one item, the target proposition was: "The welfare system should be drastically cut back in size." In the first part of the AET used by Stanovich and West (1997)—the prior belief section—participants indicated their degree of agreement with a series of twenty-three target propositions such as this on a four-point scale: strongly agree (scored as 4), agree (3), disagree (2), strongly disagree (1). The AET prior opinion items varied greatly in the degree to which the sample as a whole endorsed them—from a low of 1.79 for the item "It is more dangerous to travel by air than by car" to a high of 3.64 for the item "Seat belts should always be worn when traveling in a car."

After completing several other questionnaires and tasks, the participants complete the second part of the AET. The instructions introduced the subjects to a fictitious individual, Dale, whose arguments they were to evaluate. Each of the 23 items on the second part of the instrument began with Dale stating a belief about an issue. The 23 beliefs were identical to the target propositions that the subjects had rated their degree of agreement with on the prior belief part of the instrument (e.g., "The welfare system should be drastically cut back in size"). Dale then provides a justification for the belief. For example, "The welfare system should be drastically reduced in size because welfare recipients take advantage of the system and buy expensive foods with their food stamps." A critic then presents an argument to counter this justification. For example, "Ninety-five percent of welfare recipients use their food stamps to obtain the bare essentials for their families." The subject is told to assume that the facts in the counterargument are correct. Finally, Dale attempts to rebut the counterargument, for example, "Many people who are on welfare are lazy and don't want to work for a living." The subject is told to evaluate the strength of Dale's *rebuttal* to the critic's argument. The subject then evaluates the strength of the rebuttal on a four-point scale: very strong (coded as 4), strong (3), weak (2), very weak (1).

The analysis of performance on the AET required that the subjects' evaluations of argument quality be compared to some objective standard. We employed a summary measure of eight experts' evaluations of these rebuttals as an operationally defined standard of argument quality. Specifically, three full-time faculty members of the Department of Philosophy at the University of Washington, three full-time faculty members of the Department of Philosophy at the University of California, Berkeley, and the two authors of the scale (Stanovich & West, 1997) judged the strength of the rebuttals. The median correlation between the judgments of the eight experts was 0.74. Although the problems were devised by the two authors, the median correlations between the authors' judgments and those of the external experts were reasonably high (0.78 and 0.73, respectively) and roughly equal to the median correlation among the judgments of the six external experts themselves (0.73). Thus, the median of the eight experts' judgments of the rebuttal quality served as the objective index of argument quality for each item.

As an example, on the above item, the median of the experts' ratings of the rebuttal was 1.5 (between weak and very weak). The mean rating given for the item by the subjects was 1.93 (weak) although the participants' mean prior belief score indicated a neutral opinion (2.64).

Stanovich and West (1997) provide many indications of the validity of the experts' ratings—for example, the experts were vastly more consistent among themselves in their evaluation of the items than were the subjects. Another consideration was that subjects of higher cognitive ability were more likely to agree with the *experts* in their judgments of argument quality than with their fellow subjects of lower cognitive ability.

We examined individual differences in subjects' relative reliance on objective argument quality and prior belief by running separate regression analyses on *each subject's* responses. That is, we constructed a separate multiple regression equation for each subject. In the Stanovich and West (1997) study, the subject's evaluations of argument quality served as the criterion variable in each of 349 separate regression analyses. The 23 evaluation scores were regressed simultaneously on both the 23 argument quality scores and the 23 prior opinion scores. Thus, we conducted a total of 349 individual regression analyses (one for each subject). For each subject, these analyses resulted in two beta weights—one for argument quality and one for prior belief. The former beta weight—an indication of the degree of reliance on argument quality independent of prior belief—is the primary indicator of one's ability to evaluate arguments independent of one's beliefs.

The mean standardized beta weight for argument quality was 0.330 (SD = 0.222). These values varied widely—from a low of –0.489 to a high of 0.751. Only 30 of 349 subjects had beta weights less than zero. Across the 349 regressions, the mean beta weight for prior belief was 0.151 (SD = 0.218). Although low, this mean was significantly different from zero ($t(348) = 12.93$, $p < 0.001$). Nevertheless, these values also varied widely—from a low of –0.406 to a high of 0.662. Thus, individuals vary substantially in their reliance on argument quality and prior belief when evaluating the rebuttal arguments.

The items from our 1997 study (somewhat revised) form the basis of the Argument Evaluation subtest of the CART (see the appendix for sample items). Several of the items have undergone small wording changes as we have studied them through the years, but they are not substantially different from what we used in our original studies (Sá et al., 1999; Stanovich & West, 1997, 1998c). The instructions for deriving the raw beta weights are repeated in the appendix. In the appendix, the beta weight for experts' rating is translated into a CART score of 0 to 5. The translation of the beta weight into CART points is based on the results of RT60 rather than our earlier work, because the mean beta weight for argument quality (expert's opinion) in RT60 was 0.199, substantially lower than that observed in our earlier work (0.330), perhaps because of slight historical changes in some of the issues involved and the arguments surrounding them.

Stanovich and West (1997) demonstrated that there is considerable variability in the argument quality beta weight and that this variability has an orderly distribution (see their figure 1). In their study, this argument evaluation measure displayed a 0.35 correlation with SAT total scores and a 0.29 correlation with an actively open-minded thinking measure similar to that described in chapter 11 (the correlations in Stanovich & West, 1998c, were similar: 0.33 and 0.23; and those in Sá et al., 1999, were a little higher: 0.45 and 0.44).[8] Our newest study that included the Argument Evaluation subtest, RT60, produced findings similar to these correlations obtained in earlier studies. Other laboratories have also done convergent work with our argument evaluation items (Oyer et al., 2012; Svedholm & Lindeman, 2013; Thompson & Evans, 2012).

The analytic logic we used in designing the Argument Evaluation subtest was intended to provide a means of separating prior belief from the assessment of how well an individual can track argument quality. The beta weight for argument quality (experts' rating) in our analyses captures a quintessential aspect of critical thought: the ability to evaluate the quality of an argument independent of one's feelings and personal biases about the proposition at issue. It is true that our measure, in requiring a separate regression analysis for each subject, is somewhat more complicated to score than other measures of the avoidance of myside thinking. Nonetheless, it does have several advantages. First of all, although 23 different issues is still a modest number as far as generalization is concerned, it is substantially larger than the number of different issues that are usually aggregated in myside bias studies in the literature. When only one or two issues are used, as in Baron (1995) and Stanovich and West (2008a), it becomes unclear whether the individual differences obtained are issue specific or are generalizable across different opinion domains. In our 1997 data, we assessed the reliability of the measure by correlating the regression weights obtained from the odd items with the regression weights obtained from the even items and found a disappointing reliability coefficient of only 0.35. It was somewhat higher (0.51) in RT60. Nevertheless, despite the instability of beta weights estimated from so few items, we were able to consistently observe the correlations with individual difference variables (correlations in the 0.25 to 0.40 range).

It should be noted that the myside-bias difference score in other studies (e.g., Stanovich & West, 2008a) is not analogous to the regression coefficient for argument quality (experts' rating) in our argument evaluation subtest. A high score on the myside-bias difference score in Stanovich and West (2008a) is a direct indicator of the magnitude of the bias, whereas a low score for beta weight in our paradigm may come about in a variety of ways. High correlations of subject evaluations with prior opinions do serve to reduce the beta weight for argument quality in our analyses. However, the beta weight for argument quality in our paradigm will also be low if a subject is simply a poor argument evaluator.

Our task is a measure of the ability to reason in situations where prior beliefs may be interfering (that is, reasoning in the face of potentially interfering prior beliefs). Our statistical measure (the critical beta weight) combines context-free reasoning ability with the ability to ignore prior opinion. It is thus a more complex index (and clearly less desirable than a direct myside-bias index for some purposes). Nevertheless, it involves the decontextualization skills that rational thinking theorists have defined as critical to the ability to separate evidence from opinion (see Stanovich, 1999, chapter 6, for a discussion; see also Stanovich, 2003, 2004).

Avoiding Overconfidence: The Knowledge Calibration Subtest

In the Probabilistic and Statistical Reasoning subtest (chapter 5), we described tests of various probabilistic reasoning tendencies. In this section, we will discuss a subtest of the CART that measures an aspect of probabilistic reasoning not covered in that subtest: how people calibrate their degree of knowledge with probability estimates.

Psychologists have done numerous studies using the so-called knowledge calibration paradigm (Fischhoff, Slovic, & Lichtenstein, 1977; Griffin

& Tversky, 1992; Hilton et al., 2011; Yates, Lee, & Bush, 1997). In this paradigm, the subject makes a large set of probability judgments. Of course, a single probability judgment by itself is impossible to evaluate. However, a large set of such judgments can be evaluated because, collectively, the set must adhere to certain statistical criteria.

For example, if the weather forecaster says there is a 90 percent chance of rain tomorrow and tomorrow turns out to be sunny and hot, there may be nothing wrong with that particular judgment. It just happened to be unexpectedly sunny on that particular day. However, if you found out that on half of the days the weatherperson said there was a 90 percent chance of rain and it did not rain, then you would be justified in seriously questioning the accuracy of weather reports from this outlet. You accept that the weatherperson does not know on which 10 percent of the days it will not rain, but overall you expect that if, across the years, the weatherperson has predicted "90 percent chance of rain" on 50 different days, then on about 45 of them it will have rained.

The most popular method for the assessment of knowledge calibration proceeds in exactly the same way as we evaluate the weatherperson. People answer multiple choice or true/false questions and, for each item, provide a confidence judgment indicating their subjective probability that their answer is correct. Epistemic rationality is optimized only when one-to-one calibration is achieved—that the set of items assigned a subjective probability of 0.70 should be answered correctly 70 percent of the time, that the set of items assigned a subjective probability of 0.80 should be answered correctly 80 percent of the time, and so on. This is what is meant by good knowledge calibration. It means, in a sense, that a person must know what he knows, and what he does *not* know, as well. If such close calibration is not achieved, then a person is not epistemically rational because his or her beliefs do not map onto the world in an important way.

The standard finding across a wide variety of knowledge calibration experiments has been one of overconfidence.[9] Subjective probability estimates are consistently higher than the obtained percentage correct. The overconfidence effect in knowledge calibration is thought to derive at least in part from our tendency to fix on the first answer that comes to mind, to then assume "ownership" of that answer, and to cut mental costs by then privileging that answer in subsequent evidence search and evaluation. The evidence retrieved for each of the response alternatives forms the basis for the confidence judgments, but the subject remains unaware that the recruitment of evidence was biased—that evidence was recruited only for the favored alternative. As a result, subjects end up with too much confidence in their answers. Thus, as figure 4.1 indicates, overconfidence in knowledge calibration groups with effects such as anchoring and framing in being an instance of miserly cognition that results from serial associative cognition with a focal bias.[10]

Overconfidence effects have been found in perceptual and motor domains as well as in knowledge calibration paradigms (Baranski & Petrusic, 1994, 1995; Mamassian, 2008; West & Stanovich, 1997). They are not just laboratory phenomena, but have been found in a variety of real-life domains such as the prediction of sports outcomes (Ronis & Yates, 1987), prediction of one's own behavior or life outcomes (Hoch, 1985; Vallone, Griffin, Lin, & Ross, 1990), and economic and medical forecasts (Åstebro, Jeffrey, & Adomdza, 2007; Braun & Yaniv, 1992; Groopman, 2007; Tetlock, 2005). Overconfidence is manifest in the so-called planning fallacy (see Buehler, Griffin, & Ross, 2002)—the fact that we often underestimate the time it will take to complete projects in the future (for example, to complete an honors thesis, to complete this year's tax forms, to finish a construction project).

Many different effects have been lumped under the overconfidence rubric, perhaps, in some cases, without complete theoretical justification (Griffin & Brenner, 2004; Hahn & Harris, 2014; Hilton et al., 2011; Klayman et al., 1999; Mata, Ferreira, & Sherman, 2013; Shepperd et al., 2013). For example, the relation of overconfidence in knowledge calibration to what Moore and Healy (2008) call "overplacement" is a matter of theoretical dispute (Hilton et al., 2011). Nevertheless, overconfidence in knowledge calibration is an important rational thinking skill in its own right (Kahneman, 2011) and has been related to outcome variables such as financial decisions (Biais et al., 2005; Camerer & Lovallo, 1999; Hilton et al., 2011; Russo & Schoemaker, 1992).[11]

The Knowledge Calibration subtest of the CART (see the appendix for sample items) has two parts: one a typical two-choice knowledge calibration paradigm and the other a confidence interval paradigm (see Klayman et al., 1999, for a comparison of these two methods). Part 1 of this subtest consists of our version of the much-used two-choice knowledge calibration paradigm. The methods and analyses used in this task were similar to those used in the classic literature on knowledge calibration (Lichtenstein, & Fischhoff, 1977, 1980; Ronis & Yates, 1987; Yates et al., 1989). Subjects answered 36 general knowledge questions in a two-choice format. Questions were either drawn from or inspired by books similar to Zahler and Zahler's (1988) *Test Your Cultural Literacy*. An example of a typical item (for which the correct answer is b) is:

What was the initial purpose of the League of Nations?
a. It was intended to combat the growing threat of Germany and Italy before World War I.
b. It was intended to preserve peace after World War I.

After answering each question, subjects indicated their degree of confidence in their answer on the following scale:

100 percent chance that I answered correctly (I am certain)
90 percent chance that I answered correctly
80 percent chance that I answered correctly
70 percent chance that I answered correctly
60 percent chance that I answered correctly
50 percent chance that I answered correctly (I was just guessing)

There are several different indices of knowledge calibration performance. Yates et al. (1989) and Ronis and Yates (1987) should be consulted for discussions of the computational and conceptual details of these indices (for example, calibration-in-the-small and discrimination). The most common index—and computationally the simplest—is used to score the Knowledge Calibration subtest of the CART. Termed the measure of over/underconfidence by Lichtenstein and Fischhoff (1977) and bias by Yates et al. (1989), it is simply the mean percentage confidence judgment minus the mean percentage correct.

Despite its mathematical simplicity, this difference score index is conceptually somewhat complex. Moore and Healy (2008) have discussed how it conflates what they call *overestimation* with what they call *overprecision*. Parker and Stone (2014) have shown how this overconfidence bias difference score confounds confidence and knowledge. Stanovich (1999, pp. 118–120) used a knowledge matching procedure to address this problem in a study of individual differences in overconfidence. Parker and Stone (2014) have employed a more elegant regression procedure, but their procedure requires a reference sample to construct the regression equation and thus is an inelegant solution for an assessment device such as ours. However, in various simulations, Parker and Stone (2014) demonstrate that when confidence and knowledge are positively correlated, the difference score and the regression beta weight converge on similar correlations with a third individual difference variable. Because our confidence/knowledge correlations are substantial (0.41 in Experiment 4 of Stanovich & West, 1998c; 0.38 in Turk5; and 0.44 in RT60), and for ease of computation, we used the numerically simpler difference score as our index in the CART.

We updated stimuli and methods from our earlier investigation (Stanovich & West, 1998c) in a study of 200 subjects on the Amazon Mechanical Turk (Turk5). The mean confidence rating for these 36 items was 73.8 percent (SD = 9.5) and the mean percentage correct was 64.7 percent (SD = 12.7), yielding a substantial overconfidence effect of 9.1 percent (SD = 12.5), significantly different from zero ($t(199) = 10.23$, $p < 0.001$). The positive sign of the mean score indicates that the sample as a whole displayed overconfidence, the standard finding with items of this type. An overconfidence bias was displayed by 152 (76.0 percent) of the 200 subjects completing this task. The split-half reliability (odd-even; Spearman-Brown corrected) of the raw difference score was 0.68.

At the end of the 36 questions the subject was asked to make an aggregate estimate of the number of items answered correctly. The mean estimate in Turk5 was 23.1 (SD = 3.8) items. The actual mean number of items answered correctly was 23.3 (SD = 4.5), indicating that the sample did not show an overconfidence bias overall when estimating the aggregate performance. This finding is consistent with previous research showing that aggregate estimates are often less overconfident than item-by-item estimates (Gigerenzer, Hoffrage, & Kleinbolting, 1991; Griffin & Tversky, 1992; Sniezek & Buckley, 1991). However, unlike previous experimentation where aggregate estimates still showed some level of overconfidence bias (Soll & Klayman, 2004), albeit lower than that on an item-by-item basis, our study showed no overconfidence bias at all. Nevertheless, we retain the aggregate judgment difference score as a measure of the avoidance of overconfidence because it showed a very high correlation (0.75) with the overconfidence bias measured on an item-by-item basis.

Part 2 of our knowledge calibration subtest used a confidence interval method that is less common than the two-choice procedure but is still often employed (Hilton et al., 2011) and has a considerable history in the literature (Russo & Schoemaker, 1992; Lichtenstein, Fischhoff, & Phillips, 1982). Overconfidence effects are often larger in studies using the interval method than in studies using the two-choice method (Klayman et al., 1999; Soll & Klayman, 2004). The appendix presents the instructions for soliciting 90 percent confidence intervals for the quantities. The index of the ability to avoid overconfidence using this paradigm is the number of confidence intervals that contained the correct value—the number of "hits" achieved in constructing the confidence intervals (Soll & Klayman, 2004). This index is a direct measure of the avoidance of overconfidence bias.

The same study of 200 subjects on the Amazon Mechanical Turk (Turk5) indicated that the number of hits on this 15-item measure had a reasonably high reliability (Cronbach's alpha of 0.69). The mean number of hits was 6.2 (SD = 2.8), representing a hit rate of 41 percent, which is very typical of experiments using interval estimation with instructions like ours (Soll & Klayman, 2004). Of course, the expected number of hits for a confidence interval of that type would be 13.5, indicating that the sample displayed substantial overconfidence on this measure. Only 14.0 percent of the sample had 10 or more confidence intervals that contained the correct value.

The results from the two overconfidence tasks that were completed together in this study contrasted with those of Hilton et al. (2011), who found no correlation between overconfidence bias in the two-choice paradigm and the number of confidence intervals in the interval estimation method that contained the correct value.[12] We found a correlation of 0.30 ($p < 0.001$) between the number of hits in the interval measure and the overconfidence difference score (multiplied by –1) in the two-choice method.

In this study, subjects also made an aggregate estimate after the Part 2 interval estimation (see the appendix). Specifically, they were asked: "Out of all 15 fill-in-blank questions you just answered, for how many of the 15 questions do you think the answer will turn out to be within the interval you gave?" The mean aggregate estimate was 8.1 (SD = 3.3). Because the actual number of hits was only 6.2, the aggregate estimates in the interval task, unlike those in the two-choice task, displayed considerable overconfidence. The mean overconfidence effect of 1.9 (SD = 3.6) was significantly different from zero ($t(199) = 7.76$, $p < 0.001$). Like the two-choice aggregate measure, this aggregate measure was multiplied by –1 for consistent direction of correlations.

The aggregate estimation measure from the interval method displayed a correlation of 0.47 with the item-by-item interval overconfidence measure (number of hits). Furthermore, the aggregate estimation measure from the interval method displayed a statistically significant correlation of 0.32 with the two-choice aggregate overconfidence measure. The aggregate estimation measure from the interval method displayed a statistically significant correlation of 0.32 with the item-by-item two-choice overconfidence measure (all above $p < 0.001$).

The CART scoring system for the Knowledge Calibration subtest is indicated in the appendix. A maximum of two points are allocated to the item-by-item measure of overconfidence on the two-choice method of Part 1. A maximum of one point is allocated to the aggregate overconfidence measure from the two-choice method. A maximum of two points is allocated to the avoiding overconfidence index from the confidence interval method of Part 2. And finally, a maximum of one point is allocated to the aggregate estimation index from the interval estimation method.

Despite the substantial reliabilities of both of our measures of knowledge calibration, and despite the prominence of overconfidence in discussions of cognitive biases (Kahneman, 2011; Mata et al., 2013), we have allocated only six CART points to this subtest. This is because of the ongoing debate about the conceptual interpretation of common measures of overconfidence, as well as continuing debates about the statistical complexities of these measures (Glaser, Langer, & Weber, 2013; Moore & Healy, 2008; Parker & Stone, 2014; Soll & Klayman, 2004).

The Temporal Discounting Subtest

A prudent attitude toward the future that shifts psychological focus from the "here-and-now" to consideration of future outcomes has long been central to conceptions of rationality and wisdom (Baltes & Smith, 2008; Loewenstein, Read, & Baumeister, 2003; Staudinger, Dorner, & Mickler, 2005; Sternberg, 2003; Strathman et al., 1994). Several procedures have been used to measure attitudes toward the future, including temporal discounting measures, future orientation questionnaires, and delay of gratification tasks. In chapter 11 we will discuss a Future Orientation thinking disposition subscale of the CART. However, our primary measure of prudent

Avoidance of Miserly Information Processing

decision-making tendencies will be a Temporal Discounting subtest that involves aspects of sustained override.

Recall from chapter 3 our distinction between tasks that stress detection of conflict and those that stress sustained override. The items on the Reflection versus Intuition subtest and Disjunctive Reasoning subtest discussed in chapter 7 are examples of items where the key operation is detecting that there is an alternative to the intuitively compelling response. However, there are other situations where people have no trouble with detection—with realizing that they are made up of multiple minds. In fact, the struggle between minds is almost the defining feature of these situations. They are situations where we have to resist temptation: we have to get up and make breakfast despite wanting to sleep; we have to resist an extra $3 coffee in the afternoon because we know the budget is tight this month; we are on a diet and know that our snack should be carrots and not chips; we are at a casino having promised to lose no more than $100 and are now down $107 and really should stop, but …

It is only too apparent to us in these instances that parts of our brains are at war with each other. Our natural language even has a term to designate the hard thinking that is attempting to overcome the easy thinking in these instances: willpower. Willpower is a folk term, but cognitive researchers have begun to understand it scientifically (Ainslie, 2001, 2005; Baumeister & Vohs, 2003, 2007; Loewenstein, Read, & Baumeister, 2003; Rachlin, 2000). Our colloquial notion of willpower usually refers to the ability to delay gratification or to override visceral responses prompting us to make a choice that is not in our long-term interests. The inability to properly value immediate versus delayed rewards keeps many people from maximizing their goal fulfillment. The logic of many addictions, such as alcoholism, overeating, and credit card shopping, illustrates this point. From a long-term perspective, a person definitely prefers sobriety, dieting, and keeping credit-card debt low. However, when immediately confronted with a stimulus that challenges this preference—a drink, a dessert, an item on sale—the long-term preference is trumped by the short-term desire.

Temporal discounting can be said to differ from the ratio bias and belief bias situations in being closer to so-called hot override than so-called cold override. The former refers to the override of emotions, visceral drives, or short-term temptations (by analogy to what has been called "hot" cognition in the literature; see Abelson, 1963). The latter refers to the override of overpracticed rules, Darwinian modules, or tendencies of the autonomous mind that are not necessarily linked to visceral systems (by analogy to what has been called "cold" cognition in the literature).

Both temporal discounting and delay of gratification paradigms have been used to measure this hot override tendency. These tasks typically ask participants to choose between smaller immediate rewards or substantially larger delayed rewards. Selection of the larger delayed rewards is typically scored as more optimal (Ainslie, 1975; Kirby, 1997). Mischel (Ayduk & Mischel, 2002; Mischel & Ebbesen, 1970; Mischel, Shoda, & Rodriguez, 1989; Rodriguez, Mischel, & Shoda, 1989) pioneered the study of the delay of gratification paradigm with children. As is well known, the child receives the larger reward if, after the experimenter leaves the room, the child waits until the experimenter returns and does not signal the experimenter to return by ringing a bell. If the bell is rung before the experimenter returns, the child receives only the smaller reward. The dependent variable is the amount of time that child waits before ringing the bell. Performance in the paradigm has been related to intelligence and to later educational attainment (Mischel, 2015).

There is a large literature on the extent to which adults discount monetary amounts into the future ("Would you prefer $34 now or $50 in 30 days?"; see Kirby, 2009). Many different paradigms have been used to assess how people compare a smaller reward immediately to a larger reward in the future—and how much larger the future reward has to be in order to tip the preference (Frederick, Loewenstein, & O'Donoghue, 2002; Green & Myerson, 2004; Loewenstein, Read, & Baumeister, 2003; McClure, Laibson, Loewenstein, & Cohen, 2004). Many of these paradigms yield a curve with an individual's normalized indifference points plotted against time. These paradigms generate many different parameters to summarize task performance. One widely used index of temporal discounting is based on the area under the discounting curve (AUC) that results when an individual's indifference points are plotted (Myerson, Green, & Warusawitharana, 2001). Others are derived from fitting a specific curve to the subject's discounting function. Parameters are then derived from the formulas that describe these curves. We will use a simpler scoring scheme when we describe our subtest below, one that simply operationalizes certain levels of discounting as clearly less than rational. Our index is highly correlated with all the other discounting parameters in the literature (Basile & Toplak, 2015; Myerson, Baumann, & Green, 2014).

A number of developmental studies have found that increasing age is associated with less extreme temporal discounting (Green et al., 1994; Prencipe et al., 2006; Steinberg et al., 2009) and more future orientation (Steinberg et al., 2009). Longitudinal studies have reported that positive long-term cognitive, educational, and career outcomes can be predicted from an early willingness to delay rewards (Mischel et al., 2011; Prencipe et al., 2006; Steinberg et al., 2009). Data from adults converge with these findings (Kirby, Winston, & Santiesteban, 2005). Higher intelligence is associated with a greater tendency to wait for the larger monetary reward. Shamosh and Gray (2008) meta-analyzed this literature and found that, across twenty-four different studies, the correlation between the tendency to wait for delayed larger rewards and intelligence averaged 0.23.

Although delay-discounting paradigms are laboratory tasks, research has shown that they are correlated with important outcome behaviors. For example, temporal discounting performance has been significantly associated with excessive gambling (Petry, 2001; Petry & Casarella, 1999), drug addiction and smoking (Ainslie, 2001, 2005; Kirby & Petry, 2004), financial behavior (Ashraf, Karlan, & Yin, 2004; Meier & Sprenger, 2012), prudent food-stamp usage (Shapiro, 2005), educational success (Kirby et al., 2005), and a variety of other behaviors (Chabris et al., 2008; Reimers et al., 2009).

Our Temporal Discounting subtest for the CART contains two parts. Part 1 was constructed for ease of scoring and utilized the insights from a study of different discounting methodologies by Hardisty, Thompson, Krantz, and Weber (2013). They found that a fixed-sequence titration method was very convergent with a dynamic multiple staircase method and with various matching methods of measuring discount rates. Additionally, the fixed-sequence titration method was at least as effective in predicting outcome behaviors as the other methods. Hardisty et al. (2013) concluded that "in comparison with the standard, fixed-sequence titration method, we did not find compelling advantages for the complex multiple-staircase method. ... However, the good news is that the simple titration measure, which is much more convenient to implement, remains a useful method" (p. 247).

The appendix presents the Temporal Discounting subtest of the CART. Part 1 of this subtest consists of two fixed-sequence titrations. The first is an ascending sequence and the second is a descending sequence. The first

(Temporal Discounting Staircase Increasing) involves a constant delayed reward of $100 in three months. It is compared initially with a $1 payment immediately that then increments to $2.50, $5, $7.50, $10, and then in increments of $5 up to $90. The sequence then continues through $92.50, $95, $97.50, $99, and ends with a choice of $99.50 now versus $100 in three months. We adopted a simplified scoring system for this sequence whereby we scored only those choices (20 in total) involving amounts of $90 and lower. The number of choices of the delayed reward ($100 in three months) is the raw score on this sequence. In one study of 407 subjects (RT54), the mean number of delayed, large reward choices was 16.6 (SD = 3.5), and in another study of 376 subjects (RT57), the mean number of delayed, large reward choices was 15.9 (SD = 3.7). In the appendix, the raw score on the ascending sequence is translated into CART points.

The choice of $90 and below as our cutoff for scoring this sequence was dictated by the fact that waiting for the $100 over the $90 would represent, on a simple interest basis, an annualized interest rate of 44.4 percent. Descending from there (to $85, $80, and so on) and choosing the delayed reward results in annualized interest rates of 70.6 percent, 100 percent, 133.3 percent, and so on—clearly interest rates that it would be irrational to pass up. Most economists believe that "the presence of capital markets should cause imputed discount rates to converge on the market interest rate" (Frederick et al., 2002, p. 381). That is, even in cases where we might worry that perhaps the subject is choosing the smaller immediate reward because he or she needs the money (say, $60) *now*, the subject should in fact accept the offer of $100 in three months instead. The subject could, for instance, immediately get the cash from a (bad) credit card or even a payday loan office (which are regulated in many states to charging no more than 36–40 percent annualized rates) and then pay it off in three months with the 266 percent interest rate earned by delaying and receiving $100. As Senecal et al. (2012) put it, "The rate of discounting ought to be similar to the current market interest rate. … Consider a person given a choice between receiving a smaller amount of money at a sooner time and a larger amount at a later time. If this person could borrow the smaller amount while waiting for the larger payoff, and if the larger amount is enough to repay this loan plus interest and still leave the individual with a profit, then the larger, later reward is the better choice, *regardless of when the person would want to spend the money*" (p. 568, italics in original). In light of this argument, our cutoff at 44 percent seems reasonable as an operational definition of prudent discounting.

The second fixed-sequence titration in Part 1 is a descending sequence (Temporal Discounting Staircase Decreasing, in the appendix). It involves a constant delayed reward of $2,000 in one year. It is compared initially with a $1,990 payment immediately that then decrements to $1,980, $1,950, $1,900, $1,850, $1,800, and then in decrements of $100 down to $200. The sequence then continues through $150, $100, $50, and ends with a choice of $20 now versus $2,000 in one year. We adopted a simplified scoring system for this sequence whereby we scored only those choices (18 in total) involving amounts of $1,600 and lower. The number of choices of the delayed reward ($2,000 in one year) is the raw score on this sequence. In one study of 407 subjects (RT54), the mean number of delayed, large reward choices was 14.8 (SD = 3.8), and in another study of 376 subjects (RT57), the mean number of delayed, large reward choices was 14.0 (SD = 4.2). In the appendix, the raw score on the descending sequence is translated into CART points for this subtest. The choice of $1,600 and below as our cutoff for scoring this sequence was dictated by the fact that waiting for the $2,000 over the $1,600 would represent an annualized interest rate of 25 percent. Descending from there (to $1,500, $1,400, and so on) and choosing the delayed reward results in annualized interest rates of 33.3 percent, 42.9 percent, 53.8 percent, 66.7 percent, and so on—again, clearly interest rates that it would be irrational to pass up.

A sample of Part 2 of the Temporal Discounting subtest of the CART is presented in the appendix (Temporal Discounting Mixed Questionnaire). This thirty-one-item measure asks participants to indicate the strength of their preference for either a smaller amount of money sooner or a larger amount of money later. The amounts and delays vary from item to item rather than being sequenced as in the titration method. To possibly increase the test's sensitivity, participants respond on a continuous scale rather than simply picking the smaller or larger amount.

In twenty-six of the items, the delayed larger amount corresponded to a substantial annualized increase in value, which on a simple interest basis would have resulted in value increases of between 44.7 percent and 346.7 percent if earned annually. Five of the items were filler items, where the delayed larger amount corresponded to only relatively low percentage increases in value, which on a simple interest basis would have resulted in value increases of between 1.3 percent and 3.1 percent if earned annually. These fillers were included to reduce response bias and demand characteristics favoring the large delayed reward. These items do not enter into the raw score.

The twenty-six items consisted of thirteen pairs of items. In the sooner-now condition, the sooner time was "now" for one member of a pair ("If you had a choice, would you prefer $70 now or $110 in 6 months?"), and in the sooner-lag condition, the sooner time involved a delay of 8 weeks to 12 months ("If you had a choice, would you prefer $70 in 12 months or $110 in 18 months?"). Within the thirteen pairs, the durations that distinguished the sooner from the later times was essentially identical. The thirty-one items were shuffled such that the members of each of the thirteen pairs and the five filler items were separated. This design allowed the examination of preference reversals that are sometimes called "dynamic inconsistency" (Thaler, 1981) or "common difference effects" (Loewenstein & Prelec, 2000). However, we found that scoring these dynamic inconsistencies did not increase the sensitivity or predictive validity of our subtest, so we will not discuss them further.

Each item was scored on a six-point scale, where 6 is given for very strongly preferring the delayed reward, 5 for strongly preferring the delayed reward, and on through 1 point for very strongly preferring the smaller reward given sooner. Thus, the maximum score on the twenty-six-item measure is 156 and the minimum score is 26. In one study of 407 university subjects (RT54), the mean score was 90.5 (SD = 18.7), and the reliability (Cronbach's alpha) was a very high 0.94. In another study of 376 university subjects (RT57), the mean score was 85.4 (SD = 22.0), and the reliability (Cronbach's alpha) was again 0.94. Thus, the scale is highly reliable. In the appendix, the raw score is translated into CART points. The maximum score for the entire subtest is 7 points.

In RT54, the raw score on the increasing staircase $100 task displayed a 0.54 correlation with the score on the decreasing staircase $2,000 task. These two tasks displayed correlations of 0.47 and 0.37, respectively, with Part 2: the mixed temporal discounting questionnaire. The same study contained another temporal discounting task. Subjects responded to the twenty-seven delayed-reward choices displayed in table 1 of Kirby (2009). Performance on these twenty-seven items correlated with all three of the tasks in this CART subtest: 0.67 with the mixed temporal discounting questionnaire, and 0.55 and 0.45 respectively with the increasing staircase $100 task and the decreasing staircase $2000 task.

In RT57, the patterns were very similar. The raw score on the increasing staircase $100 task displayed a 0.59 correlation with the score on the decreasing staircase $2,000 task. These two tasks displayed correlations of 0.45 and 0.40, respectively, with Part 2: the mixed temporal discounting questionnaire. As in the previous study, subjects responded to the twenty-seven delayed-reward choices displayed in table 1 of Kirby (2009). Performance on these twenty-seven items correlated with all three of the tasks in the Temporal Discounting subtest: 0.54 with the mixed temporal discounting questionnaire, and 0.49 and 0.48 respectively with the increasing staircase $100 task and the decreasing staircase $2,000 task.

Correlations with intelligence were somewhat lower than the 0.23 association obtained in the meta-analysis of Shamosh and Gray (2008). In the first study of university students discussed above (RT54), verified SAT total scores correlated 0.19 with the mixed temporal discounting questionnaire, and 0.19 and 0.25 respectively with the increasing staircase $100 task and the decreasing staircase $2,000 task. In the second study of university students discussed above (RT57), self-reported SAT total scores correlated 0.02 with the mixed temporal discounting questionnaire, and 0.10 and 0.23 respectively with the increasing staircase $100 task and the decreasing staircase $2,000 task. These rather low cognitive ability correlations involving this subtest are replicated in the study of the full-form CART that we will describe in chapter 13.

9 Probabilistic Numeracy, Financial Literacy, Sensitivity to Expected Value, and Risk Knowledge

In this chapter, we focus on a set of subtests of the CART that are more knowledge dependent than are the subtests discussed in the previous two chapters. That is, we focus on the tasks listed exclusively in the second column of table 4.1 (reproduced here as table 9.1). Although these tasks do have processing requirements, successful performance on them is much more dependent on the presence of specific declarative knowledge than is the case for the tasks from the first column of the table that were discussed in chapters 7 and 8.

The subtests of the CART discussed in this chapter focus on probabilistic numeracy, financial literacy, sensitivity to expected value, and risk knowledge. These are the tasks on the far right side of the 2 × 2 space defined in figure 4.1 (reproduced here as figure 9.1).

Although all of the tasks to be discussed in this chapter are on the far right edge of the space, indicating that they are heavily knowledge saturated, they differ somewhat in their processing requirements. The subtests range from the financial literacy measure, performance on which is most purely a measure of acquired declarative knowledge, to the probabilistic numeracy measure, which, although depending heavily on stored knowledge, does have more processing requirements than the financial literacy measure in the sense that it does require some ability to suppress alternative solutions that are not correct.

The Probabilistic Numeracy Subtest

The past decade has witnessed a veritable explosion of research on the measurement of numeracy and the relation of numeracy to other cognitive skills, as well as to practical outcomes in the real world (Cokely et al., 2012;

Table 9.1
Framework for classifying the types of rational thinking tasks and subtests on the CART.

Tasks Saturated with Processing Requirements (Detection, Sustained Override, Hypothetical Thinking)	Rational Thinking Tasks Saturated with Knowledge	Avoidance of Contaminated Mindware	Thinking Dispositions that Foster Thorough and Prudent Thought, Unbiased Thought, and Knowledge Acquisition
Probabilistic and Statistical Reasoning subtest		Superstitious Thinking subtest	Actively Open-Minded Thinking scale
Scientific Reasoning subtest		Antiscience Attitudes subtest	Deliberative Thinking scale
Avoidance of Miserly Information Processing subtests: • Reflection versus Intuition • Belief Bias Syllogisms • Ratio Bias • Disjunctive Reasoning	Probabilistic Numeracy subtest	Conspiracy Beliefs subtest	Future Orientation scale
Absence of Irrelevant Context Effects in Decision Making subtests: • Framing • Anchoring • Preference Anomalies	Financial Literacy and Economic Knowledge subtest	Dysfunctional Personal Beliefs subtest	Differentiation of Emotions scale
Avoidance of Myside Bias: • Argument Evaluation subtest	Sensitivity to Expected Value subtest		
Avoiding Overconfidence: • Knowledge Calibration subtest	Risk Knowledge subtest		
Rational Temporal Discounting			

Probabilistic Numeracy 179

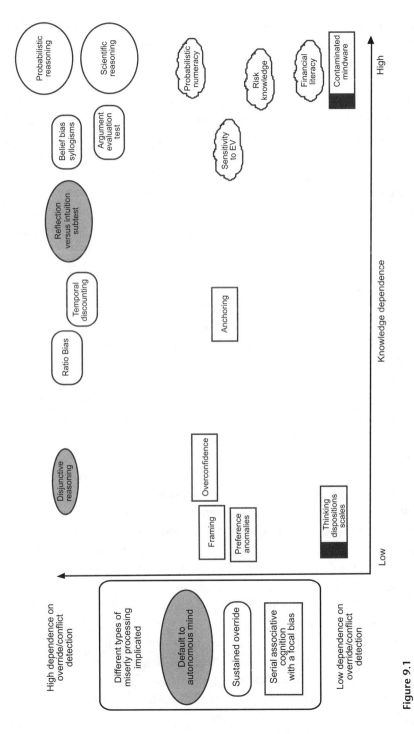

Figure 9.1
CART subsets arrayed in the process/knowledge space.

Låg et al., 2014; Liberali et al., 2012; Weller et al., 2013). Beginning with the work of Lipkus, Samsa, and Rimer (2001; see also Schwartz et al., 1997) and Peters et al. (2006), this research has mushroomed so much recently that there have been several comprehensive reviews and theoretical integrations of what has been published to date (Reyna et al., 2009; Reyna & Brust-Renck, 2015). A compelling finding has been the high level of difficulty that even educated subjects have with percentages, probabilities, and proportions (Cokely et al., 2012; Lipkus et al. 2001; Reyna & Brainerd, 2008; Reyna et al., 2009).

Several scales of numeracy have undergone psychometric evaluation and validity assessment (Cokely et al., 2012; Låg et al., 2014; Liberali et al., 2012; Schapira et al., 2012; Weller et al., 2013). Numeracy measures have been related to consumer, economic, and health decisions (Banks & Oldfield, 2007; Banks et al., 2010; Låg et al., 2014; Nelson et al., 2008; Peters, 2012; Reyna & Brainerd, 2007; Reyna et al., 2009). Levels of numeracy have been linked to other constructs that we measure in the CART, such as cognitive reflection, ratio bias, and maximizing expected utility (Cokely & Kelley, 2009; Cokely et al., 2012; Jasper et al., 2013; Klaczynski, 2014; Liberali et al., 2012; Peters et al., 2006).

What made this work so relevant to us in constructing this CART subtest was that most of the work on numeracy was not really about numeracy broadly construed, but instead was focused on what might be called "probabilistic numeracy" or "statistical numeracy." This work has focused on percentages, probabilities, and related statistical constructs (Cokely et al., 2012; Reyna & Brainerd, 2008; Weller et al., 2013). That particular characteristic of the literature seemed to make the studies of numeracy a very nice complement to the Probabilistic and Statistical Reasoning subtest, which we described in chapter 5. This subtest carries some heavy processing requirements (in terms of conflict detection and sustained override) in addition to tapping probabilistic knowledge as well. The numeracy measures in the literature complement the Probabilistic and Statistical Reasoning subtest by shifting the balance from processing components to declarative knowledge. That is, our Probabilistic and Statistical Reasoning subtest had both processing requirements and knowledge requirements, but was heavier on the former. Most measures of numeracy are heavier on the latter and lighter on the former. Numeracy tests tend to stress processing requirements less because many items do not contain strongly intuitive alternative responses.

As mentioned previously, most measures of numeracy in the cognitive psychology literature have heavily stressed knowledge of percentages and probabilities. Our Probabilistic Numeracy subtest retains this feature and uses the type of item that has been common in the literature cited above. Each of the nine items on this measure (see the appendix for sample items) is simply scored as correct or incorrect, resulting in total raw scores that vary from 0 to 9. In an unpublished study by our laboratory (RT57) involving 351 university students as subjects, the mean raw score was 4.61 (SD = 1.78). The reliability of our Probabilistic Numeracy subtest was 0.59 (Cronbach's alpha). Items varied widely in difficulty, from 14.8 percent correct to 87.7 percent correct. The Probabilistic Numeracy subtest is given a nine-point weight on the CART. That is, the CART score is equal to the raw number correct.

In the same study, the Probabilistic Numeracy subtest displayed a correlation of 0.37 with SAT total scores and a correlation of 0.47 with a ten-item version of the Reflection versus Intuition subtest of the CART. These correlations are of a similar magnitude to those found with other numeracy measures in the literature (Cokely et al., 2012; Låg et al., 2014; Liberali et al., 2012; Weller et al., 2013).

The Probabilistic Numeracy subtest displayed significant correlations with several other components of the CART that were tested in the same study. It displayed significant correlations of 0.27 with the Sensitivity to Expected Value subtest discussed later in this chapter, 0.17 with the Preference Anomalies subtest of the CART described in chapter 8, and 0.19 and 0.20 with measures of rejection of superstitious thinking and antiscientific attitudes—both measures to be described in chapter 10. Regarding the supplemental thinking dispositions assessed on the CART (see chapter 11), the Probabilistic Numeracy subtest displayed significant correlations of 0.25 with the Actively Open-Minded Thinking scale and 0.26 with the Deliberative Thinking scale.

The Financial Literacy and Economic Knowledge Subtest

As we described in chapter 2, System 1 and System 2 both have the capability to support rational behavior, but both have characteristic weaknesses that Kahneman (2011) described in his best-selling book. The bias of System 2 is laziness. The bias of System 1 is that of attribute substitution—it emits an answer from within its capabilities when it is asked a question that it (System 1) cannot strictly answer. An attribute-substituting System 1 and a lazy System 2 can combine to yield rational behavior in benign environments but can yield seriously suboptimal behavior in hostile environments.

In earlier chapters, we have briefly mentioned the idea of hostile and benign environments for Type 1 processing. A benign environment is an environment that contains useful cues that, via practice or evolutionary history, have been well represented in Type 1 subsystems. Additionally, for an environment to be classified as benign, it must not contain other individuals who will adjust their behavior to exploit those relying only on Type 1 processing. From the beginning of their research tradition, heuristics and bias researchers have often been criticized for emphasizing the errors made by the human information-processing apparatus (Christensen-Szalanski & Beach, 1984; Krueger & Funder, 2004), despite the fact that they have always emphasized that most real-world situations are benign rather than hostile.

What the critics often fail to realize is how important the relatively rarer instances of judgment in hostile environments can be. Thus, it is not assuaging to be told that many more situations in life are benign than are hostile. We cannot dismiss Type 2 thinking by saying that heuristics will get a "close enough" answer 98 percent of the time, because the 2 percent of the instances where Type 1 processing leads us seriously astray may be critical to our lives. This point is captured in an interview in *Money Magazine* with Ralph Wanger, a leading mutual fund manager. Wanger says, "The point is, 99 percent of what you do in life I classify as laundry. It's stuff that has to be done, but you don't do it better than anybody else, and it's not worth much. Once in a while, though, you do something that changes your life dramatically. You decide to get married, you have a baby. ... These rare events tend to dominate things" (quoted in Zweig, 2007, p. 102). Yet, in terms of raw numbers, these might represent only twenty to thirty decisions out of thousands that we have made throughout our lives. But the thousands are just the "laundry of life" to use Wanger's phrase. The twenty "nonlaundry" decisions are small in number and nonrecurring, and thus require Type 2 processing. As Holt (2011) notes, "If you've had 10,000 hours of training in a predictable, rapid-feedback environment—chess, firefighting, anesthesiology—then blink. In all other cases, think" (p. 17).

Given our arguments in chapters 3 and 4 about how processing requirements and knowledge bases are intertwined in tasks that require rational thinking, we must caution here that an environment will also be hostile for a decision maker if he or she lacks the requisite mindware for the decision-making situation. The knowledge base that is the focus of the subtest introduced in this section is one squarely focused on the small "nonlaundry" part of life. Modernity has subjected all citizens to the conceptual abstractions necessary to navigate the monetary and financial world. Citizens of first-world, technological societies must navigate the complexities of credit cards, car loans, pension plans, mortgages, appliance warranties, healthcare deductibles, auto insurance, life insurance, college savings plans, tax-deferred savings plans, and a myriad of other financial products and services. Unlike recurring perceptual events in the natural world which give our systems thousands of practice trials, some of these decisions in the financial domain will occur only a few times in a lifetime (e.g., pension plans, Medicare drug enrollment). There is no chance for Type 1 subsystems to receive extensive enough practice at such decisions. Instead, the way to effective decision making in the financial domain is to have learned specific knowledge relevant to each of the decisions. This is why financial literacy and economic knowledge make up an important subtest of the CART, weighted at 10 points.

For the Financial Literacy and Economic Knowledge subtest of the CART, we built on the financial literacy literature that has mushroomed in the past decade and a half (Lusardi & Mitchell, 2014). Research has shown that financial knowledge is sparse among the general population, with certain groups extremely deficient in such knowledge (Chen & Volpe, 1998; Mandell, 2009; NCEE, 2005). Higher financial literacy has been linked to positive real-life outcomes (Banks & Oldfield, 2007; Larrick, Nisbett, & Morgan, 1993), although there is still controversy about the efficacy of training in this domain and its cost effectiveness (Fernandes, Lynch, & Netemeyer, 2014; Larrick, Morgan, & Nisbett, 1990; Ross, Grossmann, & Schryer, 2014; Thaler, 2013). Nevertheless, Lusardi and Mitchell (2014) constructed a quantitative model that indicated that "it can still be socially optimal to raise financial knowledge for everyone early in life, for instance by mandating financial education in high school" (p. 9).

We accept Lusardi and Mitchell's (2014) definition of financial literacy as "people's ability to process economic information and make informed decisions about financial planning, wealth accumulation, debt, and pensions" (p. 6). In constructing our subtest, we generated items similar to those in a variety of economic literacy measures (Chen & Volpe, 1998; Klein & Buturovic, 2011; Mandell, 2009; NCEE, 2005). After pilot testing, we settled on a twenty-seven-item subtest (see the appendix for sample items). Our subscale encompasses diverse financial and economic concepts and issues such as diversification, compounding, government regulations, investment instruments, financial terminology, interest rates, supply/demand logic, pyramid schemes, government debt, risk/reward relationships, mutual funds, savings vehicles, liquidity, taxes, bonds, credit card debt, exponential growth, and sunk costs.

Each item is simply scored correct or incorrect, resulting in total raw scores that vary from 0 to 30 (one item has four subitems). In an unpublished study in our laboratory (RT57) involving 375 university students as subjects, the mean raw score was 13.3 (SD = 3.6). The reliability of our financial literacy and economic knowledge subtest was 0.53 (Cronbach's alpha). Items varied widely in difficulty, with scores ranging from 12.6 percent correct to 81.9 percent correct.

In the same study, we examined the correlation of the Financial Literacy and Economic Knowledge subtest with other measures. This subtest displayed a correlation of 0.32 with SAT total scores and a correlation of 0.41 with a ten-item version of the Reflection versus Intuition subtest. It displayed significant correlations of 0.42 with the Probabilistic Numeracy subtest described previously, and a 0.30 correlation with the Sensitivity to Expected Value subtest also discussed in this chapter (both correlations $p < 0.001$).

The Financial Literacy and Economic Knowledge subtest correlated with a rejection of Superstitious Thinking subtest (0.19, $p < 0.001$) and with the rejection of Antiscientific Attitudes subtest (0.20, $p < 0.001$), both to be described in chapter 10. It did not correlate with the rejection of Conspiracy Beliefs subtest ($r = 0.10$), described in chapter 10, or with the avoidance of Dysfunctional Personal Beliefs measure ($r = 0.05$), described in the same chapter. Regarding the supplemental thinking dispositions assessed on the CART (see chapter 11), the Financial Literacy and Economic Knowledge subtest displayed significant correlations of 0.34 with the Actively Open-Minded Thinking scale and 0.24 with the Deliberative Thinking scale (both $p < 0.001$).

These results from our study of university students are in many ways quite reasonable. One disappointing aspect of the study, however, was the very modest reliability of 0.53, given that this was a 30-point test and that we had thought it was quite well constructed. Although we had culled items a bit before using this form of the Financial Literacy subtest, it would still be possible to improve the reliability of this version of the subtest by further culling items on a psychometric basis. For example, a statistical analysis of the results of this study indicated that there were several items with negative item-test correlations. But when we looked at the nature of these items, we were given pause. Specifically, they turned out to be the type of item that one would not want to remove from a comprehensive subtest of this type.

An examination of the content of the problematic items revealed that the difficulty might not be with our subtest per se, but with applying the subtest to young university students (the mean age in our sample was 18.8, SD = 1.0) who are without much experience with finances, money, and issues beyond managing a credit card. This is also suggested by the relatively low mean score on this subtest, a score of 13.3 out of total potential score of 30. The item with the largest negative item-test correlation (–0.13) was the multiple-choice item: "Personal finance experts recommend no-load mutual funds over load funds because …" It is clear that there is no reason to expect that a nineteen-year-old would have had any experience with mutual funds. Thus, this item probably results in a good deal of guessing among that population. It is not surprising that it did not contribute to the reliability of the test (in fact, it detracted a little). But this does not mean that we would want to remove this item from a subtest of an assessment device (the CART) that is meant to be applied to a much wider population than university students. This item on no-load mutual funds would be perfectly appropriate for individuals in their mid-thirties (a very typical age range for Amazon Mechanical Turk samples), because by then it would be expected that individuals would know about savings and investment vehicles, because savings should have begun for someone in his or her mid-thirties.

It is likewise with an item that concerned knowing the difference between a tax credit and a tax deduction. Again, it is understandable that this item was not diagnostic of performance on the rest of this scale for university students. The tax situation of the average young student in our sample would not have any of the complexities that make the difference between a tax credit and a tax deduction something that would be of relevance to him or her. In contrast, subjects in their late twenties or early thirties would have begun to encounter the complexities of the tax code that would make this kind of knowledge useful.

Several more items were of this type. That is, they appear to be problematic among this university student sample, but they still concerned vital financial knowledge for people somewhat older. Therefore, although our study of university students was informative and provided us with knowledge about some important correlations with other measures (such as cognitive ability and thinking dispositions), we thought that a better assessment of the reliability of the test would come from a more diverse Amazon Mechanical Turk sample. Our Amazon Mechanical Turk study (Turk6) was completed by 391 subjects with a mean age of 33.1 (SD = 10.7). The subjects in this study did not complete the other measures that were in the study of university students, but instead filled out a brief demographics questionnaire and the Financial Literacy and Economic Knowledge subtest. The mean score in this study (16.8, SD = 4.5) was, as we predicted, substantially higher than that obtained by the university sample (13.3). The reliability of our Financial Literacy and Economic Knowledge subtest in the Mechanical Turk sample was 0.72 (Cronbach's alpha), considerably higher than that in the university sample (0.53). Items varied widely in difficulty, with scores from 16.1 percent correct to 90.0 percent correct. Exploratory factor analysis revealed a dominant factor with an eigenvalue of 2.82 that explained 61.3 percent of the variance. The second factor displayed an eigenvalue of 1.03, barely over the > 1 criterion.

Raw scores on the subtest displayed significant correlations with age ($r = 0.25$), educational level ($r = 0.22$), and sex ($r = -0.17$). Older subjects with more education performed better on this subtest than younger, less educated subjects. The mean score of males (17.6, SD = 4.8) was significantly higher than the mean score of females (15.9, SD = 4.1), $t(389) = 3.58$, $p < 0.001$. The correlation with sex was not reduced when age and education were controlled (partial $r = -0.21$). In the appendix, the raw score on the subtest is translated into the ten-point weighting that it is given on the CART.

The Sensitivity to Expected Value Subtest

To this point in the book, we have emphasized the axiomatic approach to subjective expected utility as a normative model of performance. In this approach, so-called representation theorems (Krantz, 1991; Savage, 1954) link a set of choice axioms (or choice rules, if you will) to a utility function that presumably results in an ordered and coherent set of preferences. To put it more simply and colloquially, it has been shown that if people follow the key axioms, they are behaving as if they are maximizing expected utility. Thus, direct tests of whether people are adhering to the axioms (transitivity, independence, reduction of compound lotteries, etc.) become indirect tests of whether a person is adhering to the normative model of maximizing expected utility.

There is another way to assess whether people are maximizing expected utility, however, and that is to simply present subjects with monetary choices (e.g., gambles) and to observe whether people choose the option with the highest expected value. Of course, choices between gambles have been used to study preference reversals and other preference anomalies for some time now (Birnbaum 1999; Lichtenstein & Slovic, 1971, 2006), as has been discussed in chapter 8, but the present use of such a paradigm is even simpler. Here, we are not looking at whether sets of such choices display patterns of consistency across the normative axioms, but instead we are simply examining whether the testee can recognize the higher expected value option of two alternatives. In this subtest, we are interpreting the choice of the higher expected value option as the normative choice—the one that maximizes expected utility. We adopt this stance following the work of Rabin (2000a, 2000b; see also Benjamin, Brown, & Shapiro 2013; Shapiro 2005) who has shown that, for small stakes, people should be risk neutral and choose the higher expected value option regardless of the probabilities involved.[1]

There is a small body of research linking the tendency to make high expected value choices with other components that are assessed on the CART. For example, Frederick (2005) showed that the original version of his cognitive reflection test predicted the tendency to make high expected value choices in two-choice gambles, particularly in cases where the risky choice had the higher expected value. Jasper et al. (2013) found that the tendency to make high expected value choices was related to a measure of probabilistic numeracy similar to our Probabilistic Numeracy subtest. Finally, Benjamin, Brown, and Shapiro (2013) found that cognitive ability was related to the tendency to choose the higher expected value option.

Our twenty-item measure of the Sensitivity to Expected Value subtest (see the appendix for sample items) is similar to those typically used in paradigms involving choices between gambles. Typically, the higher expected value option was at least 25 percent more valuable than the alternative. For each pair of gambles, the options have positive expected values, and none of the gambles involve losses. The range of gambles clearly falls within what the arguments of Rabin (2000a, 2000b) would lead us to consider as small stakes. The raw score on the Sensitivity to Expected Value subtest of the CART is simply the number of times out of twenty that the subject chooses the higher expected value option.

The twenty items were run in a laboratory study involving 377 university students (RT57). The mean total score was 14.8 (SD = 3.1). The reliability of the total score was moderate (Cronbach's alpha of 0.67). In the appendix, the raw score is translated into the five-point weighting that this subtest is given on the CART. In the same laboratory study, the raw total score on the Sensitivity to Expected Value subtest displayed a correlation of 0.18 with SAT total scores and a correlation of 0.30 with a ten-item version of the Reflection versus Intuition subtest. Finally, the raw total score on the Sensitivity to Expected Value subtest displayed a correlation of 0.27 with the Probabilistic Numeracy subtest of the CART.

Risk Knowledge Subtest

A long-standing finding in cognitive psychology is that although our brains contain useful mechanisms for assessing risk, sometimes people's calibration of risks in our technological society is suboptimal (Fischhoff & Kadvany, 2011; Lichtenstein, Slovic, Fischhoff, Layman, & Combs, 1978). That is, people sometimes misassess risk in their environments—sometimes worrying more about an event than its actual probability would warrant, and other times failing to be alert to risks that are much more probable (Fischhoff & Kadvany, 2011). Confirmation biases, probabilistic thinking errors, and contaminated mindware have been linked to people's misassessment of risks in their environment and to irrational public policy linked to risk (Baron, 1998, 2008; Sunstein, 2002). Perhaps even more relevant to risk misassessment, however, is the work of researchers who have studied what is termed the "vividness effect" in human memory and decision making (Li & Chapman, 2009; Slovic, 2007; Stanovich, 2009; Trout, 2008; Wang, 2009). When faced with a problem-solving or decision-making situation, people retrieve from memory the information that seems relevant to the situation at hand. The problem is that easy retrievability can be determined by factors other than frequency—a point going back to Tversky and Kahneman's (1974) classic discussion of availability. For example, it is strongly influenced by vividness and by media exposure, which often does not track the true frequency of risks (Fischhoff & Kadvany, 2011).

Misleading personal judgments based on the vividness of media-presented images are widespread. Studies have surveyed parents to see which risks to their children worried them the most (Gardner, 2008; Radford, 2005; Skenazy, 2009). Parents turned out to be most worried about their children being abducted, an event with a probability of 1 in 600,000. By contrast, the probability of their child being killed in a car crash, which the parents worried about much less, is dozens of times more likely (Gardner, 2008). Likewise, children are much more likely to drown in a swimming pool than they are to be abducted and killed by a stranger (Kalb & White, 2010). Of course, the fears of abduction are mostly a media-created worry. Car crashes, accidents (including firearm accidents), childhood obesity, and suicide at older ages are much more of a threat to our children's well-being than are events like abduction and shark attacks.

In assessing risk knowledge, we followed the classic paradigm introduced by Lichtenstein et al. (1978). The Risk Knowledge subtest (see the appendix for sample items) consists of fourteen items in which the subject chooses which of two causes of death is more likely. The total raw score on this subtest thus ranges from 0 to 14. The fourteen items were run in a laboratory study involving 182 university students (RT56). The items ranged widely in difficulty, with scores ranging from 12.1 percent correct (pneumonia versus motor vehicle accident) to 78.6 percent correct (bicycle-related versus commercial airplane crash). The mean score was 7.2 (SD = 2.5). The reliability of the total score was 0.60 (Cronbach's alpha). In the appendix, the raw score is translated into the 3-point weighting that this subtest is given on the CART.

We will present more data on the associations displayed by all of the subtests discussed above in chapter 13, when we report a study of the full-form version of the CART.

10   Contaminated Mindware

In chapter 9 we discussed subtests of the CART that assess areas of knowledge that facilitate rational thinking. While discussing the framework of the CART in chapters 3 and 4, we mentioned that knowledge becomes implicated in rationality in two different ways. In those two chapters, we illustrated that it is sometimes best to think of this in terms of failures of rationality, or in terms of thinking errors (see table 3.1). So-called mindware problems come in two types. When the knowledge bases discussed in chapter 9 are missing, we have a case of a mindware gap. A different type of mindware problem arises because not all mindware is helpful—either for attaining our goals (instrumental rationality) or for having accurate beliefs (epistemic rationality). In fact, some acquired mindware can be the direct cause of irrational actions that thwart our goals. This type of problem has been termed the "problem of contaminated mindware" (Stanovich, 2009; Stanovich, Toplak, & West, 2008).

Although there may be many specific clusters of misinformation that would support irrational thought and behavior, we tried, in this section of the CART, to assess domains with some degree of generality. In the CART, we assess several clusters of mindware that support pseudoscientific belief systems: superstitious thinking, antiscience attitudes, and conspiracy beliefs. This emphasis on pseudoscience is justified because pseudosciences are a major source of costly irrationality in society. In this chapter, we will also discuss a fourth subtest that is unrelated to the other three: the Dysfunctional Personal Beliefs subtest of the CART. It is a subtest that tests for the presence of self-defeating personal beliefs that impede people's goal attainment.

Upon examination, pseudoscientific beliefs often tend to be more costly than most people think they are. First, people tend not to think about opportunity costs. When people spend time (and money) on pseudosciences, they gain nothing and also waste time that might have been spent on more productive endeavors. Even more important, in a complex, technological society, you may be affected by pseudoscientific beliefs even if you do not share those beliefs. For example, millions of Americans in areas without fluoridation are suffering needless cavities because their neighbors are in the grip of pseudoscientific conspiracy theories about the harmful effects of fluoridation (Brody, 2012; Griffin, Regnier, Griffin, & Huntley, 2007). Some police departments have hired psychics to help with investigations even though research has shown that this practice has no effectiveness (Radford, 2010; Shaffer & Jadwiszczok, 2010). There is not a single documented case of psychic information being used to successfully find a missing person (Radford, 2009).

A clear example of how we are all hurt when pseudoscientific beliefs spread is provided by the theory (first put forth in the early 1990s and continuing to this day) that autism is connected to the early vaccination of children. This theory is false (Grant, 2011; Offit, 2008), but it has spawned an antivaccination movement. As a result, immunization rates have decreased, many more children have been hospitalized with measles than would have been otherwise, and some have died (Goldacre, 2008; Grant, 2011; Offit, 2008). Again, the lesson is that in an interconnected society, your neighbor's pseudoscientific belief might affect you even if you reject the belief yourself. Physicians are increasingly concerned about the spread of medical quackery on the Internet (Offit, 2008) and its real health costs. Many sick individuals delay getting medically appropriate treatment because they waste time chasing bogus cures. Renowned computer entrepreneur Steve Jobs ignored his doctors after being told of his pancreatic cancer and delayed surgery for nine months while he pursued unproven fruit diets, consulted a psychic, and received bogus hydrotherapy (Isaacson, 2011).

High intelligence is no inoculation against the contaminated mindware of various pseudosciences (as the case of Jobs shows). The complex mindware of pseudosciences often sounds most enticing to those of moderate to high intelligence. Search the Internet for examples of conspiracy theories, tax-evasion schemes, get-rich-quick schemes, schemes for "beating" the stock market, and procedures for winning the lottery. You will quickly see that many of them are characterized by enticing complexity. For example, many get-rich-quick schemes involve real-estate transactions that interact in a complex manner with the tax system. Many win-the-lottery books contain explanations (wrong ones!) using mathematics and probabilities. Beat-the-market stock investment advice often involves the mathematics and graphics of so-called technical analysis.

The intuition that those taken in by fraudulent investment schemes are probably not of the lowest intelligence is confirmed by the results of a study commissioned by the National Association of Securities Dealers (Consumer Fraud Research Group, 2006). The study found that 68.6 percent of the investment fraud victims had at least a BA degree. The proportion of the investment victim group with incomes over $30,000 was 74.1 percent. We can infer from these statistics that many victims of investment fraud are not of low intelligence. Likewise, the communities that were most affected by antivaccination pseudoscience were not low-SES communities, but instead were affluent, well-educated communities like Boulder, Colorado and Santa Monica, California (Offit, 2014).

Cognitive scientists have uncovered some of the reasons why intelligent people can come to have beliefs that are seriously out of kilter with reality. One explanation is in terms of so-called knowledge-projection tendencies (see Stanovich, 1999). The idea here is that in a natural ecology where most of our prior beliefs are true, processing new data through the filter of our current beliefs will lead to faster accumulation of knowledge. But the assumption here—that we are in a domain where our beliefs are largely true—is critical. To the extent that current beliefs are true, then we will assimilate further true information more rapidly. However, when the subset of beliefs that the individual is drawing on contains substantial amounts of false information, knowledge projection will *delay* the assimilation of the correct information. And herein lies the key to understanding the creationist or the Holocaust denier. The knowledge-projection tendency, efficacious in the aggregate, may have the effect of isolating certain individuals on "islands of false beliefs" from which they are unable to escape. Knowledge projection from an island of false beliefs might explain the phenomenon of otherwise intelligent people who get caught in a domain-specific web of falsity that, because of projection tendencies, they cannot escape. Such individuals often use their considerable computational power to rationalize their beliefs and to ward off the arguments of skeptics.[1]

We would, of course, expect more intelligent individuals to acquire more mindware of all types based on their superior learning abilities. This would result in them acquiring more mindware that fosters rational thought (i.e., the mindware discussed in chapter 9). However, this superior learning ability would facilitate intelligent individuals acquiring more contaminated mindware as well. Complex contaminated mindware might even require a certain level of intelligence in order to be enticing to the host. George Orwell conjectured in this vein when discussing political attitudes in the World War II era: "There is no limit to the follies that can be swallowed if one is under the influence of feelings of this kind. ... One has to belong to the intelligentsia to believe things like that: no ordinary man could be such a fool" (1968, p. 379).

Three of the four subtests we will discuss in this chapter reflect an interlocking set of beliefs surrounding pseudoscience: tendencies toward superstitious thinking, attitudes toward legitimate science, and endorsement of conspiracy beliefs. The literatures on conspiratorial ideation and attitudes toward science have tried to pull these concepts apart. That is, the literature has discussed issues such as which attitudes are causally prior to others and exactly how they are linked (Goertzel, 1994; Lewandowsky, Gignac, & Oberauer, 2013; Lewandowsky, Oberauer, & Gignac, 2013; Majima, 2015; Oliver & Wood, 2014). We will not review this literature, nor will we take a stand on which of these thinking styles is causally prior. Instead, we simply wish to assess three areas of interlocking beliefs and attitudes that have implications for avoiding contaminated mindware. Specifically, avoiding these thinking styles is an indication that a person is not likely to take irrational actions because of the presence of contaminated mindware.

In this chapter, we will first present three subtests of the CART that reflect the tendency to avoid contaminated mindware of the type contained in many of the pseudosciences mentioned above. The first subtest we will consider is the Superstitious Thinking subtest. The second subtest assesses a variety of antiscientific attitudes that might impede the critical examination of contaminated mindware. Finally, our Conspiracy Beliefs subtest probes the tendency to host contaminated mindware by assessing the tendency to endorse beliefs that are false, harmful, and unjustified because of their reliance on the action of unseen powerful forces that are empirically opaque (Oliver & Wood, 2014; Sunstein & Vermeule, 2009). As we mentioned above, we expect some degree of interrelation among these three subtests because they reflect an interacting worldview concerning how beliefs are acquired and maintained, and how various types of evidence are weighted.

The Superstitious Thinking Subtest

We have been investigating superstitious thinking measures since 1989 (Stanovich, 1989), when we began using a measure inspired by items in previous paranormal beliefs questionnaires and superstitious thinking tests (Jones, Russell, & Nickel, 1977; Tobacyk & Milford, 1983). We have obtained some support for the view argued above, that superstitious thinking is linked to attitudes toward science and strategies for dealing with evidence. For example, in a 1997 study (Stanovich & West, 1997), we found that a high superstitious thinking score was significantly associated with poor performance on the Argument Evaluation subtest, discussed in chapter 8. Furthermore, the association with avoiding myside bias on that subtest remained significant after cognitive ability had been partialled out.

Subsequent studies have reinforced this pattern. Sá, Kelley, Ho, and Stanovich (2005) studied informal reasoning abilities using a variant of the structured interview paradigm of Kuhn (1991). They found that those scoring higher in superstitious thinking were more likely to use unsophisticated forms of argumentation such as simply reiterating their original theory rather than offering evidence. Furthermore, superstitious thinking was linked to unsophisticated argumentation styles even after the variance due to cognitive ability had been parceled out.

Kokis et al. (2002) found a similar pattern with a different rational thinking task in a study of children who completed a superstitious thinking measure adapted for younger respondents. That measure was significantly negatively associated with the tendency to pay attention to noncausal base rates of the type studied by Fong et al. (1986) and discussed in chapter 5. Additionally, high scores on the superstitious thinking measure were still associated with ignoring noncausal base rates even after variation in cognitive ability had been controlled. Macpherson and Stanovich (2007) found an association between high levels of superstitious thinking and the tendency for syllogistic reasoning to be influenced by belief bias in a paradigm similar to that discussed in chapter 7.

Toplak et al. (2007) were able to link superstitious thinking to dysfunctional real-world behavior. In a study of pathological gamblers, subclinical gamblers, and a control group, they found that higher levels of superstitious thinking were significantly associated with more dysfunctional gambling behavior. Furthermore, they found that this association remained significant after controlling for cognitive ability differences between their groups.

There are twelve items on the Superstitious Thinking subtest of the CART (see the appendix for sample items). The subtest assesses belief in a variety of superstitious and paranormal ideas such as astrology, psychokinesis, luck in general, mind reading, lucky numbers and possessions, unlucky objects, horoscopes, and predicting the future. The response scale was a six-point scale with no neutral point that was scored as follows: strongly disagree (1), disagree moderately (2), disagree slightly (3), agree slightly (4), agree moderately (5), agree strongly (6).

In the actual use of this subtest, the items are intermixed with the items from the Dysfunctional Personal Beliefs subtest described in this chapter as well as some of the thinking dispositions discussed in chapter 11 (Actively Open-Minded Thinking, Deliberative Thinking, Future Orientation, Differentiation of Emotions).

The items in the Superstitious Thinking subtest were run in a laboratory study involving 377 university students (RT57). The mean total score was 32.9 (SD = 9.2). The reliability of the total score was fairly high (Cronbach's alpha of 0.81). Exploratory factor analysis revealed a dominant factor with an eigenvalue of 3.46 that explained 83.9 percent of the variance. The eigenvalue of the second factor was less than one.

In the appendix, the raw score is translated into the five-point weighting that this subtest is given on the CART. Raw scores increase in the direction of more superstitious thinking and CART scores increase in the direction of *rejection* of superstitious thinking. This will be true of the other three contaminated mindware subtests as well: higher CART scores are in the direction of *rejecting* belief in the contaminated mindware.

We will discuss the relationships with other CART tasks included in this study after we have presented two other subtests related to the tendency to acquire contaminated mindware (the Antiscience Attitudes subtest and the Conspiracy Beliefs subtest).

The Antiscience Attitudes Subtest

The Superstitious Thinking subtest just described provides a direct measure of the acquisition of contaminated mindware in a specific and highly important domain of a person's worldview. Below, we describe the Conspiracy Beliefs subtest of the CART that likewise is a direct measure of the acquisition of contaminated mindware in an important domain of an individual's worldview. In this section, we describe beliefs related to an even more important and all-encompassing domain—the domain of science itself. Attitudes toward science have been shown to relate to the acquisition of accurate scientific knowledge itself (Allum et al., 2008).

The Antiscience Attitudes subtest of the CART assesses the type of attitudes that a person takes toward scientific knowledge. Specifically, the subtest looks at whether the respondent has a tendency to eschew science and to turn instead to intuition or "gut instinct" to justify belief and action. The subtest has items that assess whether the respondent tends to rely on empirical evidence or intuition; whether the respondent believes that science is a reliable source of progress; what the respondent does when scientific facts conflict with common sense; and whether the respondent relies more on observation or on instinct.

There are thirteen items on the Antiscience Attitudes subtest of the CART (see the appendix for sample items). The response scale was a six-point scale with no neutral point, described above. The items were presented as a block and were run in a laboratory study involving 377 university students (RT57). The mean total score was 40.7 (SD = 7.6). The reliability of the total score was fairly high (Cronbach's alpha of 0.81). Exploratory factor analysis revealed a dominant factor with an eigenvalue of 3.44 that explained 87.5 percent of the variance. The eigenvalue of the second factor was less than one.

In the appendix, the raw score is translated into the five-point weighting that this subtest is given on the CART. Raw scores increase in the direction of more antiscience thinking and CART scores increase in the direction of rejection of antiscience thinking.

The Conspiracy Beliefs Subtest

Conspiracy beliefs, defined as "an effort to explain some event or practice by reference to the machinations of powerful people, who attempt to conceal their role" (Sunstein & Vermeule, 2009, p. 205), represent an important area of contaminated mindware. These beliefs, given the lack of evidence for them, appear to be remarkably prevalent, especially considering that we live in an information-saturated society. For example, depending on the survey, it appears that almost 20 percent of the North American public believes that the US government had a role in planning the 9/11 attacks (Oliver & Wood, 2014; Sunstein & Vermeule, 2009), a majority believed in a conspiracy surrounding the assassination of John F. Kennedy (Goertzel, 1994), and 15–20 percent believe that government has spread AIDS deliberately (Goertzel, 1994). Oliver and Wood (2014) summarize the results of some surveys by noting that "four nationally representative survey samples collected in 2006, 2010, and 2011 indicate that over half of the American population consistently endorse some kind of conspiratorial narrative about a current political event" (p. 953).

Not only are conspiracy beliefs prevalent, but conspiratorial ideation appears to be part of a cluster of thinking styles that interconnect superstitious behavior, antiscience attitudes, and animistic thinking (Oliver & Wood, 2014). One common finding in the literature is that conspiracy beliefs tend to go together. That is, a belief in one type of conspiracy tends to be correlated with a belief in others (Dagnall et al., 2015; Goertzel, 1994; Lewandowsky, Oberauer, & Gignac, 2013; Majima, 2015; Swami et al., 2011). This finding is so ubiquitous that one group of authors humorously titled an article "NASA Faked the Moon Landing—Therefore, Climate Science Is a Hoax" (Lewandowsky, Oberauer, & Gignac, 2013). The co-occurrence of conspiracy beliefs is so strong that it is not uncommon for people to believe in conspiracies that contradict each other (Wood, Douglas, & Sutton, 2012).

One reason that conspiracy beliefs seem to be correlated across domains is that all conspiracy theories tap common underlying psychological mechanisms. Oliver and Wood (2014) posit two innate psychological predispositions as underlying conspiratorial ideation. One is the tendency to explain events by attributing intentionality to unseen others. The second is the tendency to look for melodramatic narratives when faced with the need to explain an important event. Both of these predispositions have been discussed extensively in the literature for some time now (Banerjee & Bloom, 2014; Bloom, 2004; Boyer, 2001, 2003; Dennett, 1987, 1991; 1996; Hood, 2009; Humphrey, 1976). The strong human propensity to infer intentionality and the tendency to rely on a narrative mode of thought were two important components of what Stanovich (2003, 2004) termed a set of four "fundamental computational biases" in human cognition:

(1) the tendency to contextualize a problem with as much prior knowledge as is easily accessible, even when the problem is formal and the only solution is a content-free rule;
(2) the tendency to "socialize" problems even in situations where interpersonal cues are few;
(3) the tendency to see deliberative design and pattern in situations that lack intentional design and pattern; and
(4) the tendency toward a narrative mode of thought.

Biases 3 and 4 are the fundamental computational biases that relate to conspiratorial ideation. The former bias (3), in the case of conspiratorial ideation, is just intentionality overextended in a different way—in the case of conspiracies, to unseen entities.

It is clear why we would expect a linkage between conspiratorial ideation and the Antiscience Attitudes subtest. Scientific attitudes are System 2 products (Stanovich, 2004) that are good at uncovering System 1 defaults that should be overridden. Likewise, supernatural thinking of the type assessed in our Superstitious Thinking subtest would appear to derive from the same fundamental computational biases that Oliver and Wood (2014) posit as central to conspiratorial ideation. Thus, we would expect some positive association between all three of the subtests discussed thus far in this chapter.

There are twenty-four items on the Conspiracy Beliefs subtest of the CART (see the appendix for sample items). We drew on a large number of conspiracies studied in the literature (Goertzel, 1994; Lewandowsky, Oberauer, & Gignac, 2013; Oliver & Wood, 2014), and added a few new ones of our own. Our subtest covered a wide range of conspiratorial beliefs. Most importantly, it covered conspiracies of both the political left and political right. Unlike some previous measures, it was not just a proxy for political attitudes (see Brandt et al., 2015; Chambers, Schlenker, & Collisson, 2013; Kahan, 2013; Oliver & Wood, 2015). Some of the commonly studied conspiracies that we assessed were: the assassination of President John F. Kennedy, the 9/11 attacks, fluoridation, the moon landing, pharmaceutical industry plots, the spread of AIDS, oil industry plots, and Federal Reserve conspiracies.

The response scale that was described above was used for this subtest as well. Like the Antiscience Attitudes subtest, the items on the Conspiracy Beliefs subtest were all presented together. Five extra items were included that actually did involve collusion on the part of corporations and government. We will call these the "justified beliefs" items. Subjects who were simply adopting a strategy of strongly denying conspiracies for reasons of impression management would score unusually low on these five items.

This subtest was run in the laboratory study mentioned above that involved 377 university students (RT57). Three subjects answered all five of the justified beliefs items "disagree strongly" and also answered at least 22 of 24 target conspiracy items the same way. These three subjects were removed from subsequent analyses involving this subtest, because they seemed to have developed an unthoughtful response set for the entire subtest. The mean total score of the remaining 374 subjects on the 24 target conspiracy items was 58.6 (SD = 17.5). This represents an average score of 2.44 on each of the individual conspiracies, which is a response scale location in the middle of "disagree moderately" and "disagree slightly." The mean total score on the five justified beliefs items was 18.7 (SD = 3.7). This represents an average score of 3.73 on each of the justified beliefs items, which is a response scale location in the middle of "agree slightly" and "disagree slightly," leaning a bit more toward the former—a location substantially higher than that for the target conspiracy items, as would be expected.

The reliability of the total score on the 24 target conspiracy items was high (Cronbach's alpha of 0.92). The reliability of the total score on the five justified beliefs items was 0.69 (Cronbach's alpha). From here on, when we refer to the Conspiracy Beliefs subtest, we mean the 24 target conspiracy items only, even though in administration the five justified beliefs items were intermixed with them. In the appendix, the raw score is translated into the ten-point weighting that this subtest is given on the CART. Raw

Table 10.1
Correlations among the variables in RT57

	1	2	3	4	5	6	7
1. Conspiracy Beliefs							
2. Superstitious Thinking	.32						
3. Antiscience Attitudes	.26	.20					
4. Impression Management	–.06	–.14	–.01				
5. SAT Total	–.16	–.24	–.30	.00			
6. Reflection versus Intuition	–.14	–.19	–.23	.05	.44		
7. Financial Literacy	–.10	–.19	–.20	–.06	.32	.41	
8. Probabilistic Numeracy	–.13	–.19	–.20	–.10	.37	.47	0.42

Note: All correlations larger than .10 in absolute value are significant at the .05 level. Ns range from 348 to 377.

scores increase in the direction of more conspiracy belief and CART scores increase in the direction of rejection of conspiracy beliefs. Exploratory factor analysis of the Conspiracy Beliefs subtest revealed a dominant factor with an eigenvalue of 7.88 that explained 79.1 percent of the variance. The second factor displayed an eigenvalue (1.22) not very far over the > 1 criterion.

Table 10.1 presents the correlations between the raw score on the Conspiracy Beliefs subtest and the other variables in this study. As expected, the Conspiracy Beliefs subtest correlated ($p < 0.001$) with both the raw Superstitious Thinking subtest score ($r = 0.32$) and the raw Antiscience Attitudes subtest score ($r = 0.26$). Also as expected, the latter two subtests correlated with each other ($r = 0.20$, $p < 0.001$). Because these three subtests were self-report questionnaire measures, we were concerned about social desirability/impression management issues contributing to the variance. Therefore, in this and other studies, we have included a ten-item Impression Management scale (Cronbach's alpha = 0.72) adapted from the work of Paulhus (1991). The next line in table 10.1 indicates that impression management was not highly associated with responses on these three subtests. Two of the three correlations were not significant even with this substantial sample size, and the third was –0.14 ($p < 0.01$.); in the latter case, higher impression management was associated with rejection of superstitious thinking.

The remainder of table 10.1 presents the correlations between the three subtests discussed so far in this chapter and SAT total score, a version of the Reflection versus Intuition subtest, and the Financial Literacy subtest and the Probabilistic Numeracy subtest. The correlations varied in magnitude, but only one failed to reach statistical significance (the Conspiracy Beliefs subtest and the Financial Literacy subtest).

The demographics questionnaire in this study contained two questions on political beliefs. One concerned the respondent's vote or favored candidate in the 2012 presidential election in the United States (choices restricted to Romney and Obama). The second concerned ideological orientation and ran from conservative to liberal on a six-point scale. We examined political ideology in the context of the Conspiracy Beliefs subtest because the literature on this association is evolving. The earlier literature seemed to suggest that conspiratorial ideation was associated with the political right. However, more recent research has suggested that this was simply a function of the specific conspiracy beliefs that were studied and their distribution in the questionnaire. More balanced research and more balanced stimulus materials have suggested that conspiracy beliefs (and other tendencies, such as intolerance) are equally prevalent on the right and the left (Brandt et al., 2014; Brandt et al., 2015; Chambers et al., 2013; Kahan, 2013; Oliver & Wood, 2014).

The results from this study were consistent with the more recent work on this issue. There was no significant correlation between voting for Obama or liberalism and the score on the Conspiracy Beliefs subtest (–0.08 in both cases). No doubt this was because our questionnaire contained more items than were included in earlier research, and also because we attempted to include both right-wing and left-wing conspiracy items as well as a good number of items that spanned the political divide. We strongly achieved the latter, because only three of our twenty-four items showed a significant correlation with political ideology (belief in a SARS conspiracy, belief that climate science is a hoax, and belief that the Federal Reserve is controlled by international elites were all significantly associated with conservatism). In the aggregate, our Conspiracy Beliefs subtest was largely independent of ideology, which is a positive feature of it as a subtest in a rational thinking assessment.

Because contaminated mindware of the conspiracy-belief type is acquired over a lifetime and the university sample in RT57 had a mean age of only

18.8 years, we tested an older Amazon Mechanical Turk sample (Turk2). In addition to the Conspiracy Beliefs subtest, the subjects in the Mechanical Turk sample completed a different set of measures than were in the study of university students just discussed. The Amazon Mechanical Turk study (Turk2) was completed by 222 subjects with a mean age of 32.3 (SD = 9.8), 134 of whom were males and 88 of whom were females. Two subjects answered three or four of the justified belief items "disagree strongly" and also answered at least 22 of 24 target conspiracy items the same way. These two subjects were removed from subsequent analyses involving this subtest because they seemed to have developed an unthoughtful response set for the entire subtest. The mean total score of the remaining 220 subjects on the 24 target conspiracy items was 64.7 (23.1), significantly higher than the mean in the university sample (58.6; $t(592) = 3.65$, $p < 0.001$). The mean total score of the Mechanical Turk sample on the five justified beliefs items was 22.1 (SD = 4.8), also significantly higher than the mean in the university sample (18.7; $t(592 = 8.92$, $p < 0.001$). These differences in conspiracy beliefs between samples did not replicate in another study, RT60 (see chapter 13), perhaps because both groups there were well-compensated monetarily and both completed the entire CART.

The reliability of the Conspiracy Beliefs subtest in the Mechanical Turk sample was 0.94 (Cronbach's alpha), even higher than that in the university sample (0.92). The reliability of the total score on the five justified beliefs items was 0.76 (Cronbach's alpha), again higher than that in the university sample (0.69). Exploratory factor analysis of the Conspiracy Beliefs subtest in the Mechanical Turk sample revealed a dominant factor with an eigenvalue of 10.16 that explained 76.7 percent of the variance. The second factor displayed an eigenvalue (1.31) not very far over the > 1 criterion.

The Dysfunctional Personal Beliefs Subtest

The previous three subtests probed an interlocking set of beliefs and attitudes toward science, toward the nature of explanation, and toward the definition of evidence, as well as attitudes toward belief justification and expertise. It is not surprising that these three subtests displayed significant intercorrelations. The fourth subtest on the CART that concerns contaminated mindware taps a different domain and thus would not be expected to correlate with the other three.

There is an enormous literature in clinical psychology documenting how certain personal beliefs are associated with suboptimal psychological and behavioral outcomes (Bentz, Williamson, & Franks, 2004; Bernard & Cronan, 1999; Butler, Beck, & Cohen, 2007; Christensen, Moran, & Wiebe, 1999; Kendall et al., 2008; Leahy, Holland, & McGinn, 2012; Lindner et al., 1999; Palmer, Gilleen, & David, 2015; Terjesen, Salhany, & Sciutto, 2009). Many of these beliefs are self-defeating in that they subvert the very goals that people have, and they cause the very worry and distress that most people wish to avoid. Examples of these would be personal stances that fail to accept the necessary imperfection of human performance or that fail to acknowledge the inevitable variability in the social world's response to oneself. Extreme perfectionism is often a worldview associated with negative behavioral and psychological outcomes. An excessive concern with social acceptance is another. Problems in dealing with the inevitable uncontrollable events in life are also part of this cluster of unhealthy beliefs about one's personal world.

Drawing on this large literature relating personal worldviews to psychological outcomes, and after some pilot work, we developed the nine-item Dysfunctional Personal Beliefs subtest of the CART (see the appendix for sample items). The subtest was run in the same laboratory study involving 377 university students (RT57) that was discussed above. The mean total score was 32.5 (SD = 7.2). The reliability of the total score was fairly high (Cronbach's alpha of 0.76). Exploratory factor analysis revealed a dominant factor with an eigenvalue of 2.55 that explained 95.8 percent of the variance. The eigenvalue of the second factor was less than one. In the appendix, the raw score is translated into the five-point weighting that this subtest is given on the CART. Raw scores increase in the direction of more dysfunctional beliefs and CART scores increase in the direction of rejection of dysfunctional beliefs.

In the same laboratory study, the raw total score on the Dysfunctional Personal Beliefs subtest failed to correlate significantly with either SAT total scores (0.04) or with a ten-item version of the Reflection versus Intuition subtest (–0.10). Finally, the Dysfunctional Personal Beliefs subtest failed to correlate significantly with the three measures of contaminated mindware discussed previously: Superstitious Thinking (0.09), Antiscience Attitudes (–0.10), and Conspiracy Beliefs (–0.03). Finally, one caution in using the Dysfunctional Personal Beliefs subtest is that it did correlate significantly with the Impression Management measure discussed previously ($r = -0.29$, $p < 0.001$). Thus, the fact that it is carrying some impression management/social desirability variance must be a consideration in interpreting any relationships involving this subtest.

We will present more data on the associations displayed by all of the subtests discussed above in chapters 12 and 13 when we report studies of the short-form and full-form versions of the CART.

11 The Dispositions and Attitudes of Rationality

In this chapter, we will consider the thinking dispositions of rationality, which in the CART are supplemental measures. These thinking dispositions help to contextualize the scores on other components of the CART. However, they are not scored as part of the CART itself. Rather, we developed them to serve as useful supplements to the CART subtests. We discussed the reasons for not including these dispositions in the CART in chapter 2. As we pointed out there, and as we will discuss in more detail in this chapter, thinking disposition measures are not direct measures of rationality. Instead, they provide clues as to which underlying mechanisms are involved in suboptimal thinking. In the remainder of this chapter, we will briefly contextualize thinking dispositions in terms of broad psychological theory. We will then discuss the particular measures that we feel are tightly enough linked to rational thinking that they are useful supplements to the CART subtests—that is, measures that might help to contextualize the reasons for the level of performance on other parts of the CART.

Thinking Dispositions in Psychology

As discussed in chapter 2, the distinction between cognitive capacities and thinking dispositions is an old one in psychology, although the latter term goes under a variety of names.[1] Despite this diversity of terminology, most authors use such terms similarly—to refer to relatively stable psychological mechanisms and strategies that tend to generate characteristic behavioral tendencies and tactics (see Buss, 1991). In this chapter, we will follow many others (Ennis, 1987; Perkins, 1995) in using the term "thinking dispositions."

Cognitive capacities are the types of cognitive processes studied by information-processing researchers seeking the underlying cognitive basis of performance on IQ tests. Perceptual speed, discrimination accuracy, working memory capacity, and the efficiency of the retrieval of information stored in long-term memory are examples of cognitive capacities that underlie traditional psychometric intelligence and have been extensively investigated. Thinking dispositions, in contrast, are better viewed as cognitive styles. Rational thinking dispositions are those that relate specifically to the adequacy of belief formation and decision making. Examples of rational thinking dispositions include "the disposition to weigh new evidence against a favored belief heavily (or lightly), the disposition to spend a great deal of time (or very little) on a problem before giving up, or the disposition to weigh heavily the opinions of others in forming one's own" (Baron, 1985, p. 15).

As discussed in chapter 2, there is substantial evidence that thinking dispositions can explain variance in components of rational thinking after the variance due to cognitive ability has been controlled. This is what would be expected if, as argued in chapter 2, thinking dispositions help us to separate the reflective from the algorithmic mind. Both intelligence and thinking dispositions underlie rational thinking. Neither is a direct measure of rational thinking itself. Intelligence and thinking dispositions are in a sense diagnostics of the potential locus of the problem when a person is not rational. Again, recall the tripartite model of mind discussed in chapter 2. Figure 2.2 reminds us that rationality is a more encompassing construct than either fluid intelligence or thinking dispositions. Rationality depends on mechanisms in all the minds of the tripartite model.

The thinking dispositions scales discussed in this chapter are supplemental to the subtests of the CART, and are not primary measures of rationality themselves, because they are not maximizing concepts like the other constructs on the CART. Optimal functioning does not result from maximizing cognitive styles. Instead, rationality, plotted against most thinking dispositions, is an inverted U-shaped function. One does not maximize rationality by maximizing the reflectivity/impulsivity dimension, for example, because such a person might get lost in interminable pondering and never make a decision. One does not maximize the dimension of belief flexibility, either, because such a person might end up with a pathologically unstable personality. Reflectivity and belief flexibility are "good" cognitive styles only in the sense that most people are too low on both dimensions (Baron, 2008); most people would be more rational if they increased their degrees of reflectivity and belief flexibility. But this does not mean that either of these thinking dispositions should always be maximized.

The Thinking Dispositions That Foster Rationality

Not all of the thinking dispositions studied by psychologists relate to rationality. Also, there may be more scales measuring cognitive styles than there are true underlying constructs. This is because many thinking disposition measures, although they may be named differently by their investigators, are really measuring very similar things. We settled on four different thinking disposition scales for the CART that reflect relatively disparate domains of cognitive regulation: the Actively Open-Minded Thinking scale, the Deliberative Thinking scale, the Future Orientation scale, and the Differentiation of Emotions scale.

We have been investigating actively open-minded thinking skills for almost two decades. We were inspired by the work of Baron (1985) to operationalize this concept, and we have refined the scale and examined its correlates in several studies (Stanovich & West, 1997, 2007; see also Baron et al., 2015; Svedholm & Lindeman, 2013). Our current scale taps several interrelated aspects of open-minded thought: the avoidance of epistemological absolutism, willingness to perspective-switch, willingness to decontextualize, and the tendency to consider alternative opinions and evidence (Samuelson & Church, 2015). It has similarities to measures in the literature such as dogmatism and absolutism scales (Rokeach, 1960), need for closure measures (Kruglanski & Webster, 1996), and two dispositional factors which Schommer (1990, 1993) calls "belief in simple knowledge" and "belief in certain knowledge." Our scale is focused more generally on the issue of epistemic self-regulation (Cederblom, 1989; Goldman, 1986; Harman, 1995; Nozick, 1993; Samuelson & Church, 2015; Thagard, 1992).

Stanovich and West (1997) found that an earlier version of our scale accounted for variance in performance on the Argument Evaluation subtest, even after the variance due to SAT total scores was partialled out. Sá, West, and Stanovich (1999) replicated this finding and additionally found that a scale of actively open-minded thinking predicted the avoidance of belief bias in syllogistic reasoning, also after the variance in a composite measure of cognitive ability was partialled out. In a study of 10- to 13-year-old children Kokis et al. (2002) found that a children's actively open-minded thinking scale was also a unique predictor of the degree of belief bias in syllogistic reasoning (that is, after cognitive ability was controlled for). Actively open-minded thinking was also a unique predictor of the use of noncausal base rates in a task adapted for children.

Sá et al. (2005) found that those scoring higher in actively open-minded thinking used fewer unsophisticated forms of argumentation, such as simply reiterating their original theory, and were more likely to offer valid evidence. Furthermore, actively open-minded thinking was linked to sophisticated argumentation styles even after the variance due to cognitive ability had been partialled out. Sá and Stanovich (2001) found actively open-minded thinking to be an independent predictor of judgments of the mental content of others. Finally, one of our actively open-minded thinking scales correlated significantly with a particular myside bias measure that has little association with cognitive ability (Stanovich & West, 2008a).

In several other studies, we found that an earlier version of our actively open-minded thinking scale (in conjunction with other thinking dispositions in an amalgamated variable) predicted performance, on a variety of heuristics and biases tasks, after partialling cognitive ability. These heuristics and biases tasks include noncausal base rate tasks, hypothesis evaluation tasks, four-card selection tasks, covariation detection, gambler's fallacy, sample size problems, conjunction fallacy, Bayesian reasoning, framing problems, ratio bias, sample size problems, and probability matching (Stanovich & West, 1998c; Toplak, West, & Stanovich, 2011, 2014a; West, Toplak, & Stanovich, 2008). Other laboratories have found our actively open-minded thinking measure to predict performance on heuristics and biases tasks and other reasoning paradigms (Baron et al., 2015; Haran, Ritov, & Mellers, 2013; Heijltjes et al., 2014, 2015).

Our Deliberative Thinking scale was designed to capture the type of cognitive variance tapped by need for cognition and typical intellectual engagement measures (Cacioppo et al., 1996; Goff & Ackerman, 1992). Pilot versions of our scale and/or the original need for cognition scale were used in some of our earlier studies. Although not as potent a predictor as actively open-minded thinking, deliberative thinking displays associations with numerous heuristics and biases tasks (Kokis et al., 2002; Macpherson & Stanovich, 2007; Toplak & Stanovich, 2002; Toplak et al., 2011, 2014a;

West et al., 2008). Sometimes, especially when combined with other dispositional measures, deliberative thinking can predict variance in rational thinking after variance due to cognitive ability has been partialled out.

Our Future Orientation scale was inspired by Strathman, Gleicher, Boninger, and Edwards (1994), who studied a construct that was centered on thinking about the future. Pilot versions of our scale and/or the original Strathman et al. (1994) scale were used in some of our earlier studies. Although not as potent a predictor as actively open-minded thinking, future orientation scales display associations with numerous heuristics and biases tasks (Toplak et al., 2011, 2014a). Toplak et al. (2007) were able to link future-oriented thinking to dysfunctional real-world behavior. In a study of pathological gamblers, subclinical gamblers, and a control group, they found that low levels of future-oriented thinking were significantly associated with more dysfunctional gambling behavior even after controlling for cognitive ability differences in their sample.

Finally, the rationale for our Differentiation of Emotions scale derives from research in cognitive neuroscience that has uncovered cases of mental pathology characterized by inadequate behavioral regulation from the emotion subsystems in the autonomous mind. Examples of this are Damasio's (1994, 1996; Bechara, Damasio, Damasio, & Anderson, 1994; Eslinger & Damasio, 1985) well-known studies of patients with damage in the ventromedial prefrontal cortex. These individuals have severe difficulties in real-life decision making but do not display the impairments in sustained attention and conscious cognitive control that are characteristic of individuals with damage in dorsolateral frontal regions. Instead, they are thought to lack the emotional signals that constrain the combinatorial explosion of possible actions to a manageable number based on somatic markers stored from similar situations in the past.

The key insight here is that there are two ways in which the behavioral regulation involving the autonomous mind can go wrong. The override failures, discussed previously, are one way. In these situations, the signals shaping behavior from the autonomous mind are too pervasive and are not trumped by Type 2 processing. The second way that behavioral regulation involving the autonomous mind can go awry has the opposite properties. In this case, the automatic regulation of goals by the autonomous mind is absent, and Type 2 processing is faced with a combinatorial explosion of possibilities because the constraining function of certain autonomous modules is missing.

There is empirical evidence for rationality failures of the two different types (Toplak et al., 2010). Dorsolateral prefrontal damage has been associated with executive functioning difficulties (and/or working memory difficulties) that can be interpreted as the failure to override automatized processes. In contrast, ventromedial damage to the prefrontal cortex has been associated with problems in behavioral regulation that are accompanied by affective disruption. Difficulties of the former but not the latter kind are associated with lowered intelligence.[2]

A laboratory marker for the type of problem that Damasio had observed is the well-known Iowa Gambling Task (Bechara et al., 1994). Inferior performance on the task has been associated with many clinical syndromes of cognitive and behavioral impairment (Mukherjee & Kable, 2014; Toplak et al., 2010). Likewise, subclinical behavioral impairments have been linked to inferior performance on the task (Stanovich, Grunewald, & West, 2003).

Despite the substantial use of the Iowa Gambling Task in the literature, we decided to use an alternative method (a questionnaire method) of tapping a similar cognitive dimension—hence, our use of the Differentiation of Emotions scale. We made this choice for two reasons. First of all, the Iowa Gambling Task is lengthy and logistically difficult to administer. Second, in our own study of subclinical and pathological gamblers (Toplak et al., 2007) we found that a questionnaire method was actually better at differentiating our subgroups. In that study, we did administer the Iowa Gambling Task to a group of pathological gamblers, a group of subclinical gamblers, and a no-problem control group. However, we also administered the Alexithymia scale revised by Bagby, Parker, and Taylor (1994), designed to assess self-reported difficulties with identifying and describing feelings. In our study (Toplak et al., 2007), the Bagby et al. (1994) Alexithymia scale differentiated the three groups more strongly than did the Iowa Gambling Task. Furthermore, the Alexithymia scale (but not the Iowa Gambling Task) differentiated the subgroups after differences due to general intelligence had been controlled. Thus, over several unpublished studies, we have been developing our own Differentiation of Emotions scale that is inspired by many of the existing Alexithymia scales.

The Thinking Disposition Scales of the CART

There are thirty items on the Actively Open-Minded Thinking scale of the CART (see the appendix for sample items). The response scale was a six-point scale (with no neutral point) that was coded as follows: strongly disagree (1), disagree moderately (2), disagree slightly (3), agree slightly (4), agree moderately (5), agree strongly (6). In the actual use of this scale, the items are intermixed with the items from the other scales described in this chapter, as well as with the Dysfunctional Personal Beliefs and Superstitious Thinking subtests described in the previous chapter.

The items were run in two laboratory studies involving 238 and 377 university students, respectively (RT56 and RT57). The mean total scores on the Actively Open-Minded Thinking scale in the two studies were 122.3 (SD = 12.9) and 123.4 (SD = 11.7), respectively. The reliability of the total score was moderate (Cronbach's alpha of 0.78 and 0.72, respectively). In table 13.14 of chapter 13, the raw score is translated into a percentile rank in our comprehensive study of the full-form CART.

The sixteen items on the Deliberative Thinking scale of the CART (see the appendix for sample items) were also run in RT56 and RT57. The mean total scores on the Deliberative Thinking scale in the two studies were 63.3 (SD = 12.7) and 62.8 (SD = 12.3), respectively. The reliability of the total score was high (Cronbach's alpha of 0.92 and 0.91, respectively). In table 13.14 of Chapter 13, the raw score is translated into a percentile rank in our comprehensive study of the full-form CART.

The fourteen items on the Future Orientation scale of the CART (see the appendix for sample items) were run in the same two laboratory studies; the mean total scores in the two studies were 56.2 (SD = 8.0) and 57.2 (SD = 8.6), respectively. The reliability of the total score was fairly high (Cronbach's alpha of 0.80 and 0.82, respectively). In table 13.14 of chapter 13, the raw score is translated into a percentile rank in our comprehensive study of the full-form CART.

The fourteen items on the Differentiation of Emotions scale of the CART (see the appendix for sample items) were run in the same two laboratory studies (RT56 and RT57); the mean total scores in were 57.7 (SD = 8.6) and 56.4 (SD = 8.7), respectively. The reliability of the total score was fairly high (Cronbach's alpha of 0.79 and 0.78, respectively). In table 13.14 of chapter

Table 11.1
Correlations among the thinking dispositions in RT56 (below the diagonal) and in RT57 (above the diagonal).

	1	2	3	4	5
1. Actively Open-Minded Thinking		.46**	.30**	.21**	.02
2. Deliberative Thinking	.53**		.23**	.09	.15*
3. Future Orientation	.23**	.17*		.02	−.05
4. Differentiation of Emotions	.27**	.24**	.03		.28**
5. Impression Management	.06	.21*	−.01	.33**	

Note: $N = 238$ below the diagonal; $N = 377$ above the diagonal.
*$p < .01$; **$p < .001$

13, the raw score is translated into a percentile rank in our comprehensive study of the full-form CART.

Table 11.1 presents the intercorrelations among the thinking dispositions in the two studies. The correlations below the diagonal come from the $N = 238$ study (RT56) and the correlations above the diagonal come from the $N = 377$ study (RT57). As can be seen, the correlations from the two studies largely replicate each other. The Actively Open-Minded Thinking scale had a consistent moderate correlation with the Deliberative Thinking scale (0.53 and 0.46, respectively). It also displayed significant but smaller correlations with the Future Orientation scale and the Differentiation of Emotions scale. The Deliberative Thinking scale and Future Orientation scale had small but significant correlations with each other.

The fifth variable in the correlation matrix is the ten-item Impression Management scale, mentioned in chapter 10. The correlations indicate that two of the four thinking dispositions scales (Deliberative Thinking and Differentiation of Emotions) displayed a significant relationship with impression management. This should be taken into consideration when the results of these scales are interpreted. However, it was consistently the case that the Actively Open-Minded Thinking scale and the Future Orientation scale did not implicate impression management and/or social desirability.

Table 11.2 presents the correlations of the four thinking disposition scales with other marker variables and some CART subtests that were included in these two studies. The first three rows of the table come from the smaller study (RT56) and the remaining rows are correlations from the larger study (RT57). Across the two studies, there was a consistent trend for cognitive

Table 11.2
Correlations among the thinking dispositions and other variables in RT56 and RT57

	Actively Open-Minded Thinking Scale	Deliberative Thinking Scale	Future Orientation Scale	Differentiation of Emotions Scale
	Study RT56 ($N = 238$)			
SAT Total	.31***	.35***	.04	.10
Reflection versus Intuition (8-item pilot version)	.33***	.37***	.17**	.15*
Probabilistic Numeracy (10 item pilot version)	.35***	.30***	.16*	.21**
	Study RT57 ($N = 377$)			
SAT Total	.28***	.25***	.05	.06
Reflection versus Intuition (10-item pilot version)	.25***	.35***	.08	.06
Probabilistic Numeracy (CART version)	.25***	.26***	.12*	.09
Superstitious Thinking (12-item CART)	–.38***	–.11*	–.01	–.18***
Antiscience Attitudes	–.50***	–.27***	–.20***	–.03
Conspiracy Beliefs	–.25***	–.09	–.13*	–.15**
Dysfunctional Personal Beliefs	–.04	–.14**	.23***	–.43***

*$p < .05$; **$p < .01$; ***$p < .001$

ability (SAT total score) to correlate with the Actively Open-Minded Thinking scale and the Deliberative Thinking scale, but not with the other two thinking disposition scales. Likewise, across both studies, a pilot reflection versus intuition measure and a pilot probabilistic numeracy task both correlated more highly with Actively Open-Minded Thinking and Deliberative Thinking than with the other two thinking dispositions. The last four rows in this table present the correlations between the four thinking dispositions and the four measures of contaminated mindware discussed in chapter 10.

The Actively Open-Minded Thinking scale was a particularly good predictor of the avoidance of pseudoscience (the Superstitious Thinking subtest, the Antiscience Attitudes subtest, and the Conspiracy Beliefs subtest). In contrast, the Differentiation of Emotions scale was a particularly good predictor of Dysfunctional Personal Beliefs.

We will present more data on the associations displayed by all of the scales discussed above in chapter 13, when we report a study of the full-form version of the CART.

III Comprehensive Rational Thinking Assessment: Data and Conclusions

12 Associations among the Subtests: A Short-Form CART

In table 4.1 of chapter 4, we presented our comprehensive framework for measuring rational thinking. Unsurprisingly to anyone familiar with this literature, it took many different subtests to capture even a minimal number of the thinking skills and domains that have been studied. We will argue in chapter 14 that ours is the most comprehensive framework that has been constructed to date. Nevertheless, when it comes to actual assessment, comprehensiveness will obviously be at odds with logistical considerations. Put simply, the more comprehensive the assessment, the longer the test will take to complete and the more taxing the task of completing the test will be. In chapter 13 we will describe a study where subjects completed all of the subtests listed in table 4.1. Subjects sometimes took as long as three hours to complete the entire CART. In this chapter, we will describe a short form of the test that takes less than two hours to complete, and we will present some data on the statistical properties of the short-form test.

Table 12.1 reproduces table 4.1 in which the framework for the CART and its subtests is presented, except that table 12.1 shades the subtests of the CART that make up the short form of the test. A short form composed of these subtests can usually be completed in less than two hours by most subjects. As indicated in the table, the short-form CART includes both the Probabilistic and Statistical Reasoning and the Scientific Reasoning subtests, as both are at the heart of most definitions of rational thinking, and they both represent the core of the heuristics and biases literature.

All of the subtests that in some way tap the avoidance of miserly processing are included in the short form: the Reflection versus Intuition subtest, the Belief Bias in Syllogistic Reasoning subtest, the Ratio Bias subtest, and the Disjunctive Reasoning subtest. These four subtests are all deemed important because a key component of rational thought is to recognize when deeper processing will yield substantial increases in the optimality of responses—a central feature of all of these subtests. People can have knowledge of certain principles of rational thinking without the propensity to use them. It is just the propensity to use available thinking tools that these tasks in part tap.

The Probabilistic Numeracy subtest is included in the short-form CART not only because it is a central skill of decision making, but also because this subtest is very short and quite statistically potent for the amount of time that it takes. It is likewise with the four measures of the avoidance of contaminated mindware. All four subtests are included in the short form because they are all presented in a questionnaire format that is easily completed in a brief amount of time.

The short form of the CART uses just one of the thinking disposition scales: the Actively Open-Minded Thinking scale. This scale is not scored for points on the short-form CART for reasons discussed in chapter 11. But, as we will see in the data from a study of the short form, this subtest is quite diagnostic of performance on many of the subtests.

Table 12.1 indicates the number of CART points allocated to each of the subtests in the short form. The short form has a total of 100 points: 38 points are allocated to the crucial Probabilistic and Statistical Reasoning and Scientific Reasoning subtests; 28 points are allocated to the four subtests assessing the avoidance of miserly processing; 25 points are allocated to various forms of contaminated mindware avoidance; and 9 points are allocated to the Probabilistic Numeracy subtest. With this short form and distribution of points, a wide variety of the most critical areas of rational thinking is tapped in a very efficient way in less than two hours.

A Study of the Short-Form CART

We examined the performance of a single group of subjects on the short-form CART in an unpublished study in our laboratory (RT59) involving 372 university student volunteers as subjects (68 males and 304 females) with a mean age of 18.7 (SD = 0.9).

The study was run online using Qualtrics in a single session, supervised in a lab setting. The subtests were run as blocks except for the Scientific Reasoning subtest, which was run in three parts, and the six items of the Disjunctive Reasoning subtest, which were interspersed among other subtests.[1] The order of administration was as follows: Disjunctive Reasoning item 1

Table 12.1

Framework for classifying the types of rational thinking tasks and subtests on the CART. Short form of the test is shaded. Short-form CART points for each subtest indicated in the cell.

Tasks Saturated with Processing Requirements (Detection, Override, Hypothetical Thinking)	Rational Thinking Tasks Saturated with Knowledge	Avoidance of Contaminated Mindware	Thinking Dispositions that Foster Thorough and Prudent Thought, Unbiased Thought, and Knowledge Acquisition
Probabilistic and Statistical Reasoning subtest, 18 points		Superstitious Thinking subtest, 5 Points	Actively Open-Minded Thinking scale
Scientific Reasoning subtest, 20 points		Antiscience Attitudes subtest, 5 points	Deliberative Thinking scale
Avoidance of Miserly Information Processing: Reflection versus Intuition subtest, 10 points; Belief Bias Syllogisms, 8 points; Ratio Bias subtest, 5 points; Disjunctive Reasoning, 5 points	Probabilistic Numeracy subtest, 9 points	Conspiracy Beliefs subtest, 10 points	Future Orientation scale
Absence of Irrelevant Context Effects in Decision Making: Framing subtest; Anchoring subtest; Preference Anomalies subtest	Financial Literacy and Economic Knowledge subtest	Dysfunctional Personal Beliefs subtest, 5 points	Differentiation of Emotions scale
Avoidance of Myside Bias: Argument Evaluation subtest	Sensitivity to Expected Value subtest		
Avoiding Overconfidence: Knowledge Calibration subtest	Risk Knowledge subtest		
Rational Temporal Discounting			

of 6, Probabilistic and Statistical Reasoning subtest, Disjunctive Reasoning item 2 of 6, Belief Bias Syllogism subtest, Disjunctive Reasoning item 3 of 6, Scientific Reasoning part 1 of 3, Disjunctive Reasoning item 4 of 6, Reflection versus Intuition subtest, Scientific Reasoning part 2 of 3, Disjunctive Reasoning item 5 of 6, Ratio Bias subtest, Scientific Reasoning part 3 of 3 (Covariation Detection), Disjunctive Reasoning item 6 of 6 (the Married problem), Probabilistic Numeracy subtest, Antiscience Attitudes subtest, Conspiracy Beliefs subtest, and the Questionnaire.

The Actively Open-Minded Thinking scale, Dysfunctional Personal Beliefs subtest, and Superstitious Thinking subtest were all mixed together and presented as part of the same questionnaire. Participants were also asked to report their SAT scores as part of the demographics information. A majority of subjects finished in under 75 minutes and most finished under 100 minutes. The data were collected over 66 days in the fall of 2014.

Table 12.2 presents the means and standard deviations of the raw scores on each of the short-form subtests in addition to the mean and standard deviation of the supplemental Actively Open-Minded Thinking scale. Also presented in the table are the lowest and highest scores obtained by a subject on that subtest, as well as the skewness value for each of the raw scores. Most of the variables had modest skews, with the exception of the Reflection versus Intuition subtest, which, consistent with previous research, had a substantial positive skew.

Table 12.2

Descriptive statistics for short-form CART subtests and composite scores in RT59 ($N = 372$)

	Mean	SD	Low/High score	Skewness	Reliability
		Raw Scores			
Probabilistic Reasoning subtest	9.84	2.84	3–17	.23	.64
Scientific Reasoning subtest	9.07	3.03	2–17	.22	.58
Reflection versus Intuition subtest	2.40	2.13	0–10	1.17	.67
Syllogistic Reasoning subtest	9.85	2.37	4–16	.46	.49
Ratio Bias subtest	46.4	11.35	12–72	–.33	.85

Table 12.2 (continued)

	Mean	SD	Low/High score	Skewness	Reliability
Disjunctive Reasoning subtest	3.45	1.74	0–6	–.61	.72
Probabilistic Numeracy subtest	4.43	1.58	0–9	.09	.50
Superstitious Thinking subtest	33.17	9.73	12–58	–.11	.85
Antiscience Attitudes subtest	40.72	7.55	13–58	–.84	.82
Conspiracy Beliefs subtest	63.08	18.60	24–111	–.09	.93
Dysfunctional Personal Beliefs subtest	32.27	6.59	13–53	.26	.74
Actively Open-Minded Thinking scale	121.70	12.32	83–163	.50	.78
CART Scores of the Subtests					
Probabilistic Reasoning subtest	9.84	2.84	3–17	.23	—
Scientific Reasoning subtest	9.07	3.03	2–17	.22	—
Reflection versus Intuition subtest	2.40	2.13	0–10	1.17	—
Syllogistic Reasoning	2.04	2.12	0–8	.89	—
Ratio Bias subtest	1.96	1.63	0–5	.33	—
Disjunctive Reasoning subtest	2.54	1.58	0–5	–.38	—
Probabilistic Numeracy subtest	4.43	1.58	0–9	.09	—
Superstitious Thinking subtest	2.33	1.66	0–5	.13	—
Antiscience Attitudes subtest	2.22	1.72	0–5	.17	—
Conspiracy Beliefs subtest	4.44	3.18	0–10	.25	—
Dysfunctional Personal Beliefs subtest	2.34	1.57	0–5	.08	—
Total Short-Form CART score	43.62	12.99	16–87	.66	.76

The far right column of table 12.2 presents the reliability (Cronbach's alpha) of the raw scores on each of the subtests. Most are adequate, with the exception of the Scientific Reasoning subtest, Belief Bias in Syllogistic Reasoning subtest, and the Probabilistic Numeracy subtest, all of which had disappointing reliabilities. However, as we will see in chapter 13, in a study of the full-form test using a more diverse sample of subjects drawn in part from the Amazon Mechanical Turk and a large group of paid (rather than volunteer) university students, the reliabilities of these three subtests all increased (the syllogisms substantially). In fact, though, these three subtests all had higher reliabilities in studies prior to RT59 as well (see chapters 6, 7, and 9, where these subtests were introduced). RT59 appears to be a bit of an outlier with respect to the reliabilities of these subtests.

Farther down the table are the mean and standard deviations of the CART scores of each of the subtests. For a few subtests, the raw score and the CART score are the same: the Probabilistic and Statistical Reasoning subtest, the Scientific Reasoning subtest, the Reflection versus Intuition subtest, and the Probabilistic Numeracy subtest.[2] Finally, the bottom of the table indicates that the mean total CART score was 43.62 (SD = 12.99), that it has a moderate positive skew, and that it has a Cronbach's alpha of 0.76 (calculated by treating subtests as items with no differential weighting of CART points allocated). The lowest score achieved on the short form was 16 and the highest score achieved was 87 out of a possible 100 points.

Table 12.3 presents the associations between the CART scores on each of the subtests and the SAT total score and the raw score on the Actively Open-Minded Thinking scale (AOT). Only the Dysfunctional Personal Beliefs subtest failed to correlate with the SAT and the AOT. All of the other correlations are significant at the 0.001 level except for the 0.131 correlation, which is significant at the 0.05 level. The final three columns of the table represent the results of a multiple regression analysis in which the SAT total scores and AOT were used to predict scores on each of the short-form subtests. The last two columns represent the standardized beta weights and give an indication of whether that variable is an independent predictor of the subtest performance when the other variable is partialled out. As expected from the raw correlations, none of the coefficients for the Dysfunctional Personal Beliefs subtest was significant. However, for most of the other short-form subtests, both variables were significant independent predictors. This is theoretically interesting because it means that variance on

Table 12.3
Individual difference predictors of short-form subtest performance and composite scores in RT59

	Correlation with SAT total (N=363)	Correlation with AOT score (N=372)	Multiple R-Squared (N=363)	Standardized Beta for SAT total	Standardized Beta for AOT score
		Subtest Scores			
Probabilistic Reasoning subtest	.424	.292	.208	.375**	.175**
Scientific Reasoning subtest	.331	.360	.180	.254**	.276**
Reflection versus Intuition subtest	.426	.320	.221	.369**	.205**
Syllogistic Reasoning subtest	.316	.336	.162	.243**	.260**
Ratio Bias subtest	.187	.202	.057	.144**	.155**
Disjunctive Reasoning subtest	.196	.131	.044	.174**	.080
Probabilistic Numeracy subtest	.428	.331	.230	.365**	.225**
Superstitious Thinking subtest	.248	.461	.216	.135*	.409**
Antiscience Attitudes subtest	.179	.505	.255	.043	.492**
Conspiracy Beliefs subtest	.226	.344	.135	.143*	.301**
Dysfunctional Personal Beliefs	–.058	.049	.008	–.077	.069
		Composite Scores			
Short-Form total CART score	.496	.552	.422	.375**	.437**
CART Contaminated Mindware (4 subtests)	.257	.536	.292	.120*	.494**
CART: 7 Remaining Subtests	.511	.440	.349	.426**	.308**

All the zero-order correlations are statistically significant at the .05 level except those involving the Dysfunctional Personal Beliefs CART score. For the beta weights: * = $p < .01$; ** = $p < .001$

most of the individual rational thinking subtests of this form is explained by both cognitive ability and thinking dispositions.

The next row in table 12.3 indicates that both the SAT and the AOT had a moderate relationship with the total short-term CART score, and that both were independent predictors in a regression analysis. Perhaps surprisingly, the AOT had a higher correlation with the total short-form CART score than the SAT. However, as is clear from the subtest relationships presented at the top of the table, the AOT was a particularly potent predictor of the contaminated mindware class of subtests. We thus attempted to see whether the AOT would remain a significant independent predictor of the other parts of the short form. First, though, the next row in the table shows the correlations and multiple regression analysis for the four contaminated mindware subtests when amalgamated together. Here it can be seen that, as predicted, the AOT had a substantially higher beta weight as a predictor of this part of the CART (0.494) than did the SAT (0.120).

The final row of the table, though, contains the key analysis of interest, the one that addresses whether the AOT is a significant unique predictor of the parts of the short-form CART that do not involve the avoidance of contaminated mindware. The answer is clearly yes. Although its beta weight is lower than that for the SAT, the AOT remains a significant unique predictor of the part of the short-form CART that does not involve the avoidance of contaminated mindware subtests. The four contaminated mindware subtests of the CART display an association of 0.41 ($p < 0.001$) with the other seven subtests combined. Finally, a test for difference between dependent correlations (Cohen & Cohen, 1983) indicated that the correlation between the AOT score and the short-form CART total (0.541) was significantly higher than the 0.277 correlation between the AOT score and the SAT total score, $t(360) = 6.17$, $p < 0.001$).

The correlation of 0.496 between the CART total score and the SAT total score indicates, consistent with our discussion of our previous research in chapter 2, a moderate degree of overlap between rationality and intelligence, but also a moderate degree of nonoverlap. This relationship is contextualized further in some additional analyses not reported in table 12.3. A 65-point CART score made up of the five subtests that correlated most strongly with the SAT (Probabilistic Reasoning, Scientific Reasoning, Reflection versus Intuition, Syllogistic Reasoning, and Probabilistic Numeracy) only raised the correlation with SAT to 0.518. A 37-point CART score made up of the three highest correlates of SAT (Probabilistic Reasoning, Reflection versus Intuition, and Probabilistic Numeracy) raised the correlation with SAT to 0.533.

Table 12.4 displays the intercorrelations of all of the subtest scores on the short-form CART. Line 12 displays the correlation of the total CART score with each of the components. Line 13 displays the subtest correlation with the total score minus that subtest. The Dysfunctional Personal Beliefs subtest failed to correlate with any of the other subtests. However, 44 of the remaining 45 correlations (minus correlations with the Dysfunctional Personal Beliefs subtest) were statistically significant, with the median correlation among the remaining ten subtests being 0.25. The Cronbach's alpha for the short-form was 0.76 when all eleven subtests were included. Eliminating the Dysfunctional Personal Beliefs subtest raises Cronbach's alpha to 0.78. These alphas were calculated by treating subtests as items with no differential weighting of CART points allocated.

A perusal of table 12.4 reveals that four of the subtests had particularly strong intercorrelations with each other: the Probabilistic and Statistical Reasoning subtest, the Scientific Reasoning subtest, the Reflection versus Intuition subtest, and the Probabilistic Numeracy subtest.[3] The part–whole correlations on lines 12 and 13 confirm that these four subtests were particularly predictive of the total short-form test score, followed by the Superstitious Thinking subtest and the Belief Bias in Syllogistic Reasoning subtest.

An exploratory principal components analysis of the eleven CART subtests revealed that only the first three components displayed eigenvalues greater than 1. These three components explained 52.6 percent of the variance. Table 12.5 displays the loadings on these three components after varimax rotation.

The seven subtests not assessing the presence of contaminated mindware all loaded moderately and fairly exclusively on component 1. Three of the four measures of contaminated mindware loaded moderately and fairly exclusively on component 2. The third component was largely defined by a singleton variable, the Dysfunctional Personal Beliefs subtest, as well as a moderate loading from the Ratio Bias subtest.

Some previous investigations have found that males tend to do better in probabilistic reasoning, probabilistic numeracy, and cognitive reflection measures (Frederick, 2005; Gal & Baron, 1996; Toplak et al., 2014a; Weller et al., 2013; West & Stanovich, 2003). In chapter 5, we presented data indicating that males outperformed females on the Probabilistic and Statistical Reasoning subtest. Regarding the CART total scores in study RT59, we

Table 12.4
Correlations among the short-form CART subtests in RT59 (*N*=372)

	1	2	3	4	5	6	7	8	9	10	11
1. Probabilistic Reasoning CART score											
2. Scientific Reasoning CART score	.44										
3. Reflection versus Intuition subtest CART score	.44	.44									
4. Syllogistic Reasoning CART score	.37	.45	.42								
5. Ratio Bias CART score	.36	.26	.29	.21							
6. Disjunctive Reasoning CART score	.21	.25	.23	.21	.06						
7. Probabilistic Numeracy CART score	.43	.40	.48	.37	.21	.22					
8. Superstitious Thinking CART score	.36	.34	.28	.29	.19	.18	.28				
9. Antiscience Attitudes CART score	.24	.26	.24	.22	.22	.14	.20	.28			
10. Conspiracy Beliefs CART score	.25	.22	.13	.17	.11	.07	.19	.41	.29		
11. Dysfunctional Personal Beliefs CART score	.08	−.03	.02	−.08	.06	.00	−.01	.07	−.03	−.01	
12. Short-Form Total CART score	.72	.71	.66	.61	.47	.39	.61	.61	.50	.53	.13
13. Short-Form CART score minus subtest	.59	.57	.55	.49	.36	.28	.52	.51	.39	.32	.01

Note: All correlations larger than .10 in absolute value are significant at the .05 level.

Table 12.5
Loadings on the first three principal components after varimax rotation

	Component 1	Component 2	Component 3
1. Probabilistic and Statistical Reasoning subtest CART score	.37	—	—
2. Scientific Reasoning subtest CART score	.40	—	—
3. Reflection versus Intuition subtest CART score	.46	—	—
4. Belief Bias in Syllogistic Reasoning subtest CART score	.40	—	—
5. Ratio Bias subtest CART score	.26	—	.35
6. Disjunctive Reasoning subtest CART score	.27	—	—
7. Probabilistic Numeracy subtest CART score	.41	—	—
8. Superstitious Thinking subtest CART score	—	.51	—
9. Antiscience Attitudes subtest CART score	—	.46	—
10. Conspiracy Beliefs subtest CART score	—	.71	—
11. Dysfunctional Personal Beliefs subtest CART score	—	—	.86

Note: Loadings less than .25 have been eliminated.

found that the mean score of the 68 males (51.9, SD = 17.1) was significantly higher than the mean score of the 304 females (41.8, SD = 11.1), $t(370) = 6.10$, $p < 0.001$, Cohen's $d = 0.818$. However, the males in our sample were also higher in cognitive ability. Of those reporting SAT total scores, we found that the mean score of the 64 males (1169, SD = 124) was significantly higher than the mean score of the 299 females (1098, SD = 115), $t(361) = 4.43$, $p < 0.001$. The correlation of −0.30 between sex and CART total score was reduced to −0.21 when SAT total score was controlled but still remained significant at the 0.001 level.

Table 12.6 displays the performance of males and females across the eleven subtests of the short-form CART. From the table it can be seen that

Table 12.6
Differences between males and females on CART subscale scores from the short form in RT59

CART Subtest	Males (n = 68)		Females (n = 304)		t(370)	Cohen's d
	M	(SD)	M	(SD)		
Probabilistic Reasoning	11.53	(3.18)	9.47	(2.61)	5.64***	0.76
Scientific Reasoning	9.68	(3.78)	8.93	(2.82)	1.84	0.25
Reflection versus Intuition	3.93	(2.91)	2.06	(1.75)	6.91***	0.93
Syllogistic Reasoning	2.60	(2.50)	1.92	(2.01)	2.43*	0.33
Ratio Bias	2.60	(1.73)	1.82	(1.57)	3.64***	0.49
Disjunctive Reasoning	2.96	(1.51)	2.44	(1.58)	2.44*	0.33
Probabilistic Numeracy	5.32	(1.86)	4.23	(1.44)	5.34***	0.72
Superstitious Thinking	2.57	(1.74)	2.27	(1.64)	1.35	0.18
Antiscience Attitudes	2.63	(1.94)	2.13	(1.66)	2.18*	0.29
Conspiracy Beliefs	4.79	(3.51)	4.36	(3.10)	1.03	0.14
Dysfunctional Personal Beliefs	3.29	(1.49)	2.13	(1.51)	5.77***	0.77

Two-tailed p values: *$p < .05$; **$p < .01$; ***$p < .001$

the overall sex difference in the CART total score is strongly determined by performance differences on four particular subtests: the Probabilistic and Statistical Reasoning subtest, the Reflection versus Intuition subtest, the Probabilistic Numeracy subtest, and the Dysfunctional Personal Beliefs subtest. The first three sex differences were somewhat attenuated by taking into account the sex confound with SAT total score mentioned previously, but the last is not. However, all were still statistically significant after controlling for SAT total score.

As table 12.7 indicates, the short-form CART total score tended to rise with the subject's year in college. An analysis of variance indicated that there was an overall effect of year in college on CART total scores, F(3,

Table 12.7
Mean short-form CART total scores as a function of year in college

	Mean CART score	SD	N
Freshman	41.3	11.9	216
Sophomore	46.7	14.1	114
Junior	46.2	12.2	29
Senior	48.5	15.1	13

Table 12.8
Approximate scores corresponding to selected percentile ranks in RT59 based on the total score of the short-form CART

Percentile Rank	Approximate Score in Sample
2	21
5	25
10	29
20	33
30	36
40	39
50	41
60	45
70	49
80	54
90	61
95	67
98	76

368) = 5.58, $p < 0.001$. Scheffé post hoc comparisons indicated that only the difference between the freshman and sophomore mean was statistically significant (however, there were few junior and senior subjects for this comparison). However, SAT total scores were higher for those longer at college, perhaps due to selection effects. Nevertheless, the correlation of 0.18 between year in college and CART total score was only reduced to 0.16 when SAT total score was controlled, and it was still significant at the 0.01 level.

Table 12.8 presents the approximate raw scores on the short-form CART that correspond to selected percentile ranks in this particular sample of subjects.

13 Associations among the Subtests: The Full-Form CART

In previous chapters we have described piecemeal some highly relevant correlations between individual subtests, and in chapter 12 we described the correlations among the short-form CART subtests in one of our studies. In this chapter we describe a study that includes all of the subtests of the CART. All twenty subtests (see table 12.1) and the four supplemental thinking dispositions scales were run on the same group of subjects.

The study, labeled RT60, involved two groups of subjects, one run in our laboratory at James Madison University (hereafter the Lab sample) and the other run using the Amazon Mechanical Turk (hereafter the Turk sample). The 350 subjects in the Lab sample (109 males and 241 females) were paid $60 for their participation over two sessions (separated by two days), and the 397 subjects in the Turk sample (231 males and 166 females) were paid $50 for their participation in a single session.

The Turk and Lab studies were both run online using Qualtrics, except that the Turk study was conducted unsupervised, whereas the Lab study was conducted supervised in a university laboratory setting. The two studies were otherwise run in the same manner, including administration instructions and order of tasks.

The subtests were administered in the following order: Probabilistic and Statistical Reasoning, Reflection versus Intuition, Probabilistic Numeracy, Belief Bias Syllogisms, Knowledge Calibration, Sensitivity to Expected Value, Temporal Discounting, Framing Part 1, Argument Evaluation Test Part 1 (Prior Opinions), Anchoring, Preference Anomalies Part 1, Risk Knowledge, Cognitive Ability measures (Analogies, Antonyms, Word Checklist), Scientific Reasoning, Disjunctive Reasoning, Ratio Bias, Antiscience Attitudes, Conspiracy Beliefs, Financial Literacy and Economic Knowledge, Framing Part 2, Argument Evaluation Test Part 2 (Evaluation), Preference Anomalies

Part 2, and Questionnaire. The Risk Knowledge subtest was at the end of the first block of subtests for the Lab group.

All of the subtests were administered as a single block in a fixed order, except for the Framing, Preference Anomalies, and Argument Evaluation subtests, which were each presented in two parts separated by several subtests. Five of the six Disjunctive Reasoning subtest items were dispersed within the Scientific Reasoning subtest so that each disjunctive reasoning item was presented separately. The sixth (the Married problem) appeared after the Ratio Bias subtest. The Actively Open-Minded Thinking scale, Deliberative Thinking scale, Future Orientation scale, Differentiation of Emotions scale, Dysfunctional Personal Beliefs subtest, and Superstitious Thinking subtest were all presented together, intermixed as part of the Questionnaire.

The mean time to finish the full form of the CART was more accurately estimated in the Lab sample. A majority of subjects finished in under 135 minutes and most finished under three hours. Turk times were harder to determine because, unlike in the Lab study, the Turk subjects were not directly monitored by an experimenter. Given that the Turk sample was not supervised in a lab setting, they were given the following additional instructions: "This survey typically requires a total of about three hours to complete. However, you will have up to six hours to complete the entire survey from the time that you accept the survey on the Mechanical Turk. If you take a break while working on the survey, it is essential that you return to complete it using the same computer and within the six-hour time limit." They were also told: "Please do not look up the answers to these questions while you are working on this survey. This is very important, because we need to know how people answer these questions without looking up the answers on the web, etc."

The Turk sample was collected over twenty-two days, and the Lab sample was collected over fifty-seven days. The study was run in early 2015.

Statistical Properties of the Full-Form CART

Table 13.1 presents the means and standard deviations of the raw scores on each of the subtests in addition to the mean and standard deviation of the four supplemental thinking disposition scales. Two subtests (Knowledge

The Full-Form CART

Calibration; Temporal Discounting) have multiple parts that are scored before the CART points are assigned.

Also presented in the table are the lowest and highest scores obtained by a subject on that subtest, as well as the skewness value for each of the raw scores. The right three columns of table 13.1 present the reliability of the raw scores on each of the subtests for the Turk sample, the Lab sample, and for the total sample. The reliabilities were generally consistent with what we have reported for each subtest in the previous chapters in studies prior to RT60. The reasons for the cases of low reliabilities (Anchoring subtest, Preference Anomalies subtest) have been discussed in previous chapters. Perhaps the most disappointing reliability was that for the Probabilistic Numeracy subtest (0.57)—disappointing because this test is so tightly associated with other CART components (see chapter 12 and later in this chapter) despite its modest reliability. Had we another chance to design this subtest we would have worked hard to increase its length and reliability.

Additionally, the last three columns indicate that for many of the subtests, the reliability in the Turk sample was higher than that in the Lab sample, and the total sample reliability was between the two. Finally, note that the reliabilities for the total sample in this study were consistently higher than the reliabilities for the subtests in our study of the short-form version of the CART, RT59, described in chapter 12. That study involved university-age subjects who were not paid for their participation. Both factors—the age of the subjects and the fact that they were not compensated—might have been responsible for the lower reliabilities in that study. This is because not only were the total sample reliabilities in RT60 higher than those in RT59, but even the Lab sample in RT60 tended to show higher reliabilities than were obtained in RT59, the lab study of the short form.

The raw scores on the subtests were translated into the CART points listed in table 4.2 for each subtest. The translation process has been discussed in the chapter that explains the subtest and in the appendix containing the subtest sample items. The mean total CART score in the total sample was 75.6 (SD = 23.2) with a skewness of 0.28. The total CART score had a Cronbach's alpha of 0.86, calculated by treating subtests as items with no differential weighting of CART points allocated. The lowest score achieved on the full form was 25 and the highest score achieved was 138 out of a possible 148 points.

Table 13.1

Descriptive statistics for the raw scores on the subtests and thinking dispositions scales in RT60

	Mean (SD)	Low–High score	Skewness	Reliability: Turk Sample	Reliability: Lab Sample	Reliability: Total Sample
CART Subtests: Raw Scores						
Probabilistic Reasoning subtest	10.84 (3.42)	2–18	−.07	.77	.70	.75
Scientific Reasoning subtest	10.20 (3.42)	2–20	.12	.70	.61	.67
Reflection versus Intuition subtest	3.97 (2.82)	0–11	.54	.78	.73	.77
Syllogistic Reasoning subtest	10.92 (2.77)	1–16	−.03	.69	.61	.65
Ratio Bias subtest	48.29 (12.4)	12–72	−.31	.91	.83	.88
Disjunctive Reasoning subtest	3.19 (1.97)	0–6	−.34	.81	.75	.78
Framing subtest	−11.65 (5.84)	−41–0	−.74	.66	.64	.66
Anchoring subtest	3.59 (1.79)	0–8	.34	.54	.27	.48
Preference Anomalies subtest	5.11 (1.66)	0–10	−.03	.18	.23	.16
Argument Evaluation subtest	.199 (.260)	−.604–.924	−.25	.50	.51	.51
Knowledge Calibration, Part 1: Item Method	11.70 (10.99)	−26.9–49.4	.09	.53	.54	.55
Knowledge Calibration, Part 2: Hits, Interval Method	6.02 (3.05)	0–15	.47	.75	.71	.75
Temporal Discounting: Ascending $100	16.41 (3.58)	0–20	−1.32	.90	.88	.89
Temporal Discounting: Descending $2000 raw score	14.32 (4.33)	0–18	−1.24	.93	.93	.93

Table 13.1 (continued)

	Mean (SD)	Low–High score	Skewness	Reliability: Turk Sample	Reliability: Lab Sample	Reliability: Total Sample
Temporal Discounting: 26 Items	80.85 (30.02)	26–156	.25	.97	.95	.97
Probabilistic Numeracy subtest	5.48 (1.75)	0–9	–.17	.59	.52	.57
Financial Literacy and Economic Knowledge subtest	15.56 (4.41)	5–28	.27	.75	.53	.72
Sensitivity to Expected Value subtest	14.97 (3.38)	5–20	–.42	.82	.70	.79
Risk Knowledge subtest	8.59 (2.58)	1–14	–.37	.64	.60	.64
Superstitious Thinking subtest raw score	28.00 (10.75)	12–69	.50	.90	.86	.89
Antiscience Attitudes subtest	35.18 (10.23)	13–66	–.06	.91	.84	.90
Conspiracy Beliefs subtest	62.40 (22.23)	24–134	.37	.95	.92	.94
Dysfunctional Personal Beliefs subtest	29.63 (7.66)	10–53	.13	.84	.70	.80
Thinking Disposition Scales: Raw Scores						
Actively Open-Minded Thinking scale	129.03 (15.74)	89–169	.21	.86	.82	.85
Deliberative Thinking scale	66.75 (14.84)	19–96	–.32	.96	.90	.95
Future Orientation scale	56.40 (9.24)	19–84	–.14	.88	.78	.84
Differentiation of Emotions scale	60.28 (11.22)	29–84	.01	.90	.82	.88

Table 13.2
Differences between the Turk sample and Lab sample in RT60 in age and cognitive ability

	Turk Sample (n = 397) M (SD)		Lab Sample (n = 350) M (SD)		t(745)	Cohen's d
Age	32.4	(9.3)	20.1	(1.8)	24.40***	1.79
SAT total[a]	1231	(155)	1122	(135)	8.64***	0.77
Analogies	11.0	(3.8)	9.6	(3.3)	5.50***	0.40
Antonyms	14.6	(5.1)	10.6	(3.6)	12.23***	0.90
Word Checklist	.580	(.213)	.409	(.169)	12.06***	0.88
Cognitive Ability Composite 3	0.95	(2.69)	−1.08	(1.97)	11.60***	0.85
Cognitive Ability Composite 4[a]	1.62	(3.33)	−1.32	(2.44)	11.80***	1.05

***$p < .001$
[a] df for SAT and Cognitive Ability Composite4 = 536

In the remaining analyses, we will be dealing with the CART scores rather than the raw scores. First, though, the characteristics of the two subsamples of RT60 are presented in more detail in table 13.2. There were large demographic differences between the Turk and Lab samples, and one has already been noted above. The Turk sample was 58.2 percent male and 41.8 percent female, whereas the Lab sample was 31.1 percent male and 68.9 percent female. We will report sex differences in performance later in the chapter. Also, as noted in table 13.2, the Turk sample was over a decade older than the Lab sample and was considerably more variable in age. Whereas the Turk sample varied in age from 19 to 68, the Lab sample varied in age from 18 to only 32. Whereas 52.9 percent of the Turk sample was 30 or older, less than 1 percent of the Lab sample was 30 or older.

We collected four indicators of cognitive ability in this study. Self-reports of SAT scores were provided by 204 subjects in the Turk sample and 334 subjects in the Lab sample. Table 13.2 indicates that the Turk sample had significantly higher SAT total scores, and the effect size was a substantial 0.77. Note, however, that only half of the Turk sample provided SAT scores.

That the Turk sample was indeed higher in intelligence was confirmed by analyses of three measures of cognitive ability that all 747 subjects completed as part of the study: a 19-item analogy task, a 30-item antonym task, and a 60-item vocabulary checklist task. All three load heavily on verbal cognitive ability. The analogy task examined the ability to understand the underlying conceptual relationship between a pair of related words and to identify another pair of words that best reflected a parallel relationship. The 19 items on the analogy task were previously used items from SAT Examinations prior to March, 2002. The antonym task examined the ability to select a word or short phrase with the opposite meaning to a target word. The 30 items in the antonym task were previously used items from Graduate Record Examinations administered prior to 1995. The reliability (Cronbach's alpha) of the analogy task was 0.72 and the reliability (Cronbach's alpha) of the antonym task was 0.78.

The vocabulary checklist measure used the checklist-with-foils format, which has been shown to be a reliable and valid way of assessing individual differences in verbal cognitive ability (Anderson & Freebody, 1983; Baddeley, Emslie, & Nimmo-Smith, 1993; Scott, de Wit, & Deary, 2006; Stanovich, West, & Harrison, 1995). The stimuli for the task were 40 words (e.g., "absolution," "irksome," "purview") and 20 pronounceable nonwords (e.g., "disler," "potomite," "seblement") taken largely from the stimulus list of Zimmerman, Broder, Shaughnessy, and Underwood (1977). The words and nonwords were intermixed via alphabetization. The subjects were told that some of the letter strings were actual words and that others were not words and that their task was to read through the list of items and to put a check mark next to those that they knew were words. Scoring on the task was determined by taking the proportion of the target words that were checked and subtracting the proportion of nonword foils checked. The split-half reliability (odd/even, Spearman-Brown corrected) of the word checklist measure was 0.86.

As indicated in table 13.2 the Turk sample significantly outperformed the Lab sample on each of these three verbal cognitive ability indicators. Listed next in the table is a composite cognitive ability indicator (Cognitive Ability Composite3) composed of the sum of the z-scores on each of the three tasks (analogies, antonyms, word checklist). The Turk group outperformed the Lab sample on this measure and the effect size of the difference on this composite was 0.85. Another composite measure (Cognitive Ability Composite4) was calculated for the 538 subjects who provided a self-reported SAT score. It was composed of the sum of the z-scores on each of the three cognitive ability tasks administered in the study (analogies, antonyms, word checklist) plus the z-score of the self-reported SAT total score. The Turk sample displayed higher scores on this measure as well, with an effect size of 1.05.

The superior performance of the Turk sample extended to the CART itself, as is evident in table 13.3. At the top of the table it can be observed that the Turk sample achieved a significantly higher score on the CART than the Lab sample. The effect size of the difference was 0.59. The remainder of the table presents comparisons of the Turk sample and Lab sample performance across the other subtests of the CART, with the CART score (rather than the raw score) on each of the subtests providing the metric. The Turk sample performed significantly better on 16 of the 20 subtests. The Lab sample significantly outperformed the Turk sample on only one subtest: the Temporal Discounting subtest. On three of the subtests the two groups did not differ: the Ratio Bias subtest, the Disjunctive Reasoning subtest, and the Sensitivity to Expected Value subtest. At the bottom of table 13.3, the performance of the two groups on the four thinking dispositions is compared. The Turk sample scored significantly higher on three of the four thinking dispositions, the exception being the Future Orientation subscale.

Table 13.4 presents the correlations between three measures of cognitive ability and the CART Total score and subtest scores. The left column presents the correlations with Cognitive Ability Composite3 for the full sample of 747 subjects. The middle column presents the correlation with the SAT total score that was self-reported by 538 subjects in the full sample. The far right column presents the correlations with Cognitive Ability Composite4—the composite that included the SAT total score along with the Analogies, Antonyms, and Word Checklist tasks (Composite4 was also calculated for only 538 subjects) in the full sample.

Cognitive Ability Composite3 displayed a moderately high correlation with the CART total score and moderate correlations with most of the CART subtests except for the Preference Anomalies subtest, the Temporal Discounting subtest, the Sensitivity to Expected Value subtest, and the Dysfunctional Personal Beliefs subtest. In general, Cognitive Ability Composite3 displayed higher correlations than the SAT total score. Cognitive Ability Composite4 displayed correlations very similar to those of Cognitive Ability Composite3, although occasionally a bit higher.

Table 13.3
CART score differences between the Turk sample and Lab sample in RT60

	Turk Sample (n = 397)		Lab Sample (n = 350)		t(745)	Cohen's d
	M	(SD)	M	(SD)		
Full-Form CART total score	81.8	(24.7)	68.6	(19.1)	8.07***	.59
CART Subtest Points						
Probabilistic Reasoning	11.2	(3.6)	10.2	(3.1)	4.06***	.30
Scientific Reasoning	10.6	(3.6)	9.7	(3.1)	3.49***	.26
Reflection versus Intuition	4.67	(2.85)	3.15	(2.49)	7.70***	.56
Syllogistic Reasoning	3.39	(2.55)	2.73	(2.43)	3.59***	.26
Ratio Bias	2.28	(1.90)	2.22	(1.64)	0.52	.04
Disjunctive Reasoning	2.31	(1.82)	2.37	(1.63)	−0.47	.03
Framing	3.47	(2.02)	2.72	(1.88)	5.27***	.38
Anchoring	2.12	(.92)	1.62	(.91)	7.51***	.55
Preference Anomalies	1.62	(1.01)	1.23	(.93)	5.43***	.40
Argument Evaluation Test	2.44	(1.80)	2.10	(1.71)	2.70**	.20
Knowledge Calibration	2.52	(1.75)	1.68	(1.46)	7.04***	.52
Temporal Discounting	2.89	(2.47)	3.73	(2.15)	−4.92***	.36
Probabilistic Numeracy	5.81	(1.79)	5.11	(1.63)	5.55***	.41
Financial Literacy	5.48	(5.20)	3.49	(2.06)	10.97***	.80
Sensitivity to Expected Value	2.73	(1.66)	2.81	(1.56)	−0.66	.05
Risk Knowledge	2.24	(.91)	1.73	(.96)	7.40***	.54
Superstitious Thinking	3.72	(1.59)	2.59	(1.65)	9.53***	.70
Antiscience Attitudes	3.75	(1.76)	2.64	(1.82)	8.43***	.62
Conspiracy Beliefs	5.31	(3.65)	4.21	(3.15)	4.40***	.32
Dysfunctional Personal Beliefs	3.22	(1.76)	2.57	(1.61)	5.22***	.38
Thinking Disposition Scales: Raw Scores						
Actively Open-Minded Thinking	132.4	(16.3)	125.2	(14.1)	6.36***	.47
Deliberative Thinking	68.2	(17.2)	65.1	(11.4)	2.82**	.21
Future Orientation	56.6	(10.1)	56.2	(8.1)	0.50	.04
Differentiation of Emotions	63.8	(11.4)	56.2	(9.6)	9.81***	.72

Two-tailed p values: *p < .05; **p < .01; ***p < .001

Table 13.4
Correlations between cognitive ability and performance on the total full-form CART score and subtest CART scores in RT60

	Cognitive Ability Composite3	SAT Total	Cognitive Ability Composite4
CART total score	.687	.469	.705
Probabilistic Reasoning subtest	.505	.375	.520
Scientific Reasoning subtest	.554	.375	.571
Reflection versus Intuition subtest	.538	.485	.591
Syllogistic Reasoning subtest	.499	.338	.505
Ratio Bias subtest	.248	.135	.234
Disjunctive Reasoning subtest	.280	.256	.282
Framing subtest	.275	.202	.312
Anchoring subtest	.314	.221	.328
Preference Anomalies subtest	.126	.184	.156
Argument Evaluation subtest	.375	.224	.380
Knowledge Calibration subtest	.378	.199	.375
Temporal Discounting subtest	.061	.034	.027
Probabilistic Numeracy subtest	.473	.390	.491
Financial Literacy subtest	.625	.388	.646
Sensitivity to Expected Value subtest	.209	.113	.196
Risk Knowledge subtest	.440	.251	.424
Superstitious Thinking subtest	.472	.187	.438
Antiscience Attitudes subtest	.436	.352	.457
Conspiracy Beliefs subtest	.343	.166	.335
Dysfunctional Personal Beliefs subtest	.082	.032	.083

For Cognitive Ability Composite3 ($n = 747$)
Correlations > .075 significant at the .05 level, two tailed
Correlations > .126 significant at the .001 level, two tailed
For Cognitive Ability Composite4 and SAT ($n = 538$)
Correlations > .086 significant at the .05 level, two-tailed
Correlations > .141 significant at the .001 level, two-tailed

Table 13.5 presents the correlations between the four thinking dispositions scales and the CART Total score and subtest scores. The Actively Open-Minded Thinking scale (AOT) was by far the strongest correlate of CART performance.[1] The Deliberative Thinking scale was substantially less predictive of CART performance, but was the next highest correlate. The 0.625 correlation between the Actively Open-Minded Thinking scale and CART performance is particularly notable, given that only four of the twenty CART subtests are self-report measures like the AOT scale (the Superstitious Thinking subtest, the Antiscience Attitudes subtest, the Conspiracy Beliefs subtest, and the Dysfunctional Personal Beliefs subtest). Of the sixteen performance subtests, five had notably high correlations with the AOT: the Probabilistic and Statistical Reasoning subtest, the Scientific Reasoning subtest, the Reflection versus Intuition subtest, the Syllogistic Reasoning subtest, and the Financial Literacy and Economic Thinking subtest. A test for difference between dependent correlations (Cohen & Cohen, 1983) indicated that the correlation between the AOT score and the CART Total (0.625) was significantly higher than the 0.517 correlation between the AOT score and Cognitive Ability Composite3, $t(744) = 4.77$, $p < 0.001$).

Table 13.6 presents a series of multiple regression analyses that examine whether cognitive ability and the score on the Actively Open-Minded Thinking scale are independent predictors of CART performance. For example, the first row of this table indicates that, for the entire sample, Cognitive Ability Composite3 and the Actively Open-Minded Thinking scale score predicted 57.0 percent of the variance in the CART total score. Both Cognitive Ability Composite3 and the AOT score were independent predictors of the CART Total score. Both have substantial beta weights, although cognitive ability is the stronger predictor. The next multiple regression in the table adds the SAT total score as a third predictor in a regression equation involving the 538 subjects who self-reported an SAT score. Although SAT is a significant independent predictor in this analysis, both Cognitive Ability Composite3 and the AOT score were more potent predictors of the CART Total score than the SAT. The next two lines of this table indicate that when the same two regression analyses are run on the Turk sample only, the results were substantially the same. The next two rows of table 13.6 indicate that the same is true when these analyses were run on the lab sample only. There, in the analysis involving three predictors, SAT total score was a

Table 13.5
Correlations between the four Thinking Disposition scales and performance on the full-form total CART score and subtest scores in RT60

	Actively Open-Minded Scale	Deliberative Thinking Scale	Future Orientation Scale	Differentiation of Emotions Scale
CART total score	.625	.344	.298	.263
Probabilistic Reasoning subtest	.434	.237	.177	.110
Scientific Reasoning subtest	.502	.268	.239	.148
Reflection versus Intuition subtest	.394	.246	.150	.112
Syllogistic Reasoning subtest	.392	.245	.192	.072
Ratio Bias subtest	.250	.162	.206	.104
Disjunctive Reasoning subtest	.242	.128	.130	.023
Framing subtest	.200	.080	.111	.056
Anchoring subtest	.249	.162	.163	.208
Preference Anomalies subtest	.207	.071	.086	.055
Argument Evaluation subtest	.339	.203	.159	.183
Knowledge Calibration subtest	.286	.062	.120	.098
Temporal Discounting subtest	.077	.088	.174	-.093
Probabilistic Numeracy subtest	.350	.243	.212	.076
Financial Literacy subtest	.474	.253	.248	.270

Table 13.5 (continued)

	Actively Open-Minded Scale	Deliberative Thinking Scale	Future Orientation Scale	Differentiation of Emotions Scale
Sensitivity to Expected Value	.135	.098	.016	.004
Risk Knowledge subtest	.339	.177	.119	.167
Superstitious Thinking subtest	.549	.156	.212	.364
Antiscience Attitudes subtest	.612	.326	.240	.269
Conspiracy Beliefs subtest	.412	.152	.188	.245
Dysfunctional Personal Beliefs subtest	.162	.323	.009	.503

(N = 747)
Correlations > .075 significant at the .05 level, two-tailed
Correlations > .098 significant at the .01 level, two-tailed
Correlations > .126 significant at the .001 level, two-tailed

somewhat more potent predictor than it was in the previous analyses with the Total and Turk samples.

In chapter 12, in our discussion of the data from our study of the short-form CART (RT59), we presented data showing that the AOT was a particularly strong predictor of performance on the four subtests that assessed the ability to avoid contaminated mindware. The next regression in table 13.6 indicates that this was also true in the present study of the full form (RT60). In this regression, the AOT was a stronger predictor of performance on these four subtests (Superstitious Thinking, Antiscience Attitudes, Conspiracy Beliefs, and Dysfunctional Personal Beliefs) than was the Cognitive Ability Composite3 score (beta weight of 0.505 versus 0.218).

Because the AOT was such a strong predictor of the contaminated mindware section of the CART, we ran several analyses to examine whether it was just that part of the CART that was accounting for the predictive power of the AOT overall. In general, this was not the case. The AOT remained a significant unique predictor (albeit not as strong a predictor as cognitive ability) even on the parts of the test that did not contain self-report

Table 13.6
Individual difference predictors of performance in RT60

	Multiple R-Squared	Standardized Beta for Cognitive Ability3	Standardized Beta for AOT Score	Standardized Beta for SAT Total
CART total Score Total sample (N = 747)	.570	.496***	.367***	
CART total score Total sample (N = 538)	.582	.418***	.346***	.172***
CART total score Turk sample (n = 397)	.586	.507***	.372***	
CART total score Turk sample (n = 204)	.608	.486***	.355***	.112*
CART total score Lab sample (n = 350)	.438	.402***	.379***	
CART total score Lab sample (n = 334)	.478	.286***	.347***	.256***
Four Contaminated Mindware subtests (N = 747)	.418	.218***	.505***	
16 remaining subtests (N = 747)	.506	.526***	.280***	
Probabilistic Reasoning and Scientific Reasoning subtests (N = 747)	.424	.443***	.300***	

Table 13.6 (continued)

	Multiple R-Squared	Standardized Beta for Cognitive Ability3	Standardized Beta for AOT Score	Standardized Beta for SAT Total
10 CART Subtests Saturated with Processing Requirements (N = 747)	.395	.460***	.251***	
4 CART Subtests Saturated with Knowledge Requirements (N = 747)	.449	.545***	.200***	

*Beta weights significant at the .05 level
**Beta weights significant at the .01 level
***Beta weights significant at the .001 level

measures of contaminated mindware. For example, the next regression analysis in the table examined as a criterion variable the total score on the other sixteen subtests—that is, the sixteen subtests that did not involve self-report measures of contaminated mindware. It is apparent that the AOT remained a significant unique predictor (after cognitive ability had been controlled), although its beta weight (0.280) was, as expected, somewhat lower than it was in the total score for all twenty subtests. In short, the AOT is a significant unique predictor of CART subtests that do not involve the avoidance of contaminated mindware.

The remaining regressions in table 13.6 simply demonstrate that the AOT was a significant predictor of virtually all the components of the CART (i.e., the association between AOT and total test performance was not uniquely through the contaminated mindware subtests). For example, the next regression analysis used as a criterion variable the combined scores on two of the most important subtests of the CART: the Probabilistic and Statistical Reasoning subtest and the Scientific Reasoning subtest (this composite is discussed in more detail below). Here, the AOT achieved a significant beta weight of 0.300. The next regression analyses combined the scores on the ten subtests that were heavily saturated with processing requirements (i.e., the ten subtests that uniquely reside in the first column of table 12.1). The AOT again achieved a significant beta weight of 0.251 in this analysis. Finally, we constructed a composite variable consisting of the four subtests that were heavily saturated with knowledge requirements (the Probabilistic Numeracy subtest, Financial Literacy subtest, Sensitivity to Expected Value subtest, and Risk Knowledge subtest). The final regression analysis indicates that cognitive ability was an especially potent predictor of this subset of tasks, but that nonetheless the AOT was still a significant predictor, with a beta weight of 0.200.

Another concern regarding the AOT as a predictor is that because it is a self-report measure, it might be carrying variance attributable to impression management or social desirability. We included a ten-item Impression Management scale in RT60 (Cronbach's alpha = 0.80) adapted from the work of Paulhus (1991). The top of table 13.7 shows the correlations of this measure with the four thinking dispositions. Two of the four (Deliberation scale and the Differentiation of Emotion scale) carry significant impression management variance, but the AOT is not one of those two. These results mirror those reported in table 11.1.

Table 13.7

Correlations between the CART thinking dispositions and CART subtest scores and the ten-item Impression Management scale in RT60 ($N = 747$)

	Impression Management
Thinking Dispositions	
Actively Open-Minded Thinking scale	.05
Deliberative Thinking scale	.15***
Future Orientation scale	.04
Differentiation of Emotions scale	.35***
CART Subtests	
Probabilistic Reasoning subtest	−.08*
Scientific Reasoning subtest	−.02
Reflection versus Intuition subtest	.03
Syllogistic Reasoning subtest	−.03
Ratio Bias subtest	−.04
Disjunctive Reasoning subtest	−.04
Avoidance of Framing subtest	.09*

Table 13.7 (continued)

	Impression Management
Avoidance of Anchoring subtest	.04
Avoidance of Preference Anomalies subtest	−.04
Argument Evaluation subtest	.02
Knowledge Calibration subtest	.09*
Temporal Discounting subtest	−.08*
Probabilistic Numeracy subtest	.01
Financial Literacy subtest	.05
Sensitivity to Expected Value	−.08*
Risk Knowledge subtest	.02
Superstitious Thinking subtest	.15***
Antiscience Attitudes subtest	.03
Conspiracy Beliefs subtest	.11**
Dysfunctional Personal Beliefs subtest	.25***
CART Composite Scores	
Full-Form CART	.04
Short-Form CART	.04
Probabilistic Reasoning and Scientific Reasoning subtests	−.06
4 Contaminated Mindware subtests	.18***
4 subtests saturated with knowledge	.01
10 subtests saturated with processing requirements	.00

*correlations significant at the .05 level two tailed
**correlations significant at the .01 level two tailed
***correlations significant at the .001 level two tailed

Farther down, the table displays the correlations between Impression Management scores and the scores on the twenty CART subtests. For the most part, these correlations are close to zero, and in the few cases where they were statistically significant due to our substantial sample size they were less than 0.10 in absolute magnitude. The exception concerns the four subtests that measure the avoidance of contaminated mindware. Three of the four had significant correlations with the Impression Management scale, and all three were greater than 0.10 in absolute magnitude. So while, in theory, these three subtests of the CART might be correlating with another indicator because of impression management factors implicated in their self-report format, this does not seem to be the case with the AOT itself, because the latter itself does not correlate with Impression Management scores (a finding that was replicated in the studies reported in chapter 11).

The bottom of table 13.7 gives the correlations between Impression Management scores and the composite scores on the CART. Neither the full-form CART nor the short-form CART as a whole carries impression management variance. However, a composite score made up of the four contaminated mindware subtests does display a significant 0.18 correlation with Impression Management scores.

Table 13.8 displays the intercorrelations of all of the subtest scores on the CART. Line 22 displays the correlation of the full-form total CART score with each of the components. Line 23 displays the subtest correlation with the total score minus that subtest. As in RT59 (see chapter 12), the Dysfunctional Personal Beliefs subtest largely failed to correlate with any of the other subtests. However, 162 of the remaining 171 correlations were statistically significant, with the median correlation among the remaining nineteen subtests being 0.25. As mentioned previously, these intercorrelations resulted in a Cronbach's alpha of 0.86 for the total score when all twenty subtests (unweighted) were included. Eliminating the Dysfunctional Personal Beliefs subtest raises Cronbach's alpha to 0.87.

A perusal of table 13.8 reveals that three subtests had particularly strong intercorrelations: the Probabilistic and Statistical Reasoning subtest, the Scientific Reasoning subtest, and the Reflection versus Intuition subtest. The part–whole correlations in the last line of the table confirm that these three subtests were particularly predictive of total test score, followed by the Financial Literacy and Economic Knowledge subtest, the Probabilistic Numeracy subtest, and the Belief Bias in Syllogistic Reasoning subtest.

An exploratory principal components analysis of the twenty CART subtests revealed that a four-component solution provided the most coherent classification of the tasks. The first principal component explained 30.7 percent of the variance and the first four components explained 48.6 percent of the variance. Table 13.9 displays the loadings on these four components after varimax rotation.

Twelve subtests loaded on component 1, ten of them exclusively. Three of the four measures of the avoidance of contaminated mindware loaded on component 2. The third component was largely defined by singleton variables: the Temporal Discounting subtest and the Conspiracy Beliefs subtest. The fourth component was largely defined by the Dysfunctional Personal Beliefs subtest and the Ratio Bias subtest, as well as the Preference Anomalies subtest.

Sex Differences in CART Performance

In chapter 12, we described how in our study of the short-form test, the male subjects had higher total CART scores than the female subjects and significantly outperformed the female subjects on several specific subtests. Therefore, in our RT60 study of the full-form CART, we ran several analyses looking for sex differences. However, it is necessary to look at sex differences separately in the Turk sample and in the Lab sample of this study. The reason is that the Turk sample performed better than the Lab sample on the total test and on several specific subtests. However, the two samples have a gender imbalance going in opposite directions. As reported above, the Turk sample was 58 percent male, whereas the Lab sample was only 31 percent male. This imbalance, coupled with the overall performance differences between the Turk sample and the Lab sample, creates exactly the situation in which one might see versions of Simpson's paradox (Kievit, Frankenhuis, Waldorp, & Borsboom, 2013)—situations where what one might conclude from an aggregate analysis would not hold in either of the disaggregated analyses. Thus, table 13.10 displays the sex differences in performance separately for both samples.

The first four comparisons at the top of the table examine sex differences in cognitive ability across the two samples. In the Lab sample, the males had significantly higher scores on Cognitive Ability Composite3 and higher SAT total scores than the females. The effect size was moderate in both cases. In the Turk sample, although the males outscored the females on Cognitive Ability Composite3 and the SAT, neither of the differences was statistically significant and the effect size was small. Thus, when interpreting the rest of the results in the table, we should keep in mind the fact that in the Lab sample in particular, the males were higher in cognitive ability.

Continuing down the table, it can be seen that the total score on the entire CART full form was higher for males than for females in both samples and the mean difference corresponded to a moderate effect size of 0.52 and 0.65, respectively. These were somewhat lower effect size sizes than were displayed in the study of the short form described in chapter 12 (0.82). Moving down the table, we see displayed the sex differences for each of the

Table 13.8
Correlations among the subtest CART scores in RT60 (N = 747)

	1	2	3	4	5	6	7	8	9	10	11	12	13	14	15	16	17	18	19	20
1. Probabilistic Reasoning																				
2. Scientific Reasoning	.56																			
3. Reflection versus Intuition	.59	.58																		
4. Syllogistic Reasoning	.51	.55	.55																	
5. Ratio Bias	.34	.31	.26	.25																
6. Disjunctive Reasoning	.38	.40	.42	.36	.24															
7. Framing	.26	.35	.33	.26	.12	.22														
8. Anchoring	.26	.21	.27	.23	.09	.07	.10													
9. Preference Anomalies	.22	.22	.20	.16	.12	.11	.15	.08												
10. Argument Evaluation Subtest	.30	.39	.33	.32	.14	.19	.20	.16	.08											
11. Knowledge Calibration	.31	.31	.40	.30	.16	.18	.19	.22	.18	.23										
12. Temporal Discounting	.19	.13	.14	.18	.19	.15	.13	.01	.06	.06	.10									

Table 13.8 (continued)

	1	2	3	4	5	6	7	8	9	10	11	12	13	14	15	16	17	18	19	20
13. Probabilistic Numeracy	.58	.48	.59	.44	.28	.36	.23	.24	.17	.26	.27	.15								
14. Financial Literacy	.51	.53	.54	.45	.26	.26	.27	.30	.20	.37	.34	.09	.47							
15. Expected Value	.28	.30	.31	.30	.11	.25	.11	.11	.05	.15	.07	.11	.31	.25						
16. Risk Knowledge	.30	.37	.28	.26	.14	.16	.21	.21	.13	.19	.21	.06	.26	.39	.11					
17. Superstitious Thinking	.40	.40	.38	.31	.22	.18	.20	.25	.18	.27	.30	.04	.33	.42	.19	.28				
18. Antiscience Attitudes	.45	.44	.41	.33	.22	.23	.23	.26	.20	.30	.25	.07	.37	.43	.18	.32	.50			
19. Conspiracy Beliefs	.35	.30	.25	.24	.18	.12	.15	.13	.18	.22	.23	.10	.20	.28	.00	.23	.49	.41		
20. Dysfunctional Personal Beliefs	.08	.07	.11	.02	.05	.03	.02	.12	.08	.09	.05	−.11	.06	.15	.07	.04	.15	.10	.06	
Total CART score	.78	.78	.77	.68	.44	.50	.44	.36	.32	.49	.48	.28	.67	.72	.39	.45	.61	.63	.52	.18
Total CART score minus subtest	.71	.70	.71	.62	.37	.44	.37	.32	.28	.43	.43	.19	.62	.65	.33	.42	.56	.57	.40	.11

Note: All correlations larger than .075 in absolute value are significant at the .05 level.

Table 13.9

Loadings of the full-form CART subtests in RT60 on the first four principal components after varimax rotation

	Component 1	Component 2	Component 3	Component 4
Probabilistic Reasoning subtest	.31	—	—	—
Scientific Reasoning subtest	.31	—	—	—
Reflection Versus Intuition subtest	.32	—	—	—
Syllogistic Reasoning subtest	.28	—	—	—
Ratio Bias subtest	—	—	.23	.43
Disjunctive Reasoning subtest	.20	—	—	—
Avoidance of Framing subtest	—	—	—	—
Avoidance of Anchoring subtest	—	.23	—	—
Avoidance of Preference Anomalies subtest	—	—	.22	.35
Argument Evaluation subtest	.20	—	—	—
Knowledge Calibration subtest	.20	—	—	—
Temporal Discounting subtest	—	—	.46	—
Probabilistic Numeracy subtest	.28	—	—	—
Financial Literacy subtest	.29	—	—	—
Sensitivity to Expected Value subtest	—	—	—	—
Risk Knowledge subtest	.20	—	—	—
Superstitious Thinking subtest	.25	.34	—	—
Antiscience Attitudes subtest	.26	.25	—	—
Conspiracy Beliefs subtest	—	.37	.40	—
Dysfunctional Personal Beliefs subtest	—	—	—	.53

Note: Loadings less than .20 and negative loadings have been eliminated.

twenty subtests within each of the two samples. In thirty-eight of the forty comparisons the males outperformed the females, although this difference was not always statistically significant. There was one statistically significant comparison where females outperformed males: the Temporal Discounting subtest for the Lab sample (convergent with Dittrich & Leipold, 2014; Silverman, 2003a, 2003b). The differences favoring males were particularly sizable for certain subtests: the Probabilistic and Statistical Reasoning subtest, the Reflection versus Intuition subtest, the Practical Numeracy subtest, and the Financial Literacy and Economic Knowledge subtest. The bottom of the table shows the sex differences on the four thinking dispositions for each of the two samples. On two of the four thinking dispositions scales—the Actively Open-Minded Thinking scale and the Deliberative Thinking scale—males tended to outperform females.

Although we found a statistically significant difference between males and females in the full-form CART total score that was of moderate effect size, recall that at the top of this table it was demonstrated that in both the Turk and the Lab sample males were of higher cognitive ability (particularly in the Lab sample). We conducted several analyses to see how much the sex differences in CART performance were attenuated by cognitive ability and other factors. In the Turk sample, the correlation between sex (male = 1; female=2) and the CART Total score was –0.25 and was not attenuated (partial r = –0.30) when the scores on the Cognitive Ability Composite3 were controlled. In the Turk sample there was a significant 0.21 correlation between total CART score and years of education completed, although little correlation between sex and years of education completed (0.05). The correlation between sex and the CART Total score in this sample remained unattenuated (partial r = –0.31) when the scores on the Cognitive Ability Composite3 and years of education were both controlled.

In the Lab sample, the correlation between sex and the full-form CART total score was –0.29 and decreased to –0.25 when the scores on Cognitive Ability Composite3 were controlled, but it was still highly significant. Also in the Lab sample, there was a small but significant 0.13 correlation between total CART score and year in university, although little correlation between sex and year in university (0.05). The correlation between sex and the CART Total score in this sample remained largely unattenuated (partial r = –0.25) when the scores on the Cognitive Ability Composite3 and year in university were both controlled.

Table 13.10

Differences between males and females in the full-form version of the CART (RT60)

	Males		Females			
	M	(SD)	M	(SD)	t value	Cohen's d
Cognitive Ability Composite3—Turk	1.07	(2.61)	0.78	(2.80)	1.07	.11
Cognitive Ability Composite 3—Lab	−0.61	(2.06)	−1.29	(1.90)	3.01**	.35
SAT Total—Turk	1240	(150)	1218	(163)	1.02	.15
SAT Total—Lab	1160	(143)	1105	(127)	3.47***	.41
Full-form CART—Turk	87.1	(23.4)	74.5	(24.7)	5.16***	.52
Full-form CART—Lab	76.9	(22.3)	64.9	(16.1)	5.65***	.65
Probabilistic Reasoning—Turk	12.0	(3.5)	10.2	(3.5)	5.18***	.53
Probabilistic Reasoning—Lab	11.5	(3.3)	9.6	(2.8)	5.53***	.64
Scientific Reasoning—Turk	11.1	(3.4)	9.9	(3.8)	3.16**	.32
Scientific Reasoning—Lab	10.6	(3.4)	9.4	(2.9)	3.32**	.38
Reflection versus Intuition—Turk	5.2	(2.8)	4.0	(2.8)	4.13***	.42
Reflection versus Intuition—Lab	4.1	(3.0)	2.7	(2.1)	5.20***	.60
Syllogistic Reasoning—Turk	3.5	(2.5)	3.2	(2.8)	1.12	.11
Syllogistic Reasoning—Lab	3.3	(2.6)	2.5	(2.3)	3.01**	.35
Ratio Bias—Turk	2.6	(1.9)	1.9	(1.9)	3.38***	.34
Ratio Bias—Lab	2.7	(1.7)	2.0	(1.6)	3.75***	.43
Disjunctive Reasoning—Turk	2.4	(1.9)	2.2	(1.7)	1.47	.15
Disjunctive Reasoning—Lab	2.7	(1.6)	2.2	(1.6)	2.36*	.27
Framing—Turk	3.6	(2.0)	3.3	(2.0)	1.65	.17
Framing—Lab	2.8	(1.9)	2.7	(1.9)	0.36	.04
Anchoring—Turk	2.3	(0.9)	1.9	(0.9)	3.38***	.34
Anchoring—Lab	1.6	(0.9)	1.6	(0.9)	0.09	.01

Table 13.10 (continued)

	Males		Females			
	M	(SD)	M	(SD)	t value	Cohen's d
Preference Anomalies—Turk	1.8	(1.0)	1.4	(0.9)	3.45***	.35
Preference Anomalies—Lab	1.5	(1.0)	1.1	(0.9)	3.37***	.39
Argument Evaluation—Turk	2.6	(1.8)	2.3	(1.8)	1.61	.16
Argument Evaluation—Lab	2.4	(1.8)	2.0	(1.7)	2.24*	.26
Knowledge Calibration—Turk	2.6	(1.8)	2.4	(1.8)	0.74	.08
Knowledge Calibration—Lab	1.8	(1.5)	1.6	(1.4)	1.42	.16
Temporal Discounting—Turk	3.0	(2.5)	2.7	(2.3)	1.43	.15
Temporal Discounting—Lab	3.3	(2.2)	3.9	(2.1)	-2.23*	-.26
Probabilistic Numeracy—Turk	6.2	(1.7)	5.3	(1.8)	4.94***	.50
Probabilistic Numeracy—Lab	5.8	(1.7)	4.8	(1.5)	5.76***	.66
Financial Literacy—Turk	5.9	(2.8)	4.9	(2.7)	3.56***	.36
Financial Literacy—Lab	4.5	(2.4)	3.0	(1.7)	6.43***	.74
Expected Value—Turk	3.0	(1.7)	2.4	(1.6)	3.86***	.39
Expected Value—Lab	3.3	(1.5)	2.6	(1.5)	4.25***	.49
Risk Knowledge—Turk	2.2	(0.9)	2.2	(0.9)	0.03	.00
Risk Knowledge—Lab	1.9	(1.0)	1.7	(0.9)	1.85	.21
Superstitious Thinking—Turk	4.0	(1.4)	3.4	(1.7)	3.84***	.39
Superstitious Thinking—Lab	2.9	(1.7)	2.4	(1.6)	2.41*	.28
Antiscience Attitudes—Turk	4.1	(1.6)	3.3	(1.9)	4.35***	.44

Table 13.10 (continued)

	Males		Females			
	M	(SD)	M	(SD)	t value	Cohen's d
Antiscience Attitudes—Lab	3.1	(1.8)	2.4	(1.8)	3.30**	.38
Conspiracy Beliefs—Turk	5.6	(3.6)	4.8	(3.7)	2.16*	.22
Conspiracy Beliefs—Lab	4.0	(3.3)	4.3	(3.1)	-0.72	-.08
Dysfunctional Beliefs—Turk	3.5	(1.6)	2.8	(1.9)	4.13***	.42
Dysfunctional Beliefs—Lab	2.9	(1.6)	2.4	(1.6)	2.89**	.33
Actively Open-Minded scale—Turk	134.2	(15.9)	129.8	(16.6)	2.68**	.27
Actively Open-Minded scale—Lab	128.7	(16.3)	123.7	(12.8)	3.10**	.36
Deliberation scale—Turk	70.1	(15.8)	65.4	(18.6)	2.71**	.28
Deliberation scale—Lab	68.6	(12.5)	63.6	(10.6)	3.91***	.45
Future Orientation scale—Turk	57.3	(9.6)	55.6	(10.8)	1.67	.17
Future Orientation scale—Lab	56.0	(8.7)	56.3	(7.9)	–0.27	–.03
Differentiation of Emotions scale—Turk	63.4	(10.9)	64.4	(12.0)	–0.82	–.08
Differentiation of Emotions scale—Lab	56.6	(9.4)	56.1	(9.7)	0.44	.05

*$p < .05$; **$p < .01$; ***$p < .001$

Turk: 231 males and 166 females (df = 395) for all variables except SAT total, where there are 123 males and 81 females (df = 202).

Lab: 109 males and 241 females (df = 348) for all variables except SAT total, where there are 102 males and 232 females (df = 332).

The Full-Form versus the Short-Form CART

In the study described in chapter 12—RT59—each subject completed all eleven subtests of the short-form CART. Of course, the subjects in RT60 who completed all twenty subtests of the CART also completed the eleven short-form subtests. Therefore, each of the subjects in RT60 could also be given a score for the short-form CART. In RT60, the mean score for the Turk sample on the short form was 56.2 (SD = 17.8), which was significantly higher than the mean score of the Lab sample on the short form, 47.5 (SD = 14.2, $t(745)$ = 7.36, $p < 0.001$, Cohen's $d = 0.539$). The university sample in RT59 who completed only the short-form subtests achieved a mean score of 43.6 (SD = 13.0). The Lab sample in RT60 scored significantly higher than the similar university lab sample in RT59 ($t(720) = 3.85$, $p < 0.001$), probably because the former were paid and the latter were not.

In RT60, the correlation between the full-form score and the short-form score was quite high: 0.97 in the total sample and in both the Turk and Lab samples. For purposes of comparison, we might label the nine subtests that are in the full-form CART but are not in the short-form CART as the residual CART. In RT60, the residual CART had a correlation of 0.73 with the short-form CART in the total sample (0.74 in the Turk sample and 0.66 in the Lab sample). This correlation is based on independent subtests. The correlation of the residual CART with the full-form CART was of course a very high 0.87 in the total sample (0.87 in the Turk sample and 0.83 in the Lab sample) because one set is contained within the other.

Table 13.11 presents correlational comparisons across the full-form CART, short-form CART, and residual CART in RT60. By examining this table, we can see whether such factors as cognitive ability, study sample, sex, and thinking dispositions related similarly to the full-form and short-form CART, as well as to the residual CART. At the top of the table are the correlations involving cognitive ability: Cognitive Ability Composite3; SAT total score; and the Cognitive Ability Composite4. The correlations are displayed separately for the Turk sample and the Lab sample. Displayed next is the correlation between study sample (Turk versus Lab). The correlations here are negative (indicating that the Turk sample outperformed the Lab sample) because the variable was scored 1 for Turk and 2 for Lab. Presented next are the correlations involving the sex of the subject. The correlations here too are negative because they were scored as 1 for male and 2 for

Table 13.11
Correlation comparisons between the full-form CART (20 subtests), the short-form CART (11 subtests), and the residual CART (9 subtests) in RT60

	Full-Form CART	Short-Form CART	Residual CART
Cognitive Ability Composite3—Turk	.695	.671	.620
Cognitive Ability Composite3—Lab	.567	.546	.474
SAT Total—Turk	.313	.319	.253
SAT Total—Lab	.495	.489	.384
Cognitive Ability Composite4—Turk	.713	.699	.638
Cognitive Ability Composite4—Lab	.614	.595	.506
Sample (Turk = 1; Lab = 2)	−.283	−.260	−.280
Sex (Male = 1; Female = 2)	−.322	−.320	−.265
Actively Open-Minded Thinking scale—Turk	.628	.631	.508
Actively Open-Minded Thinking scale—Lab	.554	.568	.387
Deliberative Thinking scale—Turk	.267	.281	.191
Deliberative Thinking scale—Lab	.472	.470	.360
Future Orientation scale—Turk	.311	.296	.286
Future Orientation scale—Lab	.297	.278	.267

For Cognitive Ability Composite3 (N = 747)
Correlations > .075 significant at the .05 level, two-tailed
Correlations > .126 significant at the .001 level, two-tailed
For Cognitive Ability Composite4 and SAT (N = 538)
Correlations > .086 significant at the .05 level, two-tailed
Correlations > .141 significant at the .001 level, two-tailed

female. Finally, at the bottom of the table are the correlational relationships involving three of the thinking dispositions that were strongly related to CART performance (the fourth, the Differentiation of Emotions scale, was not). The correlations for the three thinking disposition measures are displayed for each of the two samples.

One generalization that holds throughout the entire table is that the correlations involving the full-form CART and the short-form are remarkably similar across all of these measures. The two forms show very similar associations with cognitive ability and with the thinking dispositions. The superior performance of the Turk sample over the Lab sample was fairly similar, and the performance of the male subjects was similarly elevated across these two forms. Another generalization that holds pretty much down the entire table is that the residual CART shows slightly lower correlations than either the short-form or the full-form CART.

Table 13.12 presents the approximate raw scores at selected percentile ranks for the Turk sample and Lab sample on the full-form of the CART.

Table 13.13 presents the approximate raw scores at selected percentile ranks for the Turk sample and Lab sample for the Short-Form scores, the Probabilistic and Statistical Reasoning Subtest scores, and the Scientific Reasoning Subtest scores.

Table 13.14 presents the approximate scores corresponding to selected percentile ranks in the Turk sample and the Lab sample for the four thinking dispositions scales of the CART.

Table 13.12

Approximate full-form CART total scores corresponding to selected percentile ranks in the Turk sample and the Lab sample from RT60

Percentile Rank	Approximate Score in Turk Sample	Approximate Score in Lab Sample
2	31	38
5	38	40
10	48	47
20	59	52.5
30	68	57
40	76	62
50	83	66
60	91	71.5
70	97	77
80	105	83
90	113	94
95	121	106
98	128	115.5

Table 13.13

Approximate scores corresponding to selected percentile ranks in the Turk sample and the Lab sample: Short-form scores; Probabilistic and Statistical Reasoning subtest scores; and Scientific Reasoning subtest scores from RT60

Percentile Rank	Short-Form Score in Turk Sample	Short-Form Score in Lab Sample	Prob Reasoning Score in Turk Sample	Prob Reasoning Score in Lab Sample	Scientific Reasoning Score in Turk Sample	Scientific Reasoning Score in Lab Sample
2	19	24	4	4	4	4
5	25	27	5	5	5	5
10	31	31	6	6	6	6
20	40	35	8	8	7	7
30	46	39.5	9	9	9	8
40	52	42	10	9.5	10	9
50	57	45	11	10	11	10
60	62	50	13	11	12	10.5
70	68	53.5	14	12	13	11
80	72	59	15	13	14	12
90	79	66	16	15	15	14
95	84	75	17	16	16	15
98	89	80.5	18	17	18	17

Table 13.14

Approximate scores corresponding to selected percentile ranks in the Turk sample and the Lab sample for the four thinking disposition scales (RT60)

Percentile Rank	AOT Score in Turk Sample	AOT Score in Lab Sample	Deliberative Thinking Score in Turk Sample	Deliberative Thinking Score in Lab Sample	Future Orientation Score in Turk Sample	Future Orientation Score in Lab Sample	Differentiation Emotions Score in Turk Sample	Differentiation Emotions Score in Lab Sample
2	99	101.5	27	44	36	38	39	39
5	105	105	36	48	39	45	44	41
10	110	109	42	51	44	48	48	44
20	119	113	54	56	48	49.5	53	47
30	124	116	61	59	51	52	58	51
40	128	120	66	61.5	54	53	62	53
50	132	123	71	64	57	55.5	65	56
60	137	127	75	67	59	58	68	58
70	141	132.5	78	70	62	60	71	62
80	146	137	82	75	65	63	74	64
90	155	145	90	79	70	67	78	69
95	159	151	93	88	73	69	82	73
98	164	157.5	95	93	77	73	83	77

Further Relationships within the CART Subtests and Even Shorter Forms

The CART is a long test with many complex parts. Thus, we were prompted to explore the properties of even shorter versions of the CART than the short form discussed in chapter 12. The shortest version of the test was a 38-point version constructed by combining the two subtests that reflect most centrally the logic of the tasks studied in the heuristics and biases literature: the Probabilistic and Statistical Reasoning subtest and the Scientific Reasoning subtest.

One question that might be asked of such an extremely short test is whether the 38-point two-subtest version does capture a substantial amount of the variance on the other eighteen subtests. To answer this question, consider the zero-order correlation between those two subtests and the other eighteen in table 13.8. The correlations indicated there are mostly moderate in size and almost all significant (Dysfunctional Personal Beliefs being the one exception). Second, we calculated the correlation between the 38-item two-subtest version and the 110-point version comprised of the other eighteen subtests. The correlation was a substantial and impressive 0.79.

A more refined analysis showing that the 38-point two-subtest version does capture specific variance on most of the other subtests is presented in table 13.15. There, each of the eighteen other subtests is, in turn, regressed on three predictors: Cognitive Ability Composite3, the AOT scale, and the 38-point short-form version consisting of the Probabilistic and Statistical Reasoning subtest and the Scientific Reasoning subtest. The multiple R-squared is presented in the first column to the right of the subtest's name. Those are not flagged for statistical significance because all were significant at the 0.001 level. The next three columns present the standardized beta weights for the three predictor variables: Cognitive Ability3, AOT scale, and the 38-point short-form version.

The results in that table are easy to summarize. Overall, the 38-point short-form version of the test is substantially connected to the variance in seventeen of the eighteen remaining subtests. Specifically, the 38-point version accounted for significant unique variance over and above that accounted for by cognitive ability and the AOT in seventeen of eighteen cases (as expected given previous results, the exception was the Dysfunctional Personal Beliefs subtest). Furthermore, in ten of the eighteen regression analyses, the 38-point two-subtest version had the largest beta weight. In four of the eighteen analyses, Cognitive Ability Composite3 had the highest beta weight, and in four of the eighteen analyses the AOT had the highest beta weight. Finally, in twelve of the eighteen analyses, the 38-point version had a higher beta weight than did the cognitive ability measure. All of these findings serve to indicate that the 38-point version has a substantial connection to the eighteen other subtests even after the variance due to cognitive ability and the AOT has been removed.

In the 38-point version of the CART, we have probably the shortest of the short-form tests that one could make out of our comprehensive assessment instrument. In such an extremely shortened form (that could be completed in 15 to 35 minutes by most subjects), the 38-point version loses the comprehensive coverage of the full-form CART but, as we have just shown, it does share considerable variance with the eighteen subtests that have been removed.

In contrast with the extremely shortened 38-point version, researchers may well be interested in the statistical properties of an only slightly shortened version of the CART that we might label the CART16. We formed a sixteen-subtest version of the CART by removing the four subtests that measure the avoidance of contaminated mindware. Certain researchers may prefer the test to be composed in this manner. Some investigators may view the inclusion of the contaminated mindware probes in a comprehensive assessment of rational thinking to be quite separate from the other subtests. We acknowledge that there are some large issues surrounding this component of rational thought that may well generate diverse views. Some might find the selection of the contaminated mindware domains to be controversial. Others might agree that our four subtests do assess valid domains of contaminated mindware but may feel that other domains are missing.

Aside from the content of the domains of contaminated mindware that are sampled on the CART, another property of these four subtests set them apart from the other subtests. These four subtests are self-report questionnaires rather than performance measures, in contrast with the other sixteen subtests (an issue we will revisit in the next chapter). Table 13.7 does show that three of these four contaminated mindware subtests have significant correlations with impression management (although two of the three correlations are low in absolute magnitude). Nevertheless, we acknowledge that the content coverage and the self-report nature of these subtests may

Table 13.15

Multiple regressions with each subtest regressed on three predictors: Cognitive Ability3, the AOT scale, and the Probabilistic Reasoning and Scientific Reasoning subtests combined in RT60 ($N = 747$)

	Multiple R-Squared	Standardized Beta: Cognitive Ability3	Standardized Beta: AOT Scale	Standardized Beta: Prob Reas + Scientific Reas Subtests
Reflection versus Intuition subtest	.473	.220***	−.003	.535***
Syllogistic Reasoning subtest	.392	.203***	.047	.453***
Ratio Bias subtest	.141	.020	.070	.321***
Disjunctive Reasoning subtest	.194	.023	.006	.423***
Framing subtest	.125	.111*	−.006	.280***
Anchoring subtest	.114	.212***	.089*	.094*
Preference Anomalies subtest	.072	−.073	.127**	.223***
Argument Evaluation subtest	.198	.181***	.130**	.221***
Knowledge Calibration subtest	.170	.239***	.073	.169***
Temporal Discounting subtest	.037	−.073	−.007	.231***
Probabilistic Numeracy subtest	.382	.177***	−.006	.499***
Financial Literacy subtest	.471	.388***	.117***	.294***
Sensitivity to Expected Value	.112	.039	−.066	.341***

Table 13.15 (continued)

	Multiple R-Squared	Standardized Beta: Cognitive Ability3	Standardized Beta: AOT Scale	Standardized Beta: Prob Reas + Scientific Reas Subtests
Risk Knowledge subtest	.222	.298***	.110**	.142**
Superstitious Thinking subtest	.360	.196***	.375***	.136**
Antiscience Attitudes subtest	.428	.061	.470***	.220***
Conspiracy Beliefs subtest	.207	.107*	.274***	.156***
Dysfunctional Personal Beliefs	.026	−.002	.164***	−.003

* $p < .05$; ** $p < .01$; *** $p < .001$
All multiple R-squared values are significant at the .001 level.

Table 13.16
Correlations among the CART forms in RT60 ($N = 747$)

	1	2	3	4	5
1. Full-form CART					
2. Short-form CART	.97				
3. CART16	.97	.93			
4. 38-point 2-subtest CART	.88	.89	.89		
5. Cognitive Ability3	.69	.66	.67	.60	
6. AOT scale	.62	.63	.55	.53	.52

make an investigator wary of them. Thus, we explored the properties of a sixteen-subtest version of the CART.

In table 13.16 we present a correlation matrix including four versions of the test: the full-form version of the CART, the short-form version of the CART (based on eleven subtests), the CART16 (all the subtests except the avoidance of contaminated mindware subtests), and the 38-point two-subtest version we discussed above. Also presented in the table are the correlations with Cognitive Ability Composite3 and the AOT scale. All these correlations are derived from the data of the entire sample ($N = 747$) from

RT60. The correlations indicate that nearly identical individual difference findings will emerge from the use of the full form, the short form, and the CART16. The latter two have a correlation of 0.97 with the full-form CART. Also encouraging is the fact that the ultrashort 38-point version has a correlation of 0.88 with the full-form CART in the data of RT60.

The next-to-last row in the correlation matrix indicates that cognitive ability has roughly similar relationships with all the forms. The 0.69 correlation with the full form CART is quite similar to the 0.66 correlation with the short-form CART and the 0.67 correlation with CART16. The correlation does drop to 0.60 when the 38-point two-subtest version is used. The last line of the table indicates that the relationships with the AOT scale behave in predictable ways. Specifically, the AOT has nearly identical 0.62 and 0.63 correlations with the full-form CART and the short-form CART. Predictably, the correlation is reduced to 0.55 when CART16 is used, because that version of the task excludes the four contaminated mindware subtests that tend to have the highest correlations with the AOT scale. Likewise, the 38-point version also excludes those four subtests and has a correlation with the AOT scale of 0.53.

As a whole, the results displayed in table 13.16 indicate that using alternative versions of the CART will essentially allow the investigator to track similar individual differences. The short-form eleven-subtest version and the CART16 track variance almost identical to the full-form CART. Finally, the 38-point version is an extremely efficient way to track largely the same variance (a 0.88 correlation with the full-form CART). Of course, this statement is made from a logistical or practical point of view. From a research, or theoretical, point of view, the full-form CART embodies a theoretical comprehensiveness that the other shorter forms lack. That is, the full-form CART retains high content validity as a measure of the multifarious construct of rational thinking that the shorter forms will necessarily lack.

14 The CART: Context, Caveats, and Questions

The rationale for the choice of the various subtests of the CART and for the composition of those subtests has been described in part 2. In those chapters, we discussed the specific literature relevant to the tasks included in each subtest. In the present chapter, we step back a bit from these microperspectives and take a broader look at the various decisions we have made in constructing the CART. We will discuss its overall structure compared with other related instruments in the literature, and we will contextualize the test within the broader literature on cognitive ability assessment. We will also engage in a critique of the CART, pointing out certain caveats and cautions that follow from its status as a beta (early prototype) version of a rational thinking assessment.

The CART Provides Wide Coverage of the Domain of Rational Thinking—Not the Best Measure of Any One Thinking Skill

It is important to understand that the value added by the CART resides in its status as a total assessment instrument—that is, in its components taken together. The subtests chosen, and the items chosen for the subtests, were determined by the logistics of a comprehensive assessment instrument designed to assess a large number of cognitive skills and knowledge bases. Thus, we do not view any of our subtests as necessarily the optimal measure of the particular thinking ability that it is tapping. If we had to choose the items for a two-hour test of just *that* construct, we might well choose differently. With respect to almost all of the subtests, the best measurement of each particular construct would necessitate a much longer subtest. No doubt the "best set" of stimuli would come from the empirical literature on the original versions of each construct, rather than the variants we have chosen or constructed for our test.

In short, in constructing the CART, we have opted for comprehensiveness—for coverage of a large number of thinking skills rather than the in-depth assessment of only a few. This obviously involved some trade-offs. In lengthening the test to capture more aspects of rational thinking, the logistics of such a long test necessitated that we sometimes sacrifice the number of items, and hence the reliability, within any given subtest. It is no doubt the case that if one were focusing on the particular construct of a particular subtest for research purposes, then one would construct a longer and hence more reliable subtest. For example, there exist numeracy measures that are more comprehensive and reliable than ours (see Liberali et al., 2012; Schapira et al., 2012; Weller et al., 2013), but for our Probabilistic Numeracy subtest we chose instead a very modest number of items that are quite diagnostic.

The CART Compared to Other Broad Measures of Rational Thinking

Although no other heavily researched assessment device bills itself as a measure of rational thinking, there exist some batteries of tasks that, although they go by a different name, would certainly come under the purview of what we term "rational thought" in this book. Two in particular warrant discussion because they have generated substantial research: the Decision-Making Competence Scale for Adults and the Halpern Critical Thinking Assessment.

The Decision-Making Competence Scale for Adults (A-DMC) was introduced into the literature by Parker and Fischhoff (2005) and Bruine de Bruin, Parker, and Fischhoff (2007). That measure consisted of seven subsections, or subtests, in the original versions. In subsequent work (Bruine de Bruin, Parker, & Fischhoff, 2012; Del Missier, Mäntylä, & Bruine de Bruin, 2012; Parker, Bruine de Bruin, & Fischhoff, 2015; Strough, Parker, & Bruine de Bruin, 2015; Weller et al., 2015), the number of subtests has been reduced to six by the elimination of the path independence section, which displayed modest reliability and showed poor convergence with other subtests. Parker and Fischhoff (2005) initially introduced a version of the Decision-Making Competence index for youths (Y-DMC) that contained

Context, Caveats, and Questions 271

the same seven subtests as the A-DMC, with the exception that the tasks in the subtests used items that were more suitable for young people.

Comparing the CART to the A-DMC we find that: several of the subtests of the A-DMC are also contained in the CART; several of the A-DMC subtests are actually components of some of the larger subtests in the CART; two subtests of the A-DMC are not included in the CART; and one subtest of the original A-DMC (path independence) is actually measured by the CART, but was dropped from later versions of the A-DMC. We will describe these relationships briefly.

As we will discuss below, the A-DMC is very heavy on process measurement, and it is much lighter than our battery on the measurement of areas of rational thinking that are saturated with knowledge. Finally, the A-DMC has no measure of declarative knowledge domains that contaminate thinking and can lead to irrational responses. Neither does the A-DMC measure thinking dispositions as supplementary information, in contrast with the CART. Of course, the more in-depth measurement of all of these other domains in the CART comes at the obvious cost that it is a much longer instrument to complete than is the A-DMC.

Two of the seven subtests of the A-DMC, resistance to overconfidence bias and resistance to framing effects, are measured in a very similar manner in the CART. The A-DMC measures overconfidence bias using different items and materials of course, but it uses essentially the same performance indices and logic (the CART does add the interval method). As is clear from table 4.1 in chapter 4, avoiding overconfidence is one of the heuristics and biases measured on the CART that is heavily saturated with processing requirements. The avoidance of framing effects is similarly process dependent, although as is clear from table 4.1, we view it as just one indicator of the important rational thinking characteristic of making decisions in a manner not influenced by irrelevant context.

The A-DMC measure of resistance to framing effects is similar to our measure of framing effects in several respects. However, its position in the CART is not as prominent as it is in the A-DMC. Resistance to framing effects is one of seven subsections in the original A-DMC, and its proportion of the test rises in later versions to one-sixth of the entire test. As is clear from our table 4.1, although the CART does assess framing, our measure of framing effects is one of three indicators of the ability to avoid irrelevant context effects in decision making. From that table, it can be seen that we view the ability to avoid anchoring affects and to avoid preference anomalies in the same domain of rational thinking. As table 4.2 indicates, our Framing subtest comprises only 6 of 148 points on the full-form CART, much less than the one-sixth weight it receives on the A-DMC.

Two components of the A-DMC, consistency in risk perception and resistance to sunk costs, are part of CART measurement, but are not separate subtests themselves. Instead, they are a subset of the items contained within other major subtests of the CART. So, for example, consistency in risk perception is measured in the Probabilistic Reasoning subtest of the CART. Likewise, the resistance to sunk costs is measured by three items contained within the Financial Literacy and Economic Knowledge subtest of the CART.

The path independence component of the A-DMC was dropped in later versions of that measure, but we assess decision-making competencies of a very similar sort on the Preference Anomalies subtest of the CART—one of three measures of the ability to avoid irrelevant context effects in decision making.

Two of the subscales on the A-DMC, applying decision rules and recognizing social norms, have no direct counterpart in the CART. The applying decision rules subtest of the A-DMC involves having the subject peruse a table with values assigned to the various features of a set of consumer alternatives. The subject is then given a statement that reflects a decision rule such as elimination by aspects, lexicographic, and equal weights rules. The subject must then decide which of the consumer items is preferred. This particular task seems very heavily loaded with information-processing requirements that would make it highly determined by cognitive ability. For example, the subject must evaluate the rule:

"Sally first selects the DVD players with the best Sound Quality. From the selected DVD players, she then selects the best on Picture Quality. Then, if there is still more than one left to choose from, she selects the one best on Programming Options."

Indeed, in the Bruine de Bruin et al. (2007) study, the correlation between a rather modest number of these items and the Raven's Matrices is 0.65. The correlation with the Nelson Denny reading test was 0.51. Some of the skills relevant to performance on this task may well be tapped by the Probabilistic Numeracy subtest of the CART, which also carries a good deal of cognitive ability variance. It should be noted that the instructions to participants for evaluation of the table in this task are very explicit and directive, unlike many tasks on the CART, where the probabilistic reasoning, or the scientific reasoning, or expected value calculation is not specifically instructed. In terms of the discussion in chapter 2, it is an optimal performance measure rather than a typical performance measure.

The seventh A-DMC subtest, recognizing social norms, is not represented on the CART. This subtest, testing the accuracy with which people perceive the social norms of their peers, is a measure of a particular type of declarative knowledge. Although we have studied various forms of knowledge perception and projection in our lab (Sá & Stanovich, 2001), it is not as closely integrated into the framework described in chapter 4, and into the frameworks of rationality in areas of cognitive science more generally (Evans 2014; Stanovich 2012). Furthermore, it is not one of the more reliable A-DMC subtests (see table 2 of Bruine de Bruin et al. 2007). More work would be needed on this measure before we would consider including it in the CART.

In short, as this discussion has illustrated, there is a considerable degree of overlap between the subtests of A-DMC and the CART, although the CART is by far the more comprehensive measure. Of course, the issue of time is relevant, and given constraints, the A-DMC does give reasonable coverage of many important areas of rational thinking. Nonetheless, the CART contains many additional measures of rational thinking, such as tests of cognitive reflection, belief bias in syllogistic reasoning, measures of miserly processing such as ratio bias and disjunctive reasoning, the avoidance of anchoring affects, the avoidance of myside bias, the assessment of temporal discounting, financial literacy, probabilistic numeracy, scientific reasoning, sensitivity to expected value, and more comprehensive measures of probabilistic reasoning.

The CART contains an assessment of bodies of declarative knowledge that inhibit rational thinking such as superstitious thinking, dysfunctional personal beliefs, the rejection of antiscience attitudes, and the rejection of conspiracy beliefs. Finally, the CART also provides assessment of supplementary thinking disposition measures such as the Actively Open-Minded Thinking scale, Deliberative Thinking scale, Future Orientation scale, and Differentiation of Emotions scale that are on the CART.

The Halpern Critical Thinking Assessment (HCTA; Halpern, 2008, 2010) is an instrument that, despite its name emphasizing critical thinking skills, assesses several components of rationality as we define them in the CART. It is an instrument that assesses five categories of critical thinking: hypothesis testing, likelihood and uncertainty, argument analysis, verbal reasoning, and problem solving skills. The HCTA is composed of twenty-five basic scenarios, and each scenario has two components, involving first an open-ended response followed by a forced choice response. Each response type is scored as part of the assessment. The open-ended response component is a notable aspect of the HCTA. As with the CART, performance on the HCTA has moderate associations with measures of cognitive ability (Halpern, 2008, 2010). Research on the correlates of the HCTA continues apace (Butler, 2012; Chan, Ho, & Ku, 2011).

The CART and the HCTA overlap considerably as measures of rational thinking, although the CART is a more comprehensive measure. The hypothesis-testing section of the HCTA measures many of the same concepts as the Scientific Reasoning subtest of the CART: the recognition that correlation does not imply causation, the need for control groups, the consideration of alternative hypotheses, and the necessity to consider P(D/~H).

The likelihood and uncertainty section of the HCTA touches on many of the rational thinking concepts tapped in the Probabilistic and Statistical Reasoning subtest of the CART: the weighting of base rate information, regression to the mean, sample size, and sampling. However, the CART subtest measures several additional concepts such as probability-matching tendencies, the gambler's fallacy, and consistent probability assessments.

The argument analysis section of the HCTA contains items somewhat like those popular in the informal reasoning literature (Klaczynski & Lavallee, 2005; Kuhn, 2007; Macpherson & Stanovich, 2007; Perkins, 1985; Toplak & Stanovich, 2003; Voss, Perkins, & Segal, 1991) and is most similar to the Argument Evaluation subtest of the CART. However, in the CART, the latter is used as an indicator of the avoidance of myside bias, whereas the argument analysis section of the HCTA assesses the ability to avoid a number of informal reasoning fallacies.

The verbal reasoning section of the HCTA assesses somewhat overlapping skills with the argument analysis section except that the former stresses more the verbal skills of precise interpretation of passages. There is nothing directly comparable on the CART to this section.

Finally, the decision-making and problem-solving section of the HCTA (which receives the largest weighting of the five sections) presents real-life decision-making conundrums that are scored for the adequacy of the solution. The types of problems studied are similar to those in the wisdom literature (Baltes & Smith, 2008; Grossmann et al., 2013). There is nothing comparable on the CART, although it should be noted that the interpretation of the responses on the problem-solving section of the HCTA are clearly open to a considerable amount of philosophical dispute. As will be discussed below, the normative appropriateness of responses on various heuristics and biases tasks has received wide discussion and has created substantial contention (Bishop & Trout, 2005; Cohen, 1981; Stein, 1996). It would be expected that such a discussion, and the relevant objections to the scoring principles involved, would be even greater on a measure such as the problem solving HCTA subsection.

In short, the HCTA does give reasonable coverage to many important areas of rational thinking. There is a considerable degree of overlap between the subtests of HCTA and the CART, although the CART is the more comprehensive measure. The CART contains many additional measures of rational thinking, such as tests of cognitive reflection, ratio bias, disjunctive reasoning, the avoidance of anchoring affects, the avoidance of myside bias, the assessment of temporal discounting, financial literacy, numeracy, scientific reasoning, sensitivity to expected value, and more comprehensive measures of probabilistic reasoning.

The HCTA also does not contain an assessment of bodies of declarative knowledge that inhibit rational thinking such as superstitious thinking and dysfunctional personal beliefs, the rejection of antiscience attitudes, and the rejection of conspiracy beliefs. Finally, the CART also provides assessment of the supplementary thinking disposition measures mentioned above.

In addition to the A-DMC and the Halpern instrument, there have been numerous piecemeal efforts to develop small batteries of heuristics and biases tasks. Our early (Sá et al., 1999; Stanovich & West, 1998c, 1999, Kokis et al., 2002; Toplak & Stanovich, 2002) and later (Toplak et al., 2011; Toplak et al., 2014a, 2014b; West et al., 2008) efforts have been joined by several other studies of small collections of judgment and decision-making tasks (Cokely & Kelley, 2009; Klaczynski, 2001a, 2014; Koehler & James, 2010; Liberali et al., 2012; Oechssler, Roider, & Schmitz, 2009; Strough, Karns, & Schlosnagle, 2011; Teovanović et al., 2015). Although all of these studies help to triangulate this important domain of cognition, none of these batteries in the literature come close to the comprehensiveness of the CART.

Issues Affecting Observed Relationships with the CART

The relationships that we have reported among the CART subtests and between the CART total score and other variables are most certainly affected by the degree of variability in the samples of our many studies. That variability, compared to the general population, is substantially constricted, thus attenuating any correlations that are obtained. The majority of studies that we have discussed in this volume were run using university students as subjects. Broader samples were obtained for some of the studies that involved subjects from the Amazon Mechanical Turk. Community samples of adults were used in some other studies (Toplak, West, & Stanovich, 2016). However, for the latter studies, we often did not have cognitive ability indicators, so that we could not assess the degree of variance of variables such as intelligence. Certainly, though, on many other variables such as age and socioeconomic status, the Mechanical Turk samples were more varied than those run in the university laboratories.

It is true that individuals with average and above average cognitive ability are considerably overrepresented in samples composed entirely of university students. Although our samples are no doubt constricted compared to the general population, it may be that they are much more typical of the kinds of subjects who would actually be assessed using a fully developed and standardized CART. When not using university samples, research studies will most likely employ samples from the Internet similar to the Amazon Mechanical Turk samples that we have used here. Beyond research work, however, when our prototype test is developed into a more practical assessment device in the future, the populations on which it is used will also likely be similar to the samples we have reported in this volume. As we will now discuss, two different arenas where this test might be used as a practical real-life assessment device will also have marked restriction of range.

It is likely that universities will be interested in standardized versions of the CART so that they might assess whether their educational programs are increasing the rationality of students over time. That is, we fully expect a more well-developed version of the CART to take its place alongside other measures that universities have used to operationalize the real-life effects of a college education on cognition and learning (universities already use several critical thinking tests and assessment devices like the Collegiate Learning Assessment; see Arum & Roksa, 2011). The use of such a test near or at the end of university training would provide an assessment of whether students have acquired a functional understanding of the tools of rationality that have been described in this volume. Thus, any use in this context will involve subjects much like those on which the test was developed.

Additionally, a fully developed and standardized CART might serve as an employee-assessment tool for many professional occupations such as the law and managerial decision making. Our earlier publications on the idea of a rational thinking test (Stanovich, 2009) spawned many inquiries from people in the fields of investing and finance. These inquiries were almost always from senior executives who wanted an assessment device that measured rational thought and good decision-making abilities among those they were hiring. And these individuals, of course, would be university educated individuals, by and large, who would be at least as high in cognitive ability as the students who served as subjects in the bulk of our studies. So again, although our samples compared to a general population are range restricted, when compared to the likely groups who would be taking a fully developed version of the CART, our research samples are much more representative.

Of course, the restriction of range in our research samples would serve to attenuate any observed correlations. However, another aspect of our design might well have the effect of *increasing* correlations between CART subtests and other variables, particularly with cognitive ability. This is the fact that many of the tasks from the heuristics and biases literature that we adapted for the CART were originally designed using a between-subjects logic. That is, the bias or effect that the task illustrated was a bias that was demonstrated across subjects in a between-subjects design. But such designs do not give an individual a specific score for the bias in question. Thus, many tasks in the CART (e.g., the Framing subtest, the Preference Anomalies subtest) had to be adapted into a within-subjects situation where a measure of the bias could be obtained for a particular subject. However, this might well have increased correlations with cognitive ability by signaling that an issue of consistency was at stake (high-ability subjects may be more likely to respond to the consistency cue).

In a commentary on our early research on individual differences, Kahneman (2000) pointed out that the correlations observed may well have been inflated because most of the relevant studies used within-subjects designs, which contain cues signaling the necessity of heuristic system override (Bartels, 2006; Fischhoff, Slovic, & Lichtenstein, 1979; Shafir, 1998). Stanovich and West (2008b) produced evidence that was at least somewhat consistent with this conjecture. For example, in within-subjects tests of outcome bias (Stanovich & West, 1998c) the appearance of the second item gives a pretty clear signal to the participant that there is an issue of consistency in their responses to the two different forms. The difficulty in the within-subjects version comes from the necessity of inhibiting the tendency to downgrade the decision in the negative-outcome condition, despite its having a better rationale than the positive-outcome decision.

Within-subjects framing paradigms have a similar logic. The appearance of the second problem signals that an issue of consistency is at stake. The modest cognitive ability associations that are generated by this task probably derive from lower-cognitive-ability participants who do not recognize the consistency issue or cannot suppress the attractiveness of an alternative response despite the threat to consistent responding that it represents. In contrast, there are no associations in between-subjects framing situations, as few people recognize that there is a conflict to be resolved between the framing situation presented and a potentially different response to an alternative framing.

The logic of conjunction effects problems like the Linda problem is similar (see chapter 5). Transparent, within-subjects versions are easier because they signal the conflict involved and the necessity for override. Such versions create at least modest associations with cognitive ability. In the between-subjects version, however, the association with cognitive ability is attenuated (Stanovich & West, 2008b).

The CART and the Great Rationality Debate

Researchers working in the heuristics and biases tradition tend to be so-called Meliorists (Stanovich, 1999, 2004). They assume that human reasoning is not as good as it could be, and that thinking could be improved. Thus, a Meliorist is one who feels that education and the provision of information could help make people more rational. This optimistic part of the Meliorist message derives from the fact that Meliorists see a large gap between normative models of rational responding and descriptive models of what people actually do. Over the last several decades, an alternative interpretation of the findings from the heuristics and biases research program has been championed. Contributing to this alternative interpretation have been philosophers, evolutionary psychologists, adaptationist modelers, and ecological theorists (Cohen, 1981; Cosmides & Tooby, 1996; Gigerenzer, 2007; Oaksford & Chater, 2007, 2012; Todd & Gigerenzer, 2000). They have reinterpreted the modal response (the most common response) in most of the classic heuristics and biases experiments as indicating an optimal information-processing adaptation on the part of the subjects. This group of theorists—who argue that human rationality cannot be improved but is already maximized—have been termed the "Panglossians." The Panglossian theorists often argue either that the normative model being applied is not the appropriate one because the subject's interpretation of the task is different from what the researcher assumes it is, or that the modal response in the task makes perfect sense from an evolutionary perspective. The contrasting positions of the Panglossians and Meliorists define the differing poles in what has been termed the "Great Rationality Debate" in cognitive science—the debate about how much irrationality to attribute to human cognition (Gigerenzer, 1996; Hahn & Harris, 2014; Kahneman & Tversky, 1996; Kelman, 2011; Lee, 2006; Polonioli, 2015; Samuels & Stich, 2004; Stanovich, 1999, 2004; Stanovich & West, 2000; Stein, 1996).

A reconciliation of the views of the Panglossians and Meliorists is possible, however, if we take two scientific steps. First, we must consider data patterns long ignored in the heuristics and biases literature: individual differences on rational thinking tasks. Second, we must understand the empirical patterns obtained through the lens of the modified and updated dual process theory we outlined in part 1 of this book. We have argued (Stanovich & West, 2000) that the statistical distributions of the types of goals being pursued by Type 1 and Type 2 processing are different, and that important consequences for the pursuit of rationality follow from this fact. The greater evolutionary age of *some* of the mechanisms underlying Type 1 processing accounts for why it more closely tracks ancient evolutionary goals (that is, the genes' goals), whereas Type 2 processing instantiates a more flexible goal hierarchy that is oriented toward maximizing overall goal satisfaction at the level of the whole organism. Because Type

2 processing is more attuned to the person's needs as a coherent organism than is Type 1 processing, in the minority of cases where the outputs of the two systems conflict, people will often be better off if they can accomplish a system override of the Type 1–triggered output (the full argument is contained in Stanovich, 2004).

Instances when there is a conflict between the responses primed by Type 1 and Type 2 processing are thus interpreted as reflecting conflicts between two different types of optimization—fitness maximization at the subpersonal genetic level and utility maximization at the personal level. A failure to differentiate these interests is at the heart of the disputes between researchers working in the heuristics and biases tradition and their critics in the evolutionary psychology camp. First, it certainly must be said that the evolutionary psychologists are on to something with respect to the tasks they have analyzed, because in each case the adaptive response is the *modal* response in the task—the one most subjects give. However, the work we have been describing in this book has been homing in on a data pattern relevant to this discussion—a pattern of covariation and individual differences across these tasks. We have found, as numerous previous chapters illustrate, that cognitive ability often (but not always) dissociates from the response deemed adaptive on an evolutionary analysis (Stanovich & West, 1998a, 1998c, 1999).

The evolutionary psychologists are probably correct that most Type 1 processing is evolutionarily adaptive. Nevertheless, their evolutionary interpretations do not impeach the position of the heuristics and biases researchers that the alternative response given by the minority of subjects is rational at the level of the individual. Subjects of higher analytic intelligence are simply more prone to override Type 1 processing in order to produce responses that are epistemically and instrumentally rational. This rapprochement between the two camps was introduced by Stanovich (1999) and reinforced by Kahneman and Frederick (2002). Subsequent research has only further reinforced it (see, e.g., Kelman, 2011; Samuels & Stich, 2004; Stanovich, 2004, 2011).

Independent of this theoretical rapprochement, an additional fact is embarrassing for Panglossian critics who argue that heuristics and biases researchers have used the wrong normative models (such as Cohen, 1981): people most often retrospectively endorse the Bayesian and SEU norms that they violate (Koehler & James, 2009; Shafir, 1993, 1998; Shafir & Tversky,

1995; Tversky, 1996). In introducing the collection of Amos Tversky's writings, Shafir (2003) stresses this very point: "The research showed that people's judgments often violate basic normative principles. At the same time, it showed that they exhibit sensitivity to these principles' normative appeal" (p. x). In short, when presented with a rational choice axiom that they have just violated in a choice situation, most subjects will actually endorse the axiom. That subjects endorse the strictures of rationality when presented with them explicitly suggests that they acknowledge the normative force of the axioms of rational choice. If people nevertheless make irrational choices despite consciously endorsing rational principles, this suggests that the ultimate cause of the irrational choices might reside in Type 1 processing and the miserly tendency not to override it with Type 2 processing.

The practical import for the CART of these theoretical discussions in the Great Rationality Debate is that they are essentially debates about which answer to score as correct on certain tasks in the heuristics and biases literature. For all the reasons discussed in this section, we feel secure in the scoring assumptions used for the heuristics and biases tasks in the CART. Note also that many of our subtest tasks were never the subject of controversy in any case (e.g., probabilistic numeracy, disjunctive reasoning, expected value, ratio bias, reflection versus intuition).

The CART Is Based on a Thin Rather Than a Broad Notion of Rationality

To think rationally means taking the appropriate action given one's goals and beliefs, and holding beliefs that are commensurate with available evidence—but it also means adopting appropriate goals in the first place. Instrumental rationality covers the first of these (taking the appropriate action given one's goals) and epistemic rationality covers the second (holding beliefs that are commensurate with available evidence), but the third factor (adopting appropriate goals in the first place) introduces a new issue. That issue is the distinction between a thin and broad conception of rationality. Political science theorist Jon Elster (1983) deems traditional views of instrumental rationality to be "thin" theories because the individual's goals and beliefs are accepted as they are, and evaluation centers only on whether individuals are optimally satisfying desires given beliefs. Such views do not subject to evaluation the desires and goals being maximized.

The strengths of the thin theory of instrumental rationality are well known. For example, if the conception of rationality is restricted to a thin theory, many powerful formalisms (such as the axioms of decision theory) are available to serve as normative standards for behavior. The weaknesses of the thin theory are equally well known. In not evaluating desires, a thin theory of rationality would be forced to say that Hitler was a rational person as long as he acted in accordance with the basic axioms of choice as he went about fulfilling his grotesque desires. By failing to evaluate desires, a startlingly broad range of human behavior and cognition escapes the evaluative net of the thin theory.

Developing a broad theory of rationality—one that encompasses a substantive critique of desires—has a cost, however. It means taking on some very difficult issues in philosophy and cognitive science. In previous writings (Stanovich, 2004, 2008, 2010, 2013), we have taken up some of the broader issues surrounding human judgment and decision making such as what goals are rational to pursue and when it is rational to be rational in the thin sense. We have argued that such questions implicate a kind of metarationality. Metarationality rises above the thin theory in assessing the content of the desires that are being pursued by instrumentally rational means. Despite the importance of such issues of broad rationality, they are not addressed by the CART because there are no standard paradigms for assessing metarationality skills.

The Structure of the CART

In previous chapters, we have reported data on the structure of the CART. Specifically, we have reported data on the interrelationships within the items of each of the subtests; we have reported data on the interrelations among the subtests themselves; and we have reported data on the relationships with other indicators, such as intelligence and thinking dispositions with the CART total score and the CART subtest scores.

Regarding the internal structure of each subtest, the findings are quite variable. From Table 13.1 it can be seen that many of the subtests had unproblematic reliabilities—for example, all of the contaminated mindware subtests, all of the thinking disposition scales, and several other measures as well (such as the Ratio Bias subtest, Temporal Discounting subtest, and

Sensitivity to Expected Value subtest). One subtest (Preference Anomalies) had an extremely low reliability, as we have discussed.

Two very important subtests of the CART, the Probabilistic and Statistical Reasoning subtest and the Scientific Reasoning subtest, both had somewhat modest reliabilities (0.75 and 0.67, respectively), but we do not view these as problematic. This is because the constructs measured by these two subtests are probably best viewed as "formative," rather than "reflective" (Bollen & Lennox, 1991; Jarvis et al., 2003). As Bollen and Lennox (1991) discuss, the direction of causality is from construct to measures in reflective models but from measures to construct in formative measurement models. In formative measurement models, indicators are defining characteristics of constructs, and changes in indicators cause changes in the construct. In the contrasting case of reflective models, indicators are manifestations of the construct, and changes in an indicator do not cause changes in the construct.

Furthermore, indicators in formative models do not necessarily share a common theme, but they do in the case of reflective models. In the latter, individual indicators are interchangeable and replaceable; in the former, they are not. A change in the level of an indicator in a formative model is not necessarily expected to be accompanied by changes in the levels of other indicators. By contrast, in a reflective model, a change in one indicator should be accompanied by changes in others, because the change in one reflects some kind of change in the latent construct (for lists of differences in the two models, see Bollen & Lennox, 1991; Diamantopoulis & Winklhofer, 2001; Jarvis et al., 2003; see Edwards, 2011, for a critique).

Bollen and Lennox (1991) discuss several examples of formative concepts. For example, the concept of life stress has been defined by job loss, divorce, medical problems, and the recency of a death in the family. Such a concept is formative in that the measured variables define the construct rather than the measured variables being manifestations of a unitary latent variable. A change in one indicator of the formative construct "life stress"—such as a recent death in the family—would not lead to the expectation that there would be commensurate changes in other indicators (such as divorce), as would be the case with a reflective construct. Diamantopoulis and Winklhofer (2001) cite several examples of formative constructs from the economics literature, including composite formative variables of economic welfare, human development, and quality of life. Constructs such as these have the same logic as other formative constructs in that a change in one indicator does not necessarily lead to the expectation of a change in another. The causal direction is from indicator to construct rather than from construct to indicator (as is the case with reflective models).

We believe that the Probabilistic Reasoning and Scientific Reasoning subtests of the CART are both better viewed as measuring formative constructs rather than reflective ones. We do not think that either subtest reflects a unified core ability that is reflected on many different interchangeable indicators. Instead, we think that both domains are defined largely by the disparate set of skills and tasks that make up the subtest. As formative constructs, we would not expect them to have extremely high internal reliability (Bollen & Lennox, 1991).

Even more generally, we believe that a formative approach is more appropriate for a multidimensional construct like rational thinking. Any global notion of rational thinking that is defined by the CART is surely a formative one, composed of very disparate elements that we would not necessarily expect to cohere because they are not the result of some unified latent concept of rational thought. Instead, the various components serve to define the global concept of rational thinking. However, the subtests of the CART themselves vary in that some are defined by reflective models (e.g., the Antiscientific Attitudes subtest) and others represent formative concepts (e.g., the Probabilistic Reasoning and Scientific Reasoning subtests). Perhaps some (e.g., the Probabilistic Numeracy subtest) are difficult to classify according to this dichotomous scheme, which only serves to emphasize the hybrid nature of the CART.

These points about the structure of the test reinforce the general issue that the value of the CART is somewhat independent of what future research reveals about its ultimate structure. Specifically, there is no reason to expect a g factor of rational thinking analogous to the g factor found in intelligence research. Nor should we expect some structural parallel to the Gc/Gf structure of the Cattell–Horn–Carroll theory discussed in chapter 2. Although of course the structure of the subtests and the skills measured by the CART are important research questions, the usefulness of the rational thinking concept does not depend on there being a single-factor outcome to any structural investigation—or even a simple outcome. In chapter 3, we called attention to the fact that our own earlier suggestion of an analogy between a theory of rational thinking and the CHC theory of intelligence suggests a process–content split that does not in fact carry over into the domain of rational thinking (and we have since explicitly abandoned our earlier terminology that suggested such an analogy). Heuristics and biases tasks will simply not cluster in the manner of the CHC theory because their underlying mental requirements are too diverse and too complex. We would not expect our rationality subscales to hang together in a particular predetermined structure because not only are different types of mindware involved in each but also, for many of the subtests, different types of miserliness are involved as well (default to Type 1 processing; serial associative cognition; override failure).

An oversimplified example will illustrate the point we wish to assert with respect to the CART. Imagine that highway safety researchers found that the following variables were causally associated with lifetime automobile accident frequency: braking skill; knowledge of the rules of the road; city driving skill; cornering skill; defensive driving; and a host of other relationships. In short, these skills, collectively, define a construct called "overall driver skill." Of course, we could ask these researchers whether driving skill is a g factor or whether it is really fifty separate little skills. But the point is that the outcome of the investigation of the structure of individual differences in driving skill would have no effect on the conceptual definition of what driving skill is. It may have logistical implications for measurement, however. Skills that are highly correlated might not all have to be assessed to get a good individual difference metric. But if they were all causally related to accident frequency, they would remain part of the conceptual definition of overall driver skill. If several components or measurement paradigms turn out to be highly correlated, that will make assessment more efficient and logistically easier—but it will not enhance or diminish the status of these components as aspects of overall driving skill. It is likewise with rational thinking. There is independent evidence in the literature of cognitive science that the components in table 4.1 form part of the conceptual definition of rational thought. In short, psychometric findings do not trump what cognitive scientists have found to be the conceptually essential features of rational thought and action.

The point that psychometric findings do not trump a concept that has theoretical or practical importance can be appreciated by simply looking at other emerging areas of cognitive science. Take, for example, the important area of children's developing theory of mind (ToM; Goldman, 2006; Nichols

& Stich, 2003; Wellman, 2011; Wellman & Liu, 2004). Like rationality, ToM has been found to have a variety of subcomponents (Baron-Cohen, 1995), and there is, as yet, less than complete agreement in characterizing them (Goldman, 2006; Nichols & Stich, 2003). No g factor of ToM skills has yet been identified. The many laboratory tests of the construct remain important for theoretical purposes, although they have yet to yield a psychometric gold-standard standardized test. ToM has been linked to real-world outcomes such as school learning, but the causal nature of these links has not been established, nor have the links been unambiguously shown to be independent of intelligence. Despite the nascent status of the concept, it remains an essential one in cognitive science with great promise for practical impact in educational psychology.

All of this is not to deny that it would obviously be useful to know the structure of rational thinking skills from a psychometric point of view. Our research group has contributed to clarifying that structure. However, we want to spur efforts at assessing components of rational thought, and in this early stage of the endeavor we do not want the effort to be impeded by protests that the concept cannot exist because its psychometric structure is uncertain. That structure will become clarified only once our call for greater attention to the measurement of this domain is heeded. We should not shy away from measuring something because of lack of knowledge of the full structure of its domain.

If our point here seems to have a "doth protest too much" quality, it may be because in speaking and writing about the issue of measuring rational thought we are often met with what we feel is a quite baffling question that goes, roughly, as follows: "Well, it's unlikely that you'll get a g factor of rationality isn't it? So how can you measure rational thinking?" We admit to being baffled by the implication here that we can only measure things that result in tight general factors. But for some reason the question keeps coming up in the rationality domain, perhaps because of linkages to the Aristotelian, soul-like connotations of the term "rationality" that we discussed in chapter 1. In any case, people seem to be leery about measuring it in a messy way and seem to want some pristine concept to arise from studies of rationality. They seem unsatisfied by our insistence that the concept and its measurement is bound to be multifarious and that its structure will probably end up being quite complex.

What's Missing from the CART?

Of course, our focus in this book has been on defining the CART in terms of content—in showing what we would choose to include in a comprehensive assessment of rational thinking. The CART is not, obviously, a completed instrument ready for practical use. Instead, it is a demonstration of a concept. It is clearly the case that much more psychometric work, as well as more comprehensive standardization, is needed in order to make the CART a practically usable instrument. Nevertheless, although much work remains to be done, we would argue that in this book we have taken a major step toward operationally defining what comprehensive rationality assessment would look like. In the remainder of this section, we will discuss some of the content decisions we made in constructing the CART and also touch on some domains and tasks that we omitted from the test.

As we have argued throughout, the CART is the most comprehensive device for measuring rational thinking that currently exists. Nevertheless, this does not mean that it provides an exhaustive assessment. Although our full-form measure has twenty subtests, there are still some tasks and biases from the heuristics and biases literature that have been left out. We have tried to make judicious choices in what to include and exclude from the CART. Often, our choices were dictated by the logistical constraints of aiming for a test that could be completed in two to four hours (full-form) and one to two hours (short-form). In this section, we will briefly discuss some of the inclusionary and exclusionary choices that we made, particularly the latter.

Many tests were left out of the final version of the CART because they were simply redundant with components that were easier to administer. For example, the set of tasks we used to measure the avoidance of miserly information processing (see table 4.1) could have been greatly expanded, but the four tasks that we used seem to provide a fairly wide coverage of this processing tendency. We did pilot test tasks that measured other types of attribute substitution (see Kahneman & Frederick, 2002), such as the tendency to substitute an affective reaction for more difficult task cues (studied by Slovic 2007; Slovic & Peters, 2006). We developed a task (see Stanovich & West, 2008b) based on the idea that judgments about the risks and benefits of various activities and technologies derive not from separable knowledge sources relevant to risk and benefit but from a common source—affect

(Slovic & Peters, 2006). Evidence for this conjecture derives from the finding that ratings of risk and reward are negatively correlated, both across activities within participants and across participants within activities (Finucane, Alhakami, Slovic, & Johnson, 2000). When something is rated as having high benefits it tends to be seen as having low risk, and when something is rated as having high risk it tends to be seen as having low benefits. Finucane et al. (2000) argued that such a finding is nonnormative because the risk/benefit relationship is most likely positive in a natural ecology. We found reliable individual differences in the nonnormative tendency to view risk and reward as negatively correlated (Stanovich & West, 2008b), but this task was deemed redundant given what was already included in the CART.

We made a similar decision with respect to including Wason's (1960; see Evans, in press) famous 2-4-6 task in the Scientific Reasoning subtest. In this task, the subject is told that the experimenter has a rule in mind that classifies sets of three integers (triplets) and is also told that 2-4-6 conforms to the rule. The subject is then asked to propose the next sequence of triplets and, when they do, the experimenter tells him or her whether their triplet conforms to the rule. Subjects continue proposing triplets and receiving feedback until they think they have figured out what the experimenter's rule is, at which time they should announce what they think the rule is. Although it is a famous task (see Evans, 1989, 2007a, in press; Evans & Over, 1996; Gale & Ball, 2006; Klayman & Ha, 1987; Poletiek, 2001), we omitted it from the Scientific Reasoning subtest because it is cumbersome to administer given the current format of the CART.

Some tasks were considered but ultimately rejected because we were unable to come up with a logistically tractable within-subjects version (a version that would measure the bias for an individual subject rather than as a group trend). This was a major issue in the measurement of myside bias. Although the literature is full of convincing between-subjects demonstrations of this bias (see Stanovich, West, & Toplak, 2013), we found it very difficult to turn many of the tasks in the literature into within-subjects versions appropriate for an assessment device. That is why we ended up using the Argument Evaluation subtest—a fairly old task of our own (Stanovich & West, 1997) that is probably an imperfect measure of the construct. Although it is a bit difficult to score, it turned out to be the only within-subjects measure in which we had confidence.

This logistical consideration led us to exclude several other tasks that we would have liked to have included. Often, the difficulty of constructing within-subjects versions also went along with a particular task being too long to include in a battery composed of many other tasks. We additionally chose not to use tasks that required the subject to return to the lab at a later date for the second part of a particular assessment. Had we done this with several of our tasks, the CART would have become (with its already large number of subtests) too difficult to administer in practice. Logistical considerations such as these, often combined with the issue of redundancy, were enough to dissuade us from including certain measures.

One thinking skill that is not redundant with other components of the test and that we greatly regret not including was a measure of the ability to avoid affective forecasting errors. Affective forecasting refers to our ability to predict what will make us happy in the future; people turn out to be surprisingly poor at such predictions (Gilbert, 2006; Hsee & Hastie, 2006; Kahneman, Krueger, Schkade, Schwarz, & Stone, 2006; Meyvis, Ratner, & Levav, 2010; Wilson et al., 2000; Wilson & Gilbert, 2005). A typical paradigm used to study hedonic prediction might involve asking participants to think about an impending event that will have either a desirable or undesirable outcome for them. They are asked to predict the influence that each of these outcomes would have on their overall level of happiness at a time in the future. Subsequently, once the event is known and at that future time, the participants are asked to rate their overall level of happiness.

One source of affective forecasting errors is focal bias. Researchers in the affective forecasting literature have theorized specifically about focalism interfering with hedonic predictions ("predictors pay too much attention to the central event and overlook context events," Hsee & Hastie, 2006, p. 31). Despite the importance of the affective forecasting skill, we did not include it in the CART because deriving within-subjects indices from the paradigms in that literature is difficult. Additionally, many paradigms in the literature involve some time lapse between the hedonic prediction and the target event that will cause the hedonic response, which therefore cannot be assessed in a single testing session.

Although we have done research on the bias blind spot (West, Meserve, & Stanovich, 2012), we have chosen not to include a measure of this effect in the CART. Explored in an important paper by Pronin, Lin, and Ross (2002), the bias blind spot is the label for the finding that it is relatively easy for people to recognize bias in the decisions of others, but it is difficult to detect bias in their own judgments. Pronin (2006) discussed various explanations for the bias blind spot, including naïve realism, overreliance on introspective evidence of one's own biases, and the possibility that it is a self-enhancing bias (see Scopelliti et al., 2015).

As opposed to the emphasis on domains related to social comparisons in past work on the bias blind spot, we found bias blind spots to be connected to some of the most well-known effects from the heuristics and biases literature: outcome bias, base-rate neglect, framing bias, conjunction fallacy, anchoring bias, and myside bias (West et al., 2012). Most cognitive biases in the heuristics and biases literature are negatively correlated with cognitive sophistication, whether the latter is indexed by development, by cognitive ability, or by thinking dispositions (i.e., biases decrease as cognitive sophistication increases, see Toplak et al., 2014b). This was not true for any of the bias blind spots we studied. We found that none of these bias blind spots were attenuated by measures of cognitive sophistication such as cognitive ability or thinking dispositions related to bias. If anything, a larger bias blind spot was associated with *higher* cognitive ability.

With respect to other effects in the psychology of reasoning, we have argued elsewhere that correlations with cognitive ability are a partially diagnostic pointer to the rationality of responses (Stanovich, 1999; 2004, 2011; Stanovich & West, 1999, 2000; Stanovich, West, & Toplak, 2012). If so, then these findings indicate a reinterpretation of the bias blind spot as an efficacious processing strategy rather than its more common interpretation as a processing flaw (see West et al., 2012, for the full argument). These uncertainties surrounding the bias blind spot have led us to avoid including it in the CART.

One task we have done research with but decided not to use in the CART was the Iowa Gambling Task (IGT) introduced by Bechara et al. (1994), which was discussed in chapter 11. Several groups of people with problems of behavioral regulation (gamblers, people with various forms of addiction, students with conduct disorders) perform poorly on the IGT despite having average intelligence (Brevers, Bechara, Cleeremans, & Noel, 2013; Mukherjee & Kable, 2014; Petry, Bickel, & Arnett, 1998; Stanovich, Grunewald, & West, 2003; Toplak et al., 2010; Yechiam et al., 2008).

Despite the popularity of the IGT in the literature, we decided not to include it in the CART for a variety of reasons. First of all, like some of the other excluded tasks we discussed above, the IGT takes a fairly long time to administer and involves fairly tricky administration procedures. More importantly, in a study of problem gamblers (Toplak et al., 2007), we found that a questionnaire method (an alexithymia scale) of assessing the same emotional regulation problems that the IGT does was a more potent predictor of pathological gambling than the IGT itself. The measure of alexithymia we used in that study was very similar to the Differentiation of Emotions scale in the battery of thinking dispositions that are contained in the CART.

Finally, there is the issue of whether the IGT should be viewed as an indicator of an *underlying* problem that sometimes results in irrational behavior, or whether it is a measure of rationality itself. If we attempt to view it as the latter, then the way that the IGT was constructed seems problematic. This is because the IGT was deliberately designed so that the large rewards in decks A and B will be overwhelmed by infrequent but large penalties (thus resulting in a negative expected value). As Loewenstein, Weber, Hsee, and Welch (2001) point out, it would be easy to design an experiment with the opposite payoff structure—where the risky choices had a higher payoff (Shiv, Loewenstein, Bechara, Damasio, & Damasio, 2005). Indeed, there are real-world examples of just this structure. For example, stocks are riskier than bonds on a short-term basis, but they tend to outperform bonds in the longer term. It is an open question which structure (positive expected value being associated with large variance or negative expected value being associated with large variance) is more common in the real world.

It is also an empirical question whether such a reversal of payoffs would be tracked by the same subjects who maximized expected value in the standard version. In a sense, the IGT is just a particular instance of an environment in which subjects have to implicitly calculate expected value. Because we already have an Expected Value subtest in the CART, using the IGT for this purpose seems superfluous. Alternatively, if it is viewed as a measure of emotional regulation, then it is somewhat redundant with the Differentiation of Emotions scale included in the CART's supplemental thinking dispositions scales.

In earlier taxonomies of rational thinking tasks that we have published (Stanovich, West, & Toplak, 2011a), we have included unrealistic optimism (Weinstein, 1980) and self-perception biases as domains of contaminated mindware. We did not include measures of unrealistic optimism in the final version of the CART because we think that the normative status of performance on such measures is still in dispute (Hahn & Harris, 2014), that the development of the concept itself is still undergoing refinement (Shepperd, Klein, Waters & Weinstein, 2013; Shepperd, Waters, Weinstein, & Klein, 2015), and that, in some real-world environments, at least moderate levels of unrealistic optimism appear to be efficacious (Sharot, 2011). It is quite possible that the efficaciousness of the tendency toward unrealistic optimism is an inverted U function, more like the thinking dispositions on the CART, rather than the various contaminated mindware domains that we have included. Although there are some specific experimental paradigms that clearly seem to indicate inefficacious unrealistic optimism (Simmons & Massey, 2012), these paradigms are logistically impossible for use in the CART because, like many affective forecasting paradigms, they involve measurement of the outcome of future events.

These are just a few of the tasks that we chose not to include in the CART. Clearly we cannot discuss in depth every task we decided not to include. Suffice it to say that it was most often the case that the task was either redundant with things we were already measuring, or, if it was not redundant, we deemed it too costly in terms of the amount of testing time and logistical difficulty.

Caveats on the Present State of the CART

We believe that the CART is unique in its comprehensiveness and theoretical grounding as a measure of rational thinking. Nevertheless, what we have presented in this volume is a prototype, or beta version, of such a test. It should be an excellent instrument for research use, but for other applications it will need further development. In this section, we will outline some of the caveats involved in using the test in its present form.

First, certain CART subtests depend on the Euro-American cultural context in which they were developed. For example, the Financial Literacy and Economic Knowledge subtest depends on knowledge of specific sociocultural domains (investment and banking) that vary from nation to nation. The items on this subtest concern domains that may be fairly specific to Europe and North America, particularly the latter. For example, individual retirement accounts differ in logic and name from country to country even within the Western world, as do mortgages. In short, we would not expect such a subtest to be understandable for a test-taker in Indonesia.

It is likewise with subtests such as the Superstitious Thinking subtest and Conspiracy Beliefs subtest. Both of these may contain content that would not be applicable outside Europe or North America. Indeed, some of the items may well need adaptation for anywhere outside the United States and Canada. Many superstitions and conspiracy beliefs are highly culturally specific, and thus subtests of this type would need adaptation when used outside of their environments of creation. The CART16 would be a better choice in such environments (or the 38-point two-subtest version).

In chapters 12 and 13 we reported that the CART total score on both the short-form and the full-form displays a sex difference—males tend to outperform females. In both of those chapters, we discussed how this difference was much larger on subtests that implicate numeracy skills (the Probabilistic and Statistical Reasoning subtest, the Reflection versus Intuition subtest, the Ratio Bias subtest, the Probabilistic Numeracy subtest, the Financial Literacy subtest, and the Sensitivity to Expected Value subtest), as well as the Dysfunctional Personal Beliefs subtest.[1]

As we outlined in the preface, our intent was to explore the operationalization of the concept of rational thinking and ground that operationalization in a specific assessment device. As we pointed out there, we did not select items with an attempt to minimize sex differences, nor did we let an item or a subtest's correlation with intelligence determine our choice of content. The content was derived from our reading of the empirical literature on the operationalization of rationality in cognitive science. Our purpose was to elaborate and operationalize the concept of rationality—without giving intelligence pride of place and without imposing statistical restrictions such as the absence of sex differences. Our prototype measure should be viewed as an attempt to establish concept. As the CART is developed further for other practical purposes, other issues (such as sex differences) may indeed come to the fore, but we did not want them to dominate either our conceptualization or our operational choices.

Empirically, in the studies reported in chapters 12 and 13, the CART total score correlated with intelligence at the level 0.50 (short form in RT59) and 0.69 (full form in RT60). But as we argued in the preface, we were quite prepared for that relationship to be of any magnitude. Indeed, we imagined a situation in which future research showed that a latent measure of intelligence correlated perfectly with the CART total score. We are quite prepared to live with that outcome. It would then just become an empirical fact that individual differences in rationality could be perfectly predicted from an intelligence test. One could then measure *intelligence* with *our* test!

That would probably be a positive thing in a society of the future, since our test would have much more face validity than most conventional intelligence tests, including the SAT (Frey & Detterman, 2004). Indeed the new director of the College Board, David Coleman, is portrayed as being greatly concerned that a redesigned SAT have more content validity and face validity (Balf, 2014). In the past, technical criteria determined the test items, leading to the presence of the infamous obscure vocabulary words with statistical properties that pleased the psychometricians but that turned off the public because of their lack of face validity. Coleman is described as attempting to relax these psychometric requirements in favor of content that would have credibility with the public (Balf, 2014). The CART has fewer such problems. The fact that we have included measures saturated with specific knowledge domains that are relevant to everyday life, such as financial literacy, make the CART more relevant to real life.

The CART also derives content validity from the fact that it is focused on our best measures of how well people think rationally about what to do and what is true (the instrumental and epistemic aspects of rational thought). As will be seen in chapter 15, each of the components of our test has been linked to a real-world outcome by at least some research in the literature.

Moving on to other caveats, in previous chapters, we have drawn attention to various subtests that might be contaminated by social desirability and impression management. This is not the case for most of the CART because the majority of subtests are performance subtests. Nevertheless, there are a small set of measures whose interpretation is slightly obscured by impression management issues. Specifically, the Superstitious Thinking subtest has shown significant but small (absolute magnitude of 0.15 in study RT60) correlations with impression management scales, and the Dysfunctional Personal Beliefs subtest has shown even higher correlations (0.25 in RT60). Two of the thinking disposition scales, Deliberative Thinking and the Differentiation of Emotions, also had positive correlations with impression management, particularly the latter.

Sixteen of the twenty subtests in the full-form CART are performance measures, and only four are self-report questionnaire measures. We will just repeat briefly the caveat that we mentioned in earlier chapters: the thinking dispositions are indeed self-report questionnaire measures, but they are not part of the regular CART scoring.

It is nonetheless true that the four contaminated mindware measures are indeed measured in a questionnaire format that would make them easy to prep for if the contents of the test were known in advance. They are most definitely different from the rest of the CART subtests in that respect. Nevertheless, here we are getting to the general issue of the susceptibility of the CART to coaching effects. We have already noted that the thinking dispositions measured by self-reports might be problematic in this respect. But they are not formal parts of the CART total score. Many other dimensions of the CART—such as the tendency toward miserly processing and the tendency to be affected by irrelevant context—are much less subject to the criticism that they could be coached. In fact, most of the other domains assessed on the test (probabilistic and scientific reasoning, argument evaluation, overconfidence, etc.) would likewise be less subject to nonefficacious coaching effects.

What we mean by nonefficacious coaching effects needs to be explained, and we will do so by way of an analogy to the controversy about "teaching to the test" in education. The practice of "teaching to the test" is a bad thing when it raises test performance without truly affecting the general skills being assessed. If a teacher learns that a statewide history assessment will contain many questions on the Great Depression and concentrates on that during class time to the exclusion of other important topics, then the teacher has perhaps increased the probability of a good score on this particular test, but has not facilitated more general historical knowledge. This is nonefficacious coaching. But in education, not all coaching in anticipation of assessments is of this type. Early reading skills provide the contrast to assessments of history knowledge. Imagine that teachers knew of the appearance of subtests measuring word-decoding skills (letter-sound knowledge and phonemic awareness) on a statewide early reading test and "taught to" that test. In this case, the teachers would in fact be teaching the key generalizable skills of early reading (see Hulme & Snowling, 2013; Stanovich, 2000). They would be teaching the most fundamental processes underpinning reading development. This would be *efficacious* "teaching to the test."

The point here is that the coachability of a skill does not *necessarily* undermine the rationale for including the skill in an assessment. To put it colloquially, some "coaching" is little more than providing "tricks" that lift assessment scores without really changing the underlying skill, whereas other types of coaching result in changes in the underlying skill itself. There is no reason to consider the latter type of coaching a bad thing and thus no reason to consider an assessment domain to be necessarily problematic because it is coachable. Thus, it is important to realize that many rational thinking domains may be of the latter type—that is, amenable to what we might call "virtuous coaching." If a person has been "coached" to always see the relevance of base rates, to always explore all of the disjunctive possibilities, to routinely reframe decisions, to consider sample size, to see event spaces, and to see the relevance of probabilistic thinking—then one has increased rational thinking skills and tendencies. If the existence of rational thinking assessment devices such as the CART spawns such coaching, then this can be nothing but a good thing for society.

In short, we agree that coaching people on what is considered the proper response to the contaminated mindware subtests would be easy and would represent inefficacious coaching in that it would not actually alter superstitious thinking or antiscience tendencies. But most of the rest of the CART (with the exception of the thinking dispositions, which are not part of the total score) is not subject to easy and inefficacious coaching of this type.

15 The Social and Practical Implications of a Rational Thinking Test

We believe that as a research instrument, the CART is currently the premiere comprehensive measure of rational thinking. As a practical instrument, as an assessment and selection device, it is more a proof of concept. For that purpose, it will need more development, as we described in the previous chapter.

The CART Subtests and Real-World Outcomes

As discussed in detail in chapter 14, there is a trade-off between comprehensiveness and logistics in constructing any test, and that certainly applies to the CART. More components could be assessed if the test were lengthened. We have never claimed exhaustiveness for our set of components. We do claim, though, that to be globally rational in our modern society one must have the behavioral tendencies and knowledge bases listed in table 4.1 to a sufficient degree. Our society is sometimes benign, and maximal rationality is not always necessary to navigate it well; but sometimes, in important situations, our society is hostile (more on this below). In such hostile situations, to achieve adequate degrees of instrumental rationality in our present society, the skills assessed by the CART are essential.

A reasonable amount of research has already been conducted linking rational thinking tendencies to real-life decision making (Baron, Bazerman, & Shonk, 2006; Bruine de Bruin et al., 2007; Camerer, 2000; Fenton-O'Creevy, et al., 2003; Hilton, 2003; Milkman, Rogers, & Bazerman, 2008; Parker et al., 2015; Thaler, 2015; Thaler & Sunstein, 2008). The thinking skills assessed by the CART have been linked to real-life outcomes, as is illustrated in table 15.1. In that table, for each of the paradigms and subtests of the CART, an association with a real-life outcome is indicated.

Table 15.1

Association between aspects of rational thought assessed in the CART and real-world outcomes

CART Subtest	Subtest Component or Paradigm	Source for Association with Real-World Outcome	Association with Real-World Outcome
Probabilistic and Statistical Reasoning	Proper use of base rates	Koehler, Brenner, & Griffin (2002)	In studies of medical personnel, lawyers, stockbrokers, sportswriters, economists, and meteorologists, the authors concluded that the base rate likelihood was a "major predictor of miscalibration in experts' everyday judgments" (p. 714).
	Consistent probability judgments	Kramer & Gigerenzer (2005)	Newspapers regularly confuse p(outcome B/outcome A) with p(outcome A/outcome B).
	Importance of sample size	Tversky & Kahneman (1971)	Researchers run experiments of insufficient power because of the failure to fully appreciate the importance of sample size.
	Recognizing biased and unbiased samples	Wainer (1993)	Failure to recognize the importance of selection effects leads media commentators to misinterpret the implications of state-level SAT test scores.
	Resistance to gambler's fallacy	Sundali & Croson (2006)	Actual gamblers videotaped in a casino displayed the gambler's fallacy.
	Resistance to gambler's fallacy	Xu & Harvey (2014)	An analysis of 565,915 online sports bets revealed the existence of the gambler's fallacy.

Table 15.1 (continued)

CART Subtest	Subtest Component or Paradigm	Source for Association with Real-World Outcome	Association with Real-World Outcome
	Resistance to gambler's fallacy and the hot hand	Dohmen, Falk, Huffman, Marklein, & Sunde (2009)	Failure to resist the gambler's fallacy was related to overdrawing bank accounts.
	Use of chance in explanatory frameworks; understanding random processes	Wagenaar (1988); Malkiel (2015)	Problem gamblers resist chance as an explanation of patterns they see in events; stock market investors mistakenly think that they can "beat the market" because they fail to appreciate the role of chance.
	Appreciating the limits of single-case evidence in conflict with base rates	Lilienfeld (2007); Dawes (1994); Baker, McFall, & Shoham (2009); Cunningham et al. (2009)	Clinical psychologists continue to ignore more valid actuarial evidence in favor of personal experience and so-called clinical intuition. Educators based bullying-prevention program choices on the reports of educators from other schools rather than on scientific evidence.
	Understanding regression effects	Toplak, Liu, Macpherson, Toneatto, & Stanovich (2007)	Problem gamblers were less able to recognize situations with the potential for regression effects.
	Base rate neglect	Garcia-Retamero & Hoffrage (2013); Gigerenzer, Hoffrage, & Ebert (1998)	Qualified doctors have been found to misuse base rates. AIDS counseling often ignores the implications of base rates.

Table 15.1 (continued)

CART Subtest	Subtest Component or Paradigm	Source for Association with Real-World Outcome	Association with Real-World Outcome
Scientific Reasoning	Scientific control concepts; causal variable isolation; control group necessity	Offit (2008); Pashler, McDaniel, Rohrer, & Bjork (2009)	Bogus autism treatments such as facilitated communication proliferated because of a failure to appreciate the necessity of scientific control; unproven educational fads such as "learning styles" persist because they are not subjected to test by true experimental control.
	Diagnostic hypothesis testing	Groopman (2007); Croskerry (2009a, 2009b); Lilienfeld (2007); Baker, McFall, & Shoham (2009)	Physicians and clinical psychologists fail to engage in diagnostic hypothesis testing.
	Sensitivity to $P(D/\sim H)$	Kern & Doherty (1982)	Medical students failed to choose disease symptom information that would allow calculation of a likelihood ratio.
	Diagnostic covariation judgment	Chapman & Chapman (1969)	Clinicians perceive connections between Rorschach responses and diagnoses that are not present.
	Covariation detection free of belief bias; avoidance of illusory correlations	King & Koehler (2000); Gilovich & Savitsky (2002)	Belief in pseudosciences such as graphology and astrology is fostered by the phenomenon of illusory correlation.

Table 15.1 (continued)

CART Subtest	Subtest Component or Paradigm	Source for Association with Real-World Outcome	Association with Real-World Outcome
	Difference between correlation and causation; recognizing spurious correlation	Baumeister, Campbell, Krueger, & Vohs (2003)	Educators have based many programs on the assumption that high self-esteem leads to better achievement when the direction of cause, if it exists at all, is the opposite.
	Understanding falsifiability as a context for confirmation; thinking of the alternative hypothesis	Wood, Nezworski, Lilienfeld, & Garb (2003); McHugh (2008)	Spurious associations are common in Rorschach interpretation, and pseudosciences such as recovered memory are maintained through unfalsifiable arguments.
	Appreciation of converging evidence	Begley (2007); Jordan (2007); Nijhuis (2008)	The failure to appreciate the principle of converging evidence contributes to the denial of the evidence suggesting human-caused global warming.
Reflection Versus Intuition	Impulsivity scale; Matching Familiar Figures Test	Toplak, Liu, Macpherson, Toneatto, & Stanovich (2007)	Problem gamblers are higher than control groups on measures of impulsivity and lower on measures of reflectivity.
	Frederick's Cognitive Reflection Test	Gervais (2015); Calvillo & Burgeno (2015)	Reflective answers predicted counterintuitive beliefs such as belief in evolution and economic maximizing.
	Frederick's Cognitive Reflection Test	Juanchich, Dewberry, Sirota, & Narendran (2016)	Cognitive reflection predicted fewer negative decision outcomes on an inventory of 41 life outcomes, including risk behaviors.

Table 15.1 (continued)

CART Subtest	Subtest Component or Paradigm	Source for Association with Real-World Outcome	Association with Real-World Outcome
	Frederick's Cognitive Reflection Test	Barr et al. (2015)	Lower cognitive reflection was associated with reliance on smartphones as opposed to thinking.
	Frederick's Cognitive Reflection Test	Moritz et al. (2013)	High cognitive reflection was associated with the choice of economically maximizing strategies in a market environment.
	Frederick's Cognitive Reflection Test	Graffeo, Polonio, & Bonini (2015)	Higher cognitive reflection scores associated with better consumer choices.
Belief Bias in Syllogistic Reasoning	Mock jury paradigm	Hastie & Pennington (2000)	Jurors use prior knowledge in constructing narratives of a legal case even when instructed not to do so.
	Belief bias paradigm	Toplak, West, & Stanovich (2016)	High belief bias associated with high-risk real-world choices in personal finance, gambling, and drug use.
	Belief bias paradigm	Fletcher, Marks, & Hine (2012)	High belief bias associated with endorsement of superstitious attitudes.
Ratio Bias	Size of denominator	Pinto-Prades, Martinez-Perez, & Abellán-Perpiñán (2006)	Utilities of medical risks were judged significantly higher when risk information was framed using 1,000 instead of 100 as a denominator.
	Items similar to CART subtest	Låg et al. (2014)	Avoiding ratio bias was associated with understanding medical trade-offs and medical data interpretation.

Table 15.1 (continued)

CART Subtest	Subtest Component or Paradigm	Source for Association with Real-World Outcome	Association with Real-World Outcome
Disjunctive Reasoning	Exhaustive enumeration	Johnson, Hershey, Meszaros, & Kunreuther (2000)	Flight passengers mispriced insurance for component risks by not thinking disjunctively about the components.
	Exhaustive enumeration	Redelmeier & Shafir (1995)	Making more treatment options available to a physician made it *more* likely, not less, that they would refer a patient to a specialist.
Avoidance of Anchoring	Irrelevant anchoring paradigms	Stewart (2009); Northcraft & Neale (1987)	The size of minimum payment requirements affects the size of the partial payment of credit card debt. The evaluation of home values by real estate agents is affected by the listing price.
	Archival data	Bucchianeri & Minson (2013)	Overpricing leads to higher home sales prices.
Avoidance of Framing	Framing paradigms	Hilton (2003); Camerer (2000)	Professional stock market traders overweight single-day losing positions; generally, traders sell losing positions too rarely and sell winning positions too often.
	Framing paradigms	Belton, Thomson, & Dhami (2014)	Lawyers more likely to settle cases out of court when they are in a gain frame than when they are in a lose frame.

Table 15.1 (continued)

CART Subtest	Subtest Component or Paradigm	Source for Association with Real-World Outcome	Association with Real-World Outcome
Avoidance of Preference Anomalies	Outcome bias paradigms; status quo bias; endowment effects	Johnson et al. (2000); Samuelson & Zeckhauser (1988); Johnson & Goldstein (2006); Thaler & Benartzi (2004)	Insurance purchase decisions as well as utility purchases have been shown to be influenced by status quo biases; organ donations are higher in countries with presumed consent; default values strongly affect pension investment choices.
	Preferences in line with SEU axioms	Bruine de Bruin et al. (2007)	Properly applying basic rules of choice was inversely related to a Decision Outcomes Inventory of 41 negative outcomes such as loan default and drunk driving.
	Part D drug plan switches	Abaluck & Gruber (2011)	Actual Part D Medicare choices violated the dominance principle of rational choice theory.
	Frequency versus probability formats	Slovic, Monahan, & MacGregor (2000); Wu & Weseley (2013)	Risks were judged to be higher in a frequency format (10 out 100) than in a probability format (10%); Frequency information was more effective than probability format information to discourage adolescents from cell phone use while driving.
	Dollar versus percentage formats	Nelson & Rupar (2015)	Investors assess higher risk in response to dollar-formatted disclosures than to equivalent percentage-formatted disclosures.

Social and Practical Implications

Table 15.1 (continued)

CART Subtest	Subtest Component or Paradigm	Source for Association with Real-World Outcome	Association with Real-World Outcome
	Outcome bias	Chabris (2013)	Avoiding outcome bias is a critical characteristic of elite poker players.
Argument Evaluation	Myside bias paradigm	Forsythe, Nelson, Neumann, & Wright (1992)	Investors in a political prediction stock market exchange made investment decisions that were influenced by which candidate they hoped would win.
	Recognizing the validity and invalidity of informal arguments	Watson & Glaser (2006)	This showed moderate correlations with job performance in a variety of occupations.
	Mock jury paradigm	Arkes, Shoots-Reinhard, & Mayes (2012)	Jury verdicts weighted nondiagnostic information.
	Scoring offender files	Murrie, Boccaccini, Guarnera, & Rufino (2013)	Forensic psychologists risk assessments depended on whether they were working for the prosecution or the defense.
	Otherside thinking task	Sorge, Skilling, & Toplak (2015)	Offending adolescent youth produced fewer otherside arguments regarding practical issues than did a control group.
Avoiding Overconfidence	Knowledge calibration	Bruine de Bruin et al. (2007)	Degree of overconfidence was related to a Decision Outcomes Inventory of 41 negative outcomes such as loan default and drunk driving.
	Self-report paradigms	Hilton (2003); Odean (1998); Statman, Thorley, & Vorkink (2006)	Predictions of currency exchange rates by corporate treasurers are overconfident. Financial professionals are overconfident. Overconfidence increases trading volume.

Table 15.1 (continued)

CART Subtest	Subtest Component or Paradigm	Source for Association with Real-World Outcome	Association with Real-World Outcome
Temporal Discounting	Delay of gratification paradigms; time preference; future orientation	Duckworth & Seligman (2005); Mischel, Shoda, & Rodriguez (1989)	Self-discipline measures predict school grades longitudinally, independent of intelligence. Delay of gratification measured at age four predicted college entrance exam performance later in life.
	Delay of gratification paradigm	Schlam, Wilson, Shoda, Mischel, & Ayduk (2013)	Delay of gratification measures predicted body mass index thirty years later.
	Temporal discounting of reward	Petry (2001); Kirby & Petry (2004); Basile & Toplak (2015); Meier & Sprenger (2012); Shapiro (2005); Kirby et al. (2005)	Research has revealed an association between discount rates and gambling behavior, smoking, financial behavior, prudent food-stamp usage, and educational success.
	Temporal discounting of reward	Ainslie (2001, 2005); Kirby, Petry, & Bickel (1999)	Research has revealed an association between discount rates and substance abuse.
	Temporal discounting of reward	Chao, Szrek, Pereira, & Pauly (2009)	Research has revealed an association between discount rates and longevity, as well as health.
	Temporal discounting of reward	Meier & Sprenger (2012)	Research has revealed an association between discount rates and FICO credit score.
Probabilistic Numeracy	Various numeracy scales	Reyna et al. (2009)	Low numeracy ability was associated with distorted perceptions of risks and benefits of screening, reduced medication compliance, and adverse medical outcomes.

Table 15.1 (continued)

CART Subtest	Subtest Component or Paradigm	Source for Association with Real-World Outcome	Association with Real-World Outcome
	Various numeracy scales	Banks & Oldfield (2007); Banks et al. (2010)	Numeracy was associated with retirement savings.
	Berlin numeracy test	Cokely et al. (2012)	Numeracy was associated with the evaluation of everyday risks related to consumer and medical choices.
	Various numeracy scales	Låg et al. (2014)	Numeracy ability was associated with understanding medical trade-offs and medical data interpretation even after controlling for cognitive ability.
	Lipkus scale	Graffeo, Polonio, & Bonini (2015)	Higher numeracy scores were associated with better consumer choices.
	Lipkus scale	Wood et al. (2011)	Higher numeracy scores were associated with better Medicare Part D choices even after controlling for several third variables.
Financial Literacy and Economic Knowledge	Cost-benefit reasoning; limited resource reasoning	Sunstein (2002)	Environmental and other government regulations are written inefficiently because of a lack of attention to cost-benefit reasoning.
	Recognizing opportunity costs	Larrick, Nisbett, & Morgan (1993)	Salary levels were related to the ability to recognize opportunity costs.
	Avoiding sunk costs	Bruine de Bruin et al. (2007)	Committing the sunk cost fallacy was related to a Decision Outcomes Inventory of 41 negative outcomes such as loan default and drunk driving.

Table 15.1 (continued)

CART Subtest	Subtest Component or Paradigm	Source for Association with Real-World Outcome	Association with Real-World Outcome
	Understanding externalities	Heath (2001)	Most people in the United States do not realize that, because of externalities, gasoline prices are too low rather than too high.
	Awareness of the logic of exponential growth and compounding	Paulos (2003)	People save too little, fail to recognize the pitfalls of pyramid schemes, and invest foolishly because of ignorance of the mathematics of compounding.
	Understanding commons dilemmas, zero-sum, and nonzero-sum games	Bazerman, Baron, & Shonk (2001)	Oceans are overfished and traffic jams exacerbated because of the failure to recognize commons dilemmas.
	Recognizing regression effects	Malkiel (2015)	Failure to recognize regression effects in a random walk leads stock market investors to buy high and sell low.
	Appropriate mental accounting and understanding of fungibility	Thaler (1992; 2015)	Many people have money in savings accounts while simultaneously carrying credit card debt at much higher interest rates.
	Financial misconceptions	Jarvis (2000); Valentine (1998)	The belief that reward can become decoupled from risk contributes to the proliferation of Ponzi and pyramid schemes.
Sensitivity to Expected Value	Choices among gambles	Choi, Kariv, Müller, & Silverman (2014)	Utility-maximizing choices were associated with higher net worth.
	Choices among gambles	Benjamin, Brown, & Shapiro (2013)	High expected value choices were associated with higher grades in school.

Social and Practical Implications

Table 15.1 (continued)

CART Subtest	Subtest Component or Paradigm	Source for Association with Real-World Outcome	Association with Real-World Outcome
Risk Knowledge	Accurate perception of risks and benefits	Sunstein (2002)	Hazard regulations reflect the mistaken human belief that the risks and benefits of various activities are negatively related.
	Accurate perception of risks and benefits	Sivak & Flannagan (2003); Gigerenzer (2006)	The perception of relative risks of flying and driving shifted after September 11, 2001, in a manner that cost hundreds of lives.
Superstitious Thinking	Faith in Intuition scale	Epstein, Pacini, Denes-Raj, & Heier (1996)	Small correlations were observed with depression, anxiety, and stress in nonselected samples.
	Paranormal, Superstitious Thinking, and Luck scales; illusion of control	Fenton-O'Creevy et al. (2003); Barber & Odean (2000)	A measure of illusion of control was (negatively) related to several measures of stock traders' performance, including their remuneration; personal investors trade too much and thus lower their investment returns.
	Paranormal beliefs scale	Bensley, Lilienfeld, & Powell (2014)	Paranormal beliefs predicted the number of misconceptions that people had about behavior.
Antiscience Attitudes	Alternative Medicine scale	Browne, Thomson, Rockloff, & Pennycook (2015)	Rejection of traditional therapies predicted resistance to vaccination.
	Head Over Heart scale	Toplak, Liu, Macpherson, Toneatto, & Stanovich (2007)	Problem gamblers performed worse than controls on a measure of intuition versus reflective thinking.
	Epistemological understanding	Sinatra & Pintrich (2003)	Epistemological beliefs are related to the efficiency of learning.

Table 15.1 (continued)

CART Subtest	Subtest Component or Paradigm	Source for Association with Real-World Outcome	Association with Real-World Outcome
	Intuitive thinking style	Gaudiano, Brown, & Miller (2011)	Psychotherapists with intuitive thinking styles were more likely to reject science and science-backed therapies.
Conspiracy Beliefs	Conspiracy beliefs questionnaires	Beck (2008); Griffin, Regnier, Griffin, & Huntley (2007); Singh, Spencer, & Brennan (2007)	Because of conspiratorial thinking, one-third of Americans drink unfluoridated water despite voluminous scientific evidence that fluoridation can significantly reduce tooth decay.
	Conspiracy beliefs questionnaires	Offit (2008)	Vaccination panics have been caused by beliefs in pharmaceutical company conspiracies.
Dysfunctional Personal Beliefs	Measures of irrational personal beliefs	Epstein & Meier (1989)	Small to moderate correlations were found with alcohol and drug problems.
	Measures of irrational personal beliefs	Hollon & Kendall (1980); Hollon, Kendall, & Lumry (1986)	Depressed subjects endorsed more dysfunctional personal attitudes than nondepressed subjects.
	Measures of irrational personal beliefs	Chansky & Kendall (1997); Kendall et al. (2008); Ronan, Kendall, & Rowe (1994)	Children with anxiety disorders endorsed more negative social expectations and self-statements than children without anxiety disorders.

Table 15.1 (continued)

CART Subtest	Subtest Component or Paradigm	Source for Association with Real-World Outcome	Association with Real-World Outcome
Actively Open-Minded Thinking Scale	Head over Heart scale	Toplak, Liu, Macpherson, Toneatto, & Stanovich (2007)	Problem gamblers were less likely to endorse considering evidence rather than gut feelings.
	Prompting for reasons and contemplation	Howell & Shepperd (2013)	People opted more frequently to learn their risk for type 2 diabetes or cardiovascular disease if they first considered reasons to seek or avoid this information.
Deliberative Thinking Scale	Measures of need for cognition and typical intellectual engagement	Cacioppo, Petty, Feinstein, & Jarvis (1996)	Moderate negative correlations were found with anxiety, neuroticism, and procrastination.
Future Orientation Scale	Planning; Locus of Control scales	Ameriks, Caplin, & Leahy (2003); Lefcourt (1991)	The propensity to plan has been linked to lifetime wealth accumulation; locus of control has been related to achievement in school and in sports, as well as to various health outcomes.
	Consideration of Future Consequences scale	Joireman, Sprott, & Spangenberg (2005)	Low future orientation associated with fiscal irresponsibility in the domains of impulsive buying, credit card debt, and savings.
Differentiation of Emotions Scale	Measures of alexithymia	Lumley & Roby (1995); Toplak et al. (2007)	Problem gamblers performed worse than controls on measures of alexithymia.

Although we have listed only one or two citations for each cell, there is often a much more voluminous literature documenting real-world associations. The associations are of two types. Some studies represent investigations where a laboratory measure of a bias was used as a predictor of a real-world outcome; others are reports of real-world analogues of biases that were originally discovered in the lab.

With respect to the table, though, it is important to acknowledge that in most cases we do not know whether the association between a rational thinking skill and an outcome would remain if intelligence were partialled out. Clearly more work remains to be done on tracing the exact nature of the connections—that is, whether they are causal. The sheer number of real-world connections, however, serves to highlight the importance of the rational thinking skills assessed by the CART. Critics sometimes attempt to devalue the heuristics and biases research tradition by claiming that the effects demonstrated are only obtained in laboratory tasks or that they do not predict anything in real life (Gigerenzer & Goldstein, 2011). One of the purposes of table 15.1 is to provide an easily accessible demonstration that such a claim is nonsense (see also Camerer, 2000; DellaVigna, 2009; Hilton, 2003; Kelman, 2011; Shleifer, 2012; Thaler, 2015).

The Unique Features of Rationality Assessment: CART Subtests ≠ IQ Test Components

With the construction of the CART, we have made good on our conjecture from the preface and chapter 1 that it would be possible to construct an instrument designed to assess the types of cognitive skills that have been studied for forty years in the heuristics and biases literature. It is amazing that until now we have not had a battery that comprehensively assesses these cognitive skills, given their epic influence on cognitive science. The 1974 *Science* paper by Tversky and Kahneman had, by early 2015, received over 33,000 citations according to Google Scholar. Kahneman's recent (2011) book had received over 6,000 citations by the same time. These numbers, along with the 2002 Nobel Prize to Kahneman, represent an unprecedented scientific influence. Yet until the CART and the work that preceded it (Bruine de Bruin et al., 2007; Stanovich & West, 1998c), psychologists had completely neglected to develop assessment devices for these unique cognitive skills.

Of course, researchers have examined small sets of heuristics and biases tasks together before (Cokely & Kelley, 2009; Klaczynski, 2001a; Liberali et al., 2012; Stanovich & West, 1998c). Nevertheless, our collection is unique in its comprehensiveness. However, it is important to stress that the issue of measuring rationality goes far beyond the comprehensiveness of the heuristics and biases battery that is involved. Instrumental and epistemic rationality, as defined in this volume, both implicate important knowledge bases when their definitions are operationalized. The CART is unique in this particular respect—that is, in explicitly encompassing important declarative knowledge bases in its assessment model. Beyond the measurement of the important probabilistic reasoning tendencies and reflective reasoning tendencies that are well captured by the heuristics and biases tasks, the CART taps knowledge bases that importantly facilitate rational thought and behavior, as well as knowledge bases that importantly impede optimal responding. Heuristics and biases tasks are different from the more purely knowledge-based measures that the CART uses to provide a more comprehensive rationality assessment. Some components of the CART, such as the Financial and Economic Literacy subtest and Probabilistic Numeracy subtest, do not involve the conflict detection and processing override characteristics of a more typical heuristics and biases task. The contaminated mindware subtests (such as the Conspiracy Beliefs subtest and the Superstitious Thinking subtest) also do not have these characteristics.

The emphasis on heuristics and biases tasks (e.g., Probabilistic and Statistical Reasoning subtest) and subtests composed more purely of knowledge assessment (e.g., Financial Literacy subtest) in the CART highlights the two most important ways in which the CART is different from IQ tests. Now that the reader has seen the full scope of the CART, it is easier to appreciate the point regarding knowledge that we stressed in chapter 2: the knowledge bases assessed on the CART are domain specific (financial literacy; avoidance of conspiracy beliefs) and not like the broad-based vocabulary assessments of IQ tests.

Regarding the parts of the CART that are composed of heuristics and biases tasks, the logic of these tasks makes it possible for the CART to measure the *propensity* to use a cognitive skill in a way that IQ tests do not. In the domain of rational thinking, we are interested in individual differences in the *sensitivity* to probabilistic reasoning principles and in the *tendency* to apply scientific principles when seeking causal explanations. People can have knowledge of these principles without having the propensity to use them. Heuristics and biases tasks assess not only whether people have knowledge of the principles but also whether they have a propensity to see situations in terms of probabilities. A typical heuristics and biases task, like the items on our subtests, will pit a statistical way of viewing a problem against a nonstatistical way of viewing a problem—or a control-group way of thinking against correlational thinking—in order to see which kind of thinking dominates in the situation for the subject. People who can answer an explicit probability question on a test, or who can accurately define "control group" when asked, may not show the sensitivity to invoke these principles when their relevance to a problem is partially disguised. In contrast, the cognitive skills assessed by IQ tests are explicit ones. The respondent does not have to recognize their applicability—and does not have to overcome an intuitive response that the problem deliberately activates. On IQ tests, people are not "tempted" to engage in miserly processing due to the presence of an intuitively compelling alternative.

The results that we have obtained with the Actively Open-Minded Thinking scale (AOT) are consistent with these differences between the CART and IQ tests. Although the AOT scale is correlated with both, in RT60 and in RT59 the AOT correlated significantly more strongly with CART performance than with cognitive ability. Independent of this outcome, our AOT results indicate a startlingly tight linkage between a particular thinking disposition and rational thinking. A generic style of thought—one characterized by the cultivation of reflectiveness rather than impulsivity, the seeking and processing of information that disconfirms one's beliefs (as opposed to confirmation bias in evidence seeking), and the willingness to change one's beliefs in the face of contradictory evidence—has been linked in our data to a very comprehensive measure of rational thought. The results from the AOT show that there is a global mental attitude that pervades these tasks. It is certainly not a specific cognitive skill, but instead is best characterized as a generic mental attitude toward cognitive tasks—one of openness, full engagement, mental caution, exhaustiveness of thought, humility, and willingness to encompass new evidence.

Rationality itself might be multifarious, as we have argued, involving knowledge and process in complex and changing proportions across tasks and situations. It is interesting, though, in light of such process/knowledge complexity, that the AOT has such a high correlation with the CART total score. That is, despite the multifariousness of the rationality construct itself, a global thinking style—actively open-minded thinking—permeates almost all of the subtests on the full form (see table 13.5) enough to make it a strong correlate of the total score. The AOT remains a significant correlate of the total score on the CART once cognitive ability has been separated out (see table 13.6). The AOT findings show that there is something that we might want to call a "globally rational thinking attitude" (or "cognitive set"). The magnitude of the correlation between the AOT and the CART seems to warrant that inference. On a task-by-task microskill basis, there may be considerable variation in the subtests, but that does not mean that there is not a global attitude or set toward rational thinking that seems to reside in the proclivities we have termed actively open-minded thinking.

The Context for the High IQ–RQ Correlation

If the total score on the CART is viewed as an RQ, by analogy to an IQ, then we have found in RT60 a substantial correlation between a person's IQ and her or his RQ (0.69). However, we would argue that it is not high enough to warrant the idea that an IQ test provides a measure of rational thinking. The magnitude of the observed correlation leaves plenty of room for disassociations between intelligence and rationality. An analogy can be made to the educational field of learning disabilities, which was built upon discrepancies between achievement and intelligence[1] that are no greater than the discrepancies between rationality and intelligence that we have observed in our studies. For example, scores on reading comprehension tests and IQ tests can be correlated as high as 0.60 to 0.70 in samples of adults (Harris & Sipay, 1985; Stanovich, Cunningham, & Feeman, 1984), yet the entire field of dyslexia was built around the discrepancies from this relationship. People appear to have an extreme assumption of positive manifold—an assumption that all variables are positively correlated at the individual level—when it comes to most cognitive skills (Stanovich, 2009), and they are startled when any discrepancy occurs in cognitive abilities. The correlations that we have obtained still leave enough room for significant discrepancies between rationality and intelligence that draw the attention of the general public and that actually should be the focus of scientific inquiry as well (consider, for example, the phenomenon of smart people acting foolishly that we discussed in the preface).

The moderate-to-high RQ–IQ correlation that we have observed does not create a problem for several other reasons. To the extent that we have a comprehensive measure of cognitive skills grounded in rational thinking that also carries a substantial amount of IQ variance (and thus will correlate with most of the same things that IQ correlates with) this only adds to the important tools in the research psychologist's arsenal as well as potentially, in the future, adding a tool for practical use. To the extent that an RQ test adds incremental predictive power to an IQ test, that would be nice—but it is not a necessary condition for declaring an RQ test a useful asset. We discussed in the previous chapter how SAT test constructors are showing increasing concern that their test display face validity in the eyes of the public (Balf, 2014). A test constructed from the ground up to model both practical and research definitions of rationality would have more face validity regardless of whether it added incremental predictive validity to IQ tests. Interestingly, the sex differences that we have observed (see chapter 13) remain after control for cognitive ability variance is partialled out, indicating perhaps that the CART is picking up reliable cognitive variance that is independent of IQ. Interestingly, the procedure of removing sex differences from items in IQ tests would have the effect of obscuring variance related to rational thinking.

As we have outlined in previous chapters, rationality is a more encompassing construct than intelligence, theoretically. It is grounded in the practical idea that it is good to know what it is optimal to do and also good to know what is true. In contrast to rationality's solid theoretical grounding, the intelligence concept has become unalterably corrupted in public usage. That is, however solid is the intelligence concept in *scientific* practice, it is encumbered with assumptions and misunderstandings in how the public uses it (see Stanovich, 2009). The term "rationality" does not carry with it quite so much baggage and misinformation.

The term "intelligence" in public usage is surrounded by misinformation and myths that obscure the solid research that actually grounds the intelligence concept. For example, despite the fact that the IQ concept is grounded in some of the longest and most comprehensive research traditions in all of psychology, the public has been convinced by many politically correct commentators that the concept is worthless. For example, many public commentators (including some who appear in the most influential and respected mainstream media outlets) promulgate the view that

IQ tests do not measure anything important, or that there are many different kinds of intelligence, or that all people are equally intelligent in their own way.

Media commentaries pushing such ideas are quite common. For example, highly respected columnist and book author Elizabeth Kolbert (2014) wrote a highly misleading essay about taking the SAT in the influential *New Yorker*, promulgating the view that the SAT is primarily self-referential and taps nothing of importance about human cognition. She ends her essay by saying that "as befits an exam named for itself, the SAT measures those skills—and really only those skills—necessary for the SATs" (p. 41). An op-ed in the *New York Times* around the same time as Kolbert's essay echoed much of the same politically correct misinformation about the SAT and other intelligence proxies by claiming that the SAT measures only memorization (Boylan, 2014).

Almost none of the critiques of intelligence tests that one sees in the media are scientifically correct. They may be politically correct, but they are not based in science. For example, it is fashionable for media intellectuals to say that intelligence has nothing to do with real life, or that the items on IQ tests are just parlor games related only to "school smarts." Decades of research in psychology contradicts this view (Deary, 2013; Geary, 2005; Hunt, 2011; Mackintosh, 1998). IQ tests measure something that is cognitively real and that does relate to real life. Intelligence test performance predicts both occupational level attained and performance within one's chosen occupation, and it does so better than any other ability, trait, or disposition (Schmidt & Hunter, 2004). The sizes of these correlations with job performance are also larger than most found in psychological research, and they are not limited to just the skill acquisition period but also include on-the-job performance, as well (Schmidt & Hunter, 2004).

In fact, the way we use the term "intelligence" in day-to-day discourse reveals that we do not think that it is so trivial after all. People are termed "bright" and "quick" and "smart" in ways that clearly indicate that we are not talking about social or emotional qualities. These terms are used often and nearly universally with positive connotations. In fact, "bright" and "quick" and "sharp" are used in general discourse to pick out precisely a quality assessed on standard IQ tests (fluid g in particular). It may not be politically correct to laud IQ at cocktail parties, but all the parents at those same cocktail parties do want that quality for their children. When their children have behavioral/cognitive difficulties, parents are much more accepting of diagnostic categories that do not have "low IQ" attached. Thus the public's attitudes about intelligence seem very confused. People value it in private, in the context of their families, but it has become unacceptable to say so in public.

But by discussing the "IQ confusion" in the media and among the public, we do not mean to let scientists off the hook here. In the preface we discussed psychology's tendency to adopt permissive conceptualizations of what intelligence is rather than empirically grounded conceptualizations. Permissive theories include in their definitions of intelligence aspects of functioning that are captured by the vernacular term "intelligence" (e.g., adaptation to the environment, showing wisdom, creativity), whether or not these aspects are measured by existing tests of intelligence. Grounded theories, in contrast, confine the concept of intelligence to the set of mental abilities measured by IQ tests. Adopting permissive definitions of the concept of intelligence serves to obscure what is missing from extant IQ tests. Instead, to highlight the missing elements in IQ tests, we have adopted a thoroughly grounded notion of the intelligence concept in this book.

We do not see why everything in human nature, cognitively speaking, has to have the label "intelligence"—particularly when there are readily existing scientific and folk concepts for some of those things (such as rationality, creativity, wisdom, critical thinking, open-minded thinking, reflectivity, sensitivity to evidence). Permissive theorists inflate the concept of intelligence—they put into the term more than what IQ tests actually measure. One very strong tendency among permissive theorists is to use adjectives to differentiate the more encompassing parts of their intelligence concept from the "IQ test part"—what Stanovich (2009) termed "MAMBIT" (the mental abilities measured by intelligence tests). So, for instance, when Sternberg (2003) discusses high practical intelligence it can be translated to mean "optimal behavior in the domain of practical affairs," or when Gardner (1999) talks about high bodily-kinesthetic intelligence he means little more than high functioning in the bodily-kinesthetic domain. The word "intelligence" then becomes superfluous; it is there merely to add status to the domain in question. The strategy seems to be something like the following: Because intelligence is valued and we want bodily-kinesthetic talent to be valued too, we'll fuse the term "intelligence" onto it in order to transfer some of the value from intelligence to bodily-kinesthetic talent.

Social and Practical Implications

Indeed, this is why educators have been so enthusiastic about the "multiple intelligences" idea. Its scientific status is irrelevant to them. They use it as a motivational tool—to show that "everyone is intelligent in some way."

However, there are unintended consequences—some of them quite ironic—of this strategy, consequences that have been insufficiently appreciated. By inflating the word "intelligence," by associating it with more and more valued mental activities and behaviors, permissive theorists will succeed in doing just the opposite of what many of them intend—cutting "the IQ test part of intelligence" down to size. If you inflate the conceptual term "intelligence" you will inflate all its close associates as well—and one hundred years of mental testing makes it a simple historical fact that the closest associate of the term "intelligence" is "the IQ test part of intelligence."

Our strategy in this book has been different from that of the permissive theorists. The IQ tests have the label "intelligence," and thus MAMBIT will always be dominant in the folk psychology of intelligence. We would argue that it is a mistake to ignore this fact. Instead, our strategy is to open up some space for rationality in the lexicon of mental competencies. We have coherent and well-operationalized concepts of rational action and belief formation. We have a coherent and well-operationalized concept of MAMBIT. No scientific purpose is served by fusing these concepts, because they are very different. To the contrary, scientific progress is made by differentiating concepts.

The strategy of the permissive theorists ends up giving us the worst of all worlds. Short shrift is given to the concept of rationality because it is not separately named, but instead tends to be conflated with and lost within a bloated concept of intelligence. As a result, no imperative is created to actually *assess* rationality, because its semantic space has been gobbled up by the broadened view of intelligence. Although most people recognize that IQ tests do not encompass all of the important mental faculties, we often act (and talk) as if we have forgotten this fact. Where else does our surprise at smart people doing foolish things come from if not from the implicit assumption that rationality and intelligence should go together?

There is in fact some justification in the inertia of the psychometric establishment regarding changes in IQ tests and in the intelligence concept itself. Traditional intelligence research is a progressive research program in the sense that philosophers of science use that term. There is every indication that work in the traditional paradigm is carving nature at its joints.

First, the field has a consensus model in the form of the Cattell–Horn–Carroll theory of intelligence. Much work has gone into uncovering the cognitive subcomponents of fluid intelligence. We now know that there is substantial overlap in the variance in Gf and the variance in measures of working memory capacity. Importantly, the computational features of working memory have also been identified during the same period. The most critical insight has been that the central cognitive function tapped by working memory tasks is cognitive decoupling—the ability to manipulate secondary representations that do not track the world in one-to-one fashion as do primary representations (Stanovich, 2011).

Cognitive decoupling appears to be the central cognitive operation accounting for individual differences in Gf, and, because of its role in simulation and hypothetical thinking, cognitive decoupling is a crucial mental capacity. Thus, traditional intelligence tests converge on something important in mental life. They represent the fruits of a scientific research program that is progressively carving nature at an appropriate and important joint. Nevertheless, cognitive decoupling as measured on these tests is still a property of the algorithmic mind that is assessed under maximal rather than typical conditions. Such measures do not assess how typical it is for a person to engage in decoupling operations. They do not assess the tendency to engage in hypothetical thinking to aid problem solving. The ability to sustain cognitive decoupling does not guarantee rationality of behavior or thought.

With our comprehensive assessment of rational thinking, we aim to draw more attention to the skills of rational thought by measuring them systematically and by examining the correlates of individual differences in these cognitive skills. Our strategy is to press the implications of a grounded view of intelligence by constructing a formal assessment device, the CART, to measure the skills that IQ testing has largely ignored. Indeed, scientifically oriented psychometricians have been remarkably incurious about the tasks discussed in this book (particularly those tasks from the heuristics and biases tradition). We say remarkably because these tasks have been ubiquitous in cognitive psychology for decades now, have captivated the public via several popular books (Ariely, 2008; Kahneman, 2011; Thaler, 2015; Thaler & Sunstein, 2008), and have won the Nobel Prize for a fellow psychologist (Daniel Kahneman).

Social and Practical Implications

Psychometricians have been focused on the most traditional of cognitive tasks that make up standard intelligence tests, and have been exceptionally reluctant to explore the cognitive landscape that is carved out by the heuristics and biases tradition. Our initial work on these individual differences was inspired by what we thought then was a startling lacuna. After twenty-plus years we have finally managed to fill the gap more comprehensively with the CART—what some will no doubt label an RQ test. We do not recoil from the moniker because we feel that the cognitive domain is at least as well demarcated by our rationality test as by an IQ test. The CART explicitly taps knowledge related to rational action and belief; an IQ test does not. The CART explicitly taps the reflective mind; an IQ test does this much less so. In earlier chapters, we theoretically grounded the CART in the practicalities of goal fulfillment and epistemic tracking of the world; IQ tests lack such theoretical grounding.

Given our earlier arguments against misinformed critics of IQ tests, no one should mistake our position by confusing it with that of what we might call "IQ deniers." To the contrary, intelligence is one of the most well-developed scientific concepts in all of psychology. But its very ubiquitousness might have blinded us to cognitive concepts that are not so muddled in folk language, such as rationality.

Cognitive Assessment for a Hostile World

In earlier chapters we described how certain features of heuristics and biases tasks are often criticized when in fact these very features are the things that make heuristics and biases tasks psychologically interesting and highly diagnostic of the dynamics of human reasoning. For example, the fact that many heuristics and biases tasks can be construed by the subject in different ways (a statistical interpretation versus a narrative interpretation, for instance) is often seen as a weakness of such tasks when in fact it is the design feature that makes the task diagnostic. In a probabilistic reasoning task from this literature, the entire *point* is to see how dominant or nondominant the statistical interpretation is over the narrative interpretation. Likewise, the fact that many such problems have an intuitively compelling wrong answer (the problems in the Reflection versus Intuition subtest, for example) is often seen as a misleading attempt to "trick" the participant.

In fact, the presence of the compelling intuitive response is precisely what makes the problem diagnostic of the propensity toward reflective thinking.

It is true that heuristics and biases problems seem more hostile than typical IQ test problems in that the latter do not contain enticing lures toward an incorrect response. Neither is the construal of an IQ test item left up to the subject. Instead, the instructions to an IQ test item attempt to remove ambiguity in a way that is not true of a heuristics and biases problem (the Linda conjunction problem would be a prime case in point). We used the term "hostile" deliberately, to call to mind our earlier discussions of hostile and benign environments for the use of Type 1 processing. A benign environment contains useful cues that can be exploited by various heuristics. Additionally, for an environment to be classified as benign, it also must contain no other individuals who will adjust their behavior to exploit those relying only on heuristics. In contrast, a hostile environment for heuristics is one in which there are no cues that are usable by heuristic processes or that contains other agents who arrange the cues for their own advantage.

IQ tests assess the algorithmic power of the mind in benign environments. The issue of rationality raises additional questions of how the *reflective* mind is used when the environment turns hostile. Our argument in this section will be that the modern world creates many environments for thinking that are hostile rather than benign. IQ tests do not pick up these hostile aspects of the cognitive environment of modernity. This is because they assess capabilities rather than propensities, unlike many of the CART subtests (Reflection versus Intuition, Probabilistic Reasoning, Scientific Reasoning), which assess propensities as much as capabilities. Instead, the demand characteristics of IQ tests pretty much guarantee that the subject will be working at maximum efficiency on a problem that is unambiguously framed by its instructions—unlike heuristics and biases tasks, where the subject must often *choose* a particular construal.

It is in this sense that the so-called artificiality of heuristics and biases tasks is a strength, not a weakness. It is a design feature, not a bug. Why? Because, as we will argue, the modern world is, in many ways, becoming hostile for individuals who rely solely on Type 1 processing. This design feature of many tasks on the CART will thus be assessing (1) something of growing importance in the modern world, and (2) something not assessed by IQ tests.

Social and Practical Implications

As noted in our discussion of the Great Rationality Debate in chapter 14, the Panglossian theorists have shown us that many reasoning errors might have an evolutionary or adaptive basis. But the Meliorist perspective on this is that the modern world is increasingly changing so as to render those responses less than instrumentally rational for an individual. In short, the requirements for rationality are becoming more stringent as modern technological societies develop. Decision scientists Hillel Einhorn and Robin Hogarth long ago made the telling point that "in a rapidly changing world it is unclear what the relevant natural ecology will be. Thus, although the laboratory may be an unfamiliar environment, lack of ability to perform well in unfamiliar situations takes on added importance" (1981, p. 82).

Critics of the abstract content of most laboratory tasks and standardized tests have been misguided on this very point. Evolutionary psychologists have singularly failed to understand the implications of Einhorn and Hogarth's warning. They regularly bemoan the "abstract" problems and tasks in the heuristics and biases literature and imply that since these tasks are not like "real life" we need not worry that people do poorly on them. The issue is that, ironically, the argument that the laboratory tasks and tests are not like "real life" is becoming less and less true. "Life," in fact, *is becoming more like the tests.*

Try arguing with your health insurer about a disallowed medical procedure, for example. In such circumstances, we invariably find out that our personal experience, our emotional responses, our Type 1 intuitions about social fairness—are all worthless. All are for naught when talking on the phone to the representative looking at a computer screen displaying a spreadsheet with a hierarchy of branching choices and conditions to be fulfilled. The social context, the idiosyncrasies of individual experience, the personal narrative—the "natural" aspects of Type 1 processing—all are abstracted away as the representatives of modern technological-based services attempt to "apply the rules."

Unfortunately, the modern world tends to create situations where the default values of evolutionarily adapted cognitive systems are not optimal. Modern technological societies continually spawn situations where humans must decontextualize information—where they must deal abstractly and in a depersonalized manner with information rather than in the context-specific way of the Type 1 processing modules discussed by evolutionary psychologists. The abstract tasks studied by the heuristics and biases researchers often accurately capture this real-life conflict. In short, the *requirements* for rationality are often more stringent in the modern world than they were in the environment of evolutionary adaptation. This puts a premium on the use of Type 2 processing capacity to override Type 1 responses. Likewise, market economies contain agents who will exploit automatic Type 1 responding for profit (better buy that "extended warranty" on a $150 electronic device!). This again puts a premium on overriding Type 1 responses that will be exploited by others in a market economy.

Of course, we do not mean to suggest that the use of heuristics always leads us astray. As previously discussed, they often give us a useful first approximation to the optimal response in a given situation, and they do so without stressing cognitive capacity. In fact, they are so useful that influential psychologists extol their advantages even to the extent of minimizing the usefulness of the formal rules of rationality (Brandstatter, Gigerenzer, & Hertwig, 2006; Gigerenzer, 2002, 2007; Todd & Gigerenzer, 2000, 2007). Most psychologists, though, while still acknowledging the usefulness of heuristics, think that this view carries things too far. The reason is that the usefulness of the heuristics that we rely on to lighten the cognitive load depends on a benign environment.

Take as an example the so-called recognition heuristic (Gigerenzer & Todd, 1999). The idea behind such "ignorance-based decision making," as it is called, is the fact that some items of a subset that are unknown can be exploited to aid decisions. In short, the yes/no recognition response can be used as an estimation cue (Goldstein & Gigerenzer, 1999, 2002; Todd & Gigerenzer, 2007). With ingenious simulations, Gigerenzer and colleagues have demonstrated how certain information environments can lead to such things as "less-is-more" effects: where those who know less about an environment can display more inferential accuracy in it. One is certainly convinced after reading material like this that the recognition heuristic is certainly efficacious in some situations. But then one immediately begins to worry when one ponders how it relates to a market environment specifically designed to exploit it. If a person were to rely solely on the recognition heuristic as he went about his day tomorrow, he could easily be led to:

1. buy a $3 coffee when in fact a $1.25 one would satisfy him perfectly well
2. eat a single snack with the amount of fat grams he should have in an entire day

3. pay the highest bank fees
4. incur credit card debt rather than pay cash
5. buy a mutual fund with a 6 percent sales charge rather than a no-load fund

None of these behaviors serves the long-term goals of most people, even though they are triggered by the most recognized stimuli in dense urban environments. However useful the recognition heuristic is in other contexts, it is problematic in these examples. The commercial environment of a modern city is not a benign environment for a cognitive miser.

The danger of such miserly tendencies and the necessity of relying on Type 2 processing in the domain of personal finance is suggested by the well-known finding that consumers of financial services overwhelmingly purchase high-cost products that underperform in terms of investment return when compared to the low-cost strategies recommended by true experts (e.g., dollar-cost averaging into no-load index mutual funds; see Bazerman, 2001). The reason is, of course, that the high-cost fee-based products and services are the ones with high immediate recognizability in the marketplace.

Experimental studies of choice indicate that errors due to insufficient monitoring of Type 1 responses are probably made all the time. Neumann and Politser (1992) describe a study in which people were asked to choose between two insurance policies. Policy A had a $400 yearly deductible and a cost of $40 per month. Policy B had no deductible and a cost of $80 a month. A number of subjects preferred policy B because of the certainty of never having to pay a deductible if an accident occurs. However, it takes nothing more than simple arithmetic to see that people choosing policy B have fallen prey to a Type 1 tendency to avoid risk and seek certainty. Even if an accident occurs, policy B can never cost less than policy A. This is because paying the full deductible ($400) plus the monthly fee for 12 months ($480) would translate into a total cost of $880 for policy A, whereas the monthly fee of policy B for 12 months amounts to $960. Thus, even if accidents cause the maximum deductible to be paid, policy A costs less. An automatic reaction triggered by a logic of "avoid the risk of large losses" biases responses against the more economical policy A.

Modern mass communication technicians have become quite skilled at exploiting Type 1 processing defaults. These defaults are exploited by advertisers, in election campaigns, and even by governments—for example, in promoting their lottery systems. "You could be the one!" blares an ad from the Ontario Lottery Commission—thereby increasing the availability of an outcome that, in the game called 6/49, has an objective probability of 1 in 14 million.

In short, if we rely solely on Type 1 processing, we literally do not have "a mind of our own." The response of the Type 1 processor is determined by the most vivid stimulus at hand, the most readily assimilated fact, or the most salient cue available. This tendency can be easily exploited by those who control the labeling, who control what is vivid, and who control the framing. To the extent that modern society increasingly requires the Type 1 computational biases to be overridden, Type 2 overrides will be more essential to personal well-being.

The Internet has created an environment where those who spend a good part of their lives on social media must vigilantly exercise their inhibitory abilities. There are by now countless examples of how a momentary emotional lapse on these media can get propagated to thousands of people in ways that have consequences for an entire lifetime (Crane, 2012; Das & Sahoo, 2011; Ronson, 2015; Solove, 2007).

Consider the Belief Bias in Syllogistic Reasoning subtest. These syllogisms may seem to be the epitome of the "toy problem" criticism of heuristics and biases tasks. However, they are in fact indexing a cognitive skill of increasing importance in modern society—the ability to reason from the information given and at least temporarily put aside what we thought before we had received new information (Stanovich, 2004). Here, we are not arguing that it is always better to ignore what you know. Obviously, most of the time we are better served by bringing to bear all the prior knowledge we can in order to solve a problem. We are simply pointing to the fact that modernity is creating more and more situations where such unnatural decontextualization is required. The science on which modern technological societies is based often requires "ignoring what we know or believe." Testing a control group when you fully expect it to underperform compared to an experimental group is a form of ignoring what you believe. Science is a way of systematically ignoring what we know, at least temporarily (during the test), so that we can recalibrate our belief after the evidence is in. Likewise, many aspects of the contemporary legal system put a premium on detaching prior belief and world knowledge from the process of evaluating evidence. Modernity increasingly requires decontextualizing in the form of

Social and Practical Implications 327

stripping away what we personally "know" by its emphasis on such characteristics as fairness, rule-following despite context, even-handedness, sanctioning of nepotism, unbiasedness, universalism, inclusiveness, and legally mandated equal treatment. That is, all of these requirements of modernity necessitate overriding personalized knowledge.

Evolutionary psychologists have tended to minimize the importance of the requirements for decontextualizing and abstraction in modern life (the "unnaturalness" of the modern world that in fact matches the "unnaturalness" of many laboratory tasks). For example, Tooby and Cosmides (1992) use the example of how our color constancy mechanisms fail under modern sodium vapor lamps; they warn that "attempting to understand color constancy mechanisms under such unnatural illumination would have been a major impediment to progress" (p. 73)—a fair enough point. But our purpose here is to stress a corollary issue. The point is that if the modern world were structured such that making color judgments under sodium lights was critical to one's well-being, then this would be troublesome for us because our evolutionary mechanisms have not naturally equipped us for this. In fact, humans in the modern world are in just this situation vis-à-vis the mechanisms needed for fully rational action in highly industrialized and bureaucratized societies.

The cognitive equivalent of the sodium vapor lamps are the probabilities we must deal with; the causation we must infer from knowledge of what might have happened; the vivid advertising examples we must ignore; the unrepresentative sample we must disregard; the favored hypothesis we must not privilege; the rule we must follow that dictates we ignore a personal relationship; the narrative we must set aside because it does not square with the facts; the pattern that we must infer is not there because we know a randomizing device is involved; the sunk cost that must not affect our judgment; the judge's instructions we must follow despite their conflict with common sense; the contract we must honor despite its negative effects on a relative; the professional decision we must make because we know it is beneficial in the aggregate even if unclear in a given case. These are all the "sodium vapor lamps" that modern society presents to our cognitive apparatus—and if evolution has not prepared us to deal with them, so much the worse for our rational behavior in the modern world.

Thus, the long-standing debate between the Panglossians and the Meliorists can be viewed as an issue of figure and ground reversal. It is possible to accept most of the conclusions of the work of Panglossian theorists while drawing completely different morals from them. For example, evolutionary psychologists want to celebrate the astonishing job that evolution did in adapting the human cognitive apparatus to the Pleistocene environment. Certainly they are right to do so. The more we understand about evolutionary mechanisms, the more awed appreciation we have for them. But at the same time, it is not inconsistent for a person to be horrified that a multimillion dollar advertising industry is in part predicated on creating stimuli that will trigger Type 1 processing heuristics that many of us will not have the disposition to override. To Meliorists, it is no great consolation that the heuristics so triggered were evolutionarily adaptive in their day.

Evolutionary psychologists have shown that some problems can be more efficiently solved if represented in a way that coincides with how various brain modules represent information ("when people are given information in a format that meshes with the way they naturally think about probability, they can be remarkably accurate" Pinker, 1997, p. 351). The Meliorist cautions, however, that the world will not always let us deal with representations that are optimally suited to our evolutionarily designed cognitive mechanisms. We are living in a technological society where we must decide which health maintenance organization to join based on abstract statistics rather than experienced frequencies; decide on what type of mortgage to purchase; figure out what type of deductible to get on our auto insurance; decide whether to trade in a car or sell it ourselves; decide whether to lease or to buy; and think about how to apportion our retirement funds—to simply list a random set of the plethora of modern-day decisions and choices. And we must make all of these decisions based on information represented in a manner for which our brains are not adapted. To reason rationally in all of these domains (to maximize our personal utility), we are going to have to deal with probabilistic information represented in nonfrequentistic terms—in representations that the evolutionary psychologists have shown are different from our adapted algorithms for dealing with frequency information. As all of the examples that we have discussed here show, increasingly, the modern world is a hostile environment for an uncritical reliance on Type 1 processing. The CART is a cognitive test designed to capture how well prepared people are to deal with that hostile world.

Now that we have the CART, we could, *in theory*, begin to assess rationality as systematically as we do IQ. If not for professional inertia and psychologists' investment in the IQ concept, we could choose tomorrow to more formally assess rational thinking skills, focus more on teaching them, and redesign our environment so that irrational thinking is not so costly. Whereas just thirty years ago we knew vastly more about intelligence than we knew about rational thinking, this imbalance has been redressed in the last few decades because of some remarkable work in behavioral decision theory, cognitive science, and related areas of psychology. In the past two decades, cognitive scientists have developed laboratory tasks and real-life performance indicators to measure rational thinking tendencies such as sensible goal prioritization, reflectivity, and the proper calibration of evidence. People have been found to differ from each other on these indicators. These indicators are structured differently from the items used on intelligence tests. We have brought this work together here by producing the first comprehensive assessment measure for rational thinking, the CART.

Appendix: Structure and Sample Items for the Subtests and Scales of the Comprehensive Assessment of Rational Thinking

© 2015 Keith E. Stanovich and Richard F. West. All rights reserved.

CART points allocated to each subtest

CART Subtest	CART Points
Probabilistic and Statistical Reasoning	18
Scientific Reasoning	20
Reflection versus Intuition Subtest	10
Belief Bias in Syllogistic Reasoning	8
Ratio Bias	5
Disjunctive Reasoning	5
Framing	6
Anchoring	3
Preference Anomalies	3
Argument Evaluation Test	5
Knowledge Calibration	6
Rational Temporal Discounting	7
Probabilistic Numeracy	9
Financial Literacy and Economic Knowledge	10
Sensitivity to Expected Value	5
Risk Knowledge	3
Rejection of Superstitious Thinking	5
Rejection of Antiscience Attitudes	5
Rejection of Conspiracy Beliefs	10
Avoidance of Dysfunctional Personal Beliefs	5
Total CART Points	148

Subtest: Probabilistic and Statistical Reasoning

CART points: 18
Full Form: yes
Short Form: yes

Structure: The eighteen items on the Probabilistic and Statistical Reasoning subtest of the CART were distributed as follows: four items assess the ability to avoid probability matching tendencies and instead choose a maximizing strategy; five items assess the ability to avoid the gambler's fallacy; four items assess the ability to properly assign probabilities to conjunctions; three items assess sensitivity to base rates; one item assesses sensitivity to sample size considerations; and one item assesses the ability to see regression to the mean as an explanation of performance changes. The CART score is simply the number of items correct.

Presentation: This subtest is presented as a block, but within the block all the types of items are intermixed except for the conjunction problems which are presented together.

Sample Items

Probability Matching Consider the following hypothetical situation: A deck with 10 cards is randomly shuffled 10 separate times. The 10 cards are composed of 7 cards with the letter "A" on the down side and 3 cards with the letter "B" on the down side. Each time the 10 cards are reshuffled, your task is to predict the letter on the down side of the top card. Imagine that you will receive $100 for each downside letter you correctly predict, and that you want to earn as much money as possible. Indicate your predictions for each of the 10 shuffles:

I would predict _____ for Shuffle #1 A or B?
I would predict _____ for Shuffle #2 A or B?
I would predict _____ for Shuffle #3 A or B?
I would predict _____ for Shuffle #4 A or B?
I would predict _____ for Shuffle #5 A or B?
I would predict _____ for Shuffle #6 A or B?
I would predict _____ for Shuffle #7 A or B?
I would predict _____ for Shuffle #8 A or B?
I would predict _____ for Shuffle #9 A or B?
I would predict _____ for Shuffle #10 A or B?

Appendix 333

Scoring for this item: Score of 1 for 10 choices of A and a score of 0 for < 10 choices of A.

Gambler's Fallacy Item When playing slot machines, people win something about 1 in every 10 times. Nancy, however, has just won on her first three plays. What are her chances of winning the next time she plays?

a. She has better than 1 chance in 10 of winning on her next play.
b. She has less than 1 chance in 10 of winning on her next play.
*c. She has a 1 chance in 10 that she will win on her next play.

Scoring for this item: Score of 1 for choice c and score of 0 for choices a and b.

Avoiding Conjunction Fallacy Problem
a. What is the probability that you will ***not*** have root canal surgery on one tooth in the ***next year***? (percent)
[enter value between 0 and 100]
b. What is the probability that you will have root canal surgery on one tooth in the ***next five years***? (percent)
[enter value between 0 and 100]
c. What is the probability that you will have root canal surgery on one tooth and another tooth extracted in the ***next five years***? (percent)
[enter value between 0 and 100]
d. What is the probability that you will have root canal surgery on one tooth in the ***next year***? (percent)
[enter value between 0 and 100]
e. What is the probability that you will *not* have root canal surgery on one tooth in the ***next five years***? (percent)
[enter value between 0 and 100]

Scoring for this item: Root canal in five years (b) *minus* root canal surgery on one tooth and another tooth extracted in five years (c) > 0 scored as 1.

Causal Base Rate Item Professor Kellan, the director of a teacher preparation program, was designing a new course in human development and needed to select a textbook for the new course. She had narrowed her decision down to one of two textbooks: one published by Pearson and the other published by McGraw. Professor Kellan belonged to several professional organizations that provided Web-based forums for its members to share information about curricular issues. Each of the forums had a textbook evaluation section, and the websites unanimously rated the McGraw textbook as the better choice in every category rated. Categories evaluated included quality of the writing, among others. Just before Professor Kellan was about to place the order for the McGraw book, however, she asked an experienced colleague for her opinion about the textbooks. Her colleague reported that she preferred the Pearson book. What do you think Professor Kellan should do?

a. She should definitely use the Pearson textbook.
b. She should probably use the Pearson textbook.
*c. She should probably use the McGraw textbook.
*d. She should definitely use the McGraw textbook.

Scoring for this item: Score of 1 for either c or d, and score of 0 for a or b.

Regression to the Mean Item After the first three weeks of the Little League baseball season in Wichita, Kansas, the adult managers begin to post the top batting averages. Typically, after the first three weeks, several batters often have averages over .500 (i.e., they have gotten hits in over 50 percent of their at bats). However, no batter in Wichita Little League history has ever averaged over .500 at the end of the season. Why do you think this is?

a. When a batter is hot early in the season, the pitchers know it and concentrate on getting him out more later in the season.
b. Pitchers tend to get better over the course of a season, so every batter's average goes down.
*c. A player's high average at the beginning of the season may be just luck. The longer season provides a more realistic test of a batter's skill.
d. A batter who has such a hot streak at the beginning of the season is under a lot of stress to maintain his performance record. Such stress adversely affects his playing.
e. Over the season, opposing coaches devise strategies to get the best hitters out.

Scoring for this item: Score of 1 for choice c and score of 0 for all other choices.

Appendix

Subtest: Scientific Reasoning

CART points: 20
Full Form: yes
Short Form: yes

Structure: There are seventeen single-point items to this subtest, and another three points are determined from performance on a twenty-five-item measure of covariation detection. The seventeen single-point items are distributed as follows: four items that tap falsification tendencies in the four-card selection task (two deontic and two nondeontic); two items that tap knowledge of the logic of converging evidence; three items that tap the tendency to avoid drawing causal inferences from correlation evidence; five items that tap the tendency to accurately assess the likelihood ratio by processing $P(D/\sim H)$; and three problems that tap the tendency to use control-group reasoning. The remaining three points on the subtest are derived from a twenty-five-item measure of covariation detection ability.

Presentation: This subtest is presented as a block, with some items grouped by type and others not. The converging evidence items are presented together, as are the control group reasoning items and the twenty-five-item measure of covariation detection ability. The correlation/causation items and the selection task items are interspersed throughout the subtest (the two deontic items appearing before the two nondeontic items). The items tapping likelihood ratio processing were presented in two groups.

Sample Items
Falsifiability Tendencies—Selection Task Item
[nondeontic item]

Each of the tickets below has a destination on one side and a mode of travel on the other side. Here is a rule: "If 'Baltimore' is on one side of the ticket, then 'plane' is on the other side of the ticket." Your task is to decide which tickets you would need to turn over in order to find out whether the rule is being violated.

Destination: Baltimore		Destination: Washington, DC		Mode of Travel: Train		Mode of Travel: Plane	
Turn	Do not Turn	Turn	Do not Turn	Turn	Do not Turn	Turn	Do not Turn
○	○	○	○	○	○	○	○

Scoring for this item: Calculate Pollard and Evans (1987) Logic Index. Number of correct cards turned minus number of incorrect turns:

$P + NQ - NP - Q$

P (Baltimore); NP (Washington); NQ (train); Q (plane)

Logic index > 0 scored as 1

Logic index ≤ 0 scored as 0

Converging Evidence Item Alice had been experiencing unpleasant digestive problems. She hypothesized that she had developed an allergy to either milk, eggs, wheat, nuts, or shellfish, because she never experienced any digestive problems on days when she did not eat any of these items.

In an effort to determine which one of these foods was the source of her digestive problems, Alice conducted two tests. Each test started with a day in which Alice drank water but ate no food, and was followed by a day in which she ate only a single meal that varied in which of the suspected food items it contained.

Here is what Alice ate during her meal on the second day of each test:

	Milk	Eggs	Wheat	Nuts	Shellfish	Digestive Problems
Test 1	yes	yes	yes	no	no	yes
Test 2	no	no	yes	yes	yes	yes

How likely is it that Alice's digestive problems resulted from eating:

milk _____ percent likely

eggs _____ percent likely

wheat _____ percent likely

nuts _____ percent likely

shellfish _____ percent likely

[percentages must add to 100]

Scoring for this item:

Estimates for wheat ≥ 80 = Score of 1

Estimates for wheat < 80 = Score of 0

Appendix

Limits of Correlational Relationships Item Researchers have found that teenagers who smoke cigarettes tend to have lower IQ scores than teenagers who do not smoke cigarettes.

This finding means that preventing teenagers from smoking would tend to raise their IQ scores:

a. Yes
*b. No
*c. You cannot tell

Scoring for this item: Choice b and c scored as 1, and choice of a scored as 0.

Diagnostic Hypothesis Testing
Processing P(D/~H) of the Likelihood Ratio [Sample Item 1] Imagine you are a special education teacher. Talia, a student of yours, has clubbing of her fingers (significant thickening of the end of her fingers). What information would you want in order to estimate the probability that Talia has "Fustis Digitus Syndrome"? Below are 4 pieces of information that may or may not be relevant to determining the probability. Please indicate all of the pieces of information that are necessary to determine the probability, but only those pieces of information that are necessary to do so.

a. % of people without Fustis Digitus Syndrome who have clubbing of their fingers
b. % of people with Fustis Digitus Syndrome
c. % of people without Fustis Digitus Syndrome
d. % of people with Fustis Digitus Syndrome who have clubbing of their fingers

a = [DNH]; b = [H]; c = [NH]; d = [DH]

Scoring for this item: Score 1 point for choosing both parts of the likelihood ratio (a and d) only *or* in addition to any other cards, and score 0 otherwise.

Diagnostic Hypothesis Testing
Processing P(D/~H) of the Likelihood Ratio [Sample Item 2] Imagine yourself meeting Calvin Dean. Your task is to assess the probability that he is an accountant based on some information that you will be given. This will be done in two steps. At each step you will get some information that you may or may not find useful in making your assessment. After each piece of information you will be asked to assess the probability that Calvin Dean is an accountant. In doing so, consider all the information you have received to that point if you consider it to be relevant.

Your probability assessments should be numbers between 0 and 1 that express your degree of belief. 1 means "I am absolutely certain that he is an accountant. 0 means "I am absolutely certain he is not an accountant." 0.65 means "The chances are 65 out of 100 that he is an accountant," and so forth. You can use any number between 0 and 1, for example, 0.15, 0.95, etc.

Step 1: You are told that Calvin Dean attended a party in which 30 male accountants and 70 male doctors took part, 100 people all together.

Question: What do you think the probability is that Calvin Dean is an accountant? ____

Step 2: You are told that Calvin Dean is a member of the Kiwanis Club. 75% of the male accountants at the above mentioned party were members of the Kiwanis Club. 95% of the male doctors at the party were members of the Kiwanis Club.

Question: What do you think the probability is that Calvin Dean is an accountant? ____

Scoring for this item:

Step 1 correct answer (0.30) is not scored. The Bayesian posterior after Step 2 is 0.253.

Step 1 minus Step 2 difference ≤ 0 scored as 0

Step 1 minus Step 2 difference > 0 scored as 1

Step 2 answers of zero are *also* scored as 0

Control Group Reasoning Centerville High School has had an unpopular principal for the past 2 years. She is a friend of the superintendent and had no previous experience as a school administrator when she was appointed. The superintendent has recently defended the principal in public by announcing that in the time since she became principal, truancy rates at the high school have decreased by 12%.

Which of the following pieces of evidence would most refute the superintendent's claim and instead show that the principal may not be doing a good job? Choose the *best answer*:

a. An independent survey of the teachers in the state where Centerville High School is located shows that 40% more truants are reported by survey respondents than is reported in official school records.

b. Common sense indicates that there is little a principal can do to lower truancy rates. These are for the most part due to social and economic conditions beyond the control of the schools.
*c. The truancy rates of the two cities closest to Centerville in location and size have decreased by 18% in the same period.
d. The superintendent has been discovered to have business contacts with people who are known to be involved in providing security services to the school district.

Scoring for this item: Choice c is scored as 1 and other choices are scored as 0.

Covariation Detection—25 items of the following type A researcher is interested in the relationship between self-esteem and leadership qualities. Imagine that this researcher sampled 450 individuals and found that:

100 people with high self-esteem were high in leadership qualities
50 people with high self-esteem were low in leadership qualities
100 people with low self-esteem were high in leadership qualities
200 people with low self-esteem were low in leadership qualities

[$\Delta p = 0.333$]

You could classify the data in the following way:

	High in leadership qualities	Low in leadership qualities
High self-esteem	100	50
Low self-esteem	100	200

Please judge the nature and extent of the relationship between self-esteem and leadership qualities in these data.

−10 −9 −8 −7 −6 −5 −4 −3 −2 −1 0 +1 +2 +3 +4 +5 +6 +7 +8 +9 +10
strong negative association   no association   strong positive association

The covariation detection task has a maximum of 3 points, scored as follows:

Correlation of evaluation with Δp across the 25 items:
< 0.10 scored as 0
≥ 0.10 and < 0.45 scored as 1
≥ 0.45 and < 0.75 scored as 2
≥ 0.75 scored as 3

Subtest: Reflection versus Intuition

CART points: 10
Full Form: yes
Short Form: yes
Structure: There are eleven items on this subtest that yield a maximum of 10 points.
Presentation: Items are presented as a block.

Sample Items

If it takes one minute to make each cut, how long will it take to cut a 25-foot wooden plank into 25 equal pieces? [correct answer = 24 minutes; intuitive response = 25 minutes] Score of 1 point for 24 minutes and 0 for all other responses.

The number of bacteria in a container doubles each hour. If it takes 32 hours to completely fill the container, how many hours would it take for the bacteria to fill half of the container? [correct answer = 31 hours; intuitive response = 16 hours] Score of 1 point for 31 hours and 0 for all other responses.

CART Scoring

The CART score on this task is equal to the number of items answered correctly, to a maximum of 10.

Subtest: Belief Bias in Syllogistic Reasoning

CART points: 8
Full Form: yes
Short Form: yes
Structure: There are sixteen items in this subtest, eight where the believability of the conclusion and logical validity are aligned (the consistent items) and eight where the believability of the conclusion and logical validity are in conflict (the inconsistent items).
Presentation: Items are presented as a block.

Appendix

Instructions and Sample Items

In the following problems, you will be given two premises *which you must assume are true*. You must decide whether the conclusion *necessarily follows logically* from the premises. It is important that you assume the premises to be true and ignore whether the conclusion is factually correct. Rate the conclusion only in terms of whether it *necessarily* follows.

Premises
Premise 1: All flowers are carbitops.
Premise 2: All tulips are carbitops.

Conclusion
All tulips are flowers.

a. Conclusion necessarily follows from premises.
*b. Conclusion does not necessarily follow from premises.
[item type: Inconsistent AAA2]

Premises
Premise 1: All nuts are bictodes.
Premise 2: No rocks are bictodes.

Conclusion
No rocks are nuts.

*a. Conclusion necessarily follows from premises.
b. Conclusion does not necessarily follow from premises.
[item type: Consistent AEE2]

CART Scoring

Raw scores on this task range from 0 to 16. Guessing yields a score of 8. A pure validity bias and a pure believability bias result in scores of 8. Raw scores are translated into CART points as follows:

A raw score of 16 is scored as 8 CART points.
A raw score of 15 is scored as 7 CART points.
A raw score of 14 is scored as 6 CART points.
A raw score of 13 is scored as 5 CART points.
A raw score of 12 is scored as 4 CART points.

A raw score of 11 is scored as 3 CART points.
A raw score of 10 is scored as 2 CART points.
A raw score of 9 is scored as 1 CART point.
A raw score of 8 or less is scored as 0 CART points.

Subtest: Ratio Bias

CART points: 5
Full Form: yes
Short Form: yes
Structure: There are fifteen items in this subtest: twelve scored items and three filler items (unscored) that are intermixed.
Presentation: Items are presented as a block.

Sample Items
The Marble Game: [Scored test item]   Assume that you are presented with two trays of black and white marbles (pictured below). The small tray contains 5 marbles. The large tray contains 100 marbles. The marbles inside each tray will be randomly mixed up, and you must draw out a single marble from one of the trays without looking. If you draw a black marble you win $5.

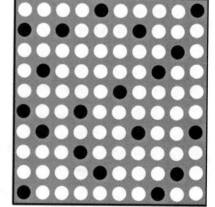

(1 black and 4 white)   (19 black and 81 white)

In a real situation, which tray would you prefer to select a marble from?
a. Strongly prefer the small tray
b. Moderately prefer the small tray
c. Slightly prefer the small tray
d. Slightly prefer the large tray
e. Moderately prefer the large tray
f. Strongly prefer the large tray

The Marble Game: [Filler item]   Assume that you are presented with two trays of black and white marbles (pictured below). The small tray contains **10** marbles. The large tray contains **100** marbles. The marbles inside each tray will be randomly mixed up, and you must draw out a single marble from one of the trays without looking. If you draw a black marble you win $5.

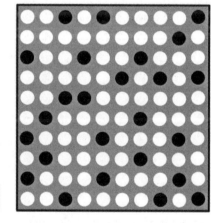

(1 black and 9 white)   (25 black and 75 white)

In a real situation, which tray would you prefer to select a marble from?
a. Strongly prefer the small tray
b. Moderately prefer the small tray
c. Slightly prefer the small tray
d. Slightly prefer the large tray
e. Moderately prefer the large tray
f. Strongly prefer the large tray

CART Scoring

The filler items, items number 2, 6, and 10 are not scored. The remaining items are scored as:

6 = Strongly prefer the small tray
5 = Moderately prefer the small tray
4 = Slightly prefer the small tray
3 = Slightly prefer the large tray
2 = Moderately prefer the large tray
1 = Strongly prefer the large tray

The twelve items are then summed to form a composite score. CART scoring is then as follows:

Summed composite scores > 61 are scored as 5 points.
Summed composite scores > 55 and ≤ 61 are scored as 4 points.
Summed composite scores > 50 to ≤ 55 are scored as 3 points.
Summed composite scores > 45 and ≤ 50 are scored as 2 points.
Summed composite scores > 40 and ≤ 45 are scored as 1 point.
Summed composite scores of 40 or less are scored as 0 points.

Subtest: Disjunctive Reasoning

CART points: 5
Full Form: yes
Short Form: yes
Structure: There are six items on this subtest, with the "Ann is married or unmarried" problem discussed in the text being the last of the six items to appear.
Presentation: These items are not presented in a block, as is the case with many other subtests of the CART. Instead, the items on this subtest are dispersed throughout the CART or throughout one of the longer subtests such as Scientific Reasoning.

Sample Item

A food warehouse ships boxes of fresh and frozen strawberries. Assume that your job is to make sure these boxes are stacked properly. A box of fresh strawberries should not touch a box of frozen strawberries, because direct contact with the colder box will cause the fresh strawberries to spoil. You find a stack of three boxes of strawberries, where the top box contains fresh strawberries and the bottom box contains frozen strawberries. However, the middle box of strawberries has no label, and, thus contains either fresh or frozen strawberries. Is a box of fresh strawberries touching a box of frozen strawberries?

Fresh Strawberries
?
Frozen Strawberries

*a. Yes
b. No
c. Cannot be determined

CART Scoring

Raw scores on this task range from 0 to 6. Raw scores are translated into CART points as follows:

A raw score of 6 is scored as 5 CART points.
A raw score of 5 is scored as 4 CART points.
A raw score of 4 is scored as 3 CART points.
A raw score of 3 is scored as 2 CART points.
A raw score of 2 is scored as 1 CART point.
Raw scores of 0 or 1 are scored as 0 CART points.

Subtest: Avoidance of Framing

CART points: 6
Full Form: yes
Short Form: no
Structure: The Avoidance of Framing subtest of the CART comprises eleven pairs of items. Seven of the pairs are designed to assess risky choice framing and four of the pairs are designed to assess attribute framing.
Presentation: The twenty-two items are presented in two blocks of eleven items—with many other subtests intervening between the two blocks. Each block of eleven items contains one item from each pair. When the positive frame of an attribute framing pair appears in one block, the negative frame appears in the other block. Likewise, when the gain frame of a risky-choice pair appears in one block, the loss frame appears

in the other block. Positive frames and gain frames appear roughly equally in each of the two blocks.

Sample Item Pair

Part a. [Attribute framing—positive] You are planning an upcoming trip that requires that you to take a flight. You are evaluating possible airlines and a particular airline that you are evaluating reports that their flights are on time 88% of the time.

How favorable do you find this particular airline?

___ 1. Very favorable.
___ 2. Favorable.
___ 3. Slightly favorable.
___ 4. Slightly unfavorable.
___ 5. Unfavorable.
___ 6. Very unfavorable.

Part b. [Attribute framing—negative] You are planning an upcoming trip that requires that you to take a flight. You are evaluating possible airlines and a particular airline that you are evaluating reports that their flights are late 12% of the time. How favorable do you find this particular airline?

How favorable do you find this particular airline?

___ 1. Very favorable.
___ 2. Favorable.
___ 3. Slightly favorable.
___ 4. Slightly unfavorable.
___ 5. Unfavorable.
___ 6. Very unfavorable.

CART Scoring

For each pair, the problems were scored by subtracting their negative frame ratings (loss in risky choice and negative frame in attribute framing) from their corresponding positive frame ratings (gain in risky choice and positive frame in attribute framing). The standard framing effects observed most commonly in the literature will yield negative values. Reverse framing effects will yield positive values. These positive values were all multiplied by –1 (+1 transformed to –1, +2 transformed to –2, etc.). The scores on the eleven items are then summed. The resulting negative composite sums are scored as follows:

Summed composite scores of 0 to –5 are scored as 6 points
Summed composite scores of –6 to –8 are scored as 5 points
Summed composite scores of –9 to –10 are scored as 4 points
Summed composite scores of –11 to –12 are scored as 3 points
Summed composite scores of –13 to –14 are scored as 2 points
Summed composite scores of –15 to –18 are scored as 1 point
Summed composite scores of –19 or less are scored as 0 points

Subtest: Avoidance of Anchoring

CART points: 3
Full Form: yes
Short Form: no
Structure: There are eight items in this subtest, each with a part a and part b that are presented together. There are four low anchors and four high anchors in the eight items.
Presentation: Items are presented as a block.

Sample Items
Correct answer is in brackets.

Part a. Is the distance from San Francisco to Hawaii more than 500 miles? [Low anchor]
a. Yes
b. No

Part b. What do you think the distance from San Francisco to Hawaii is? ___ miles [2387 miles]   Scoring for this item:
Score of 1 point if estimate is 1000 or more and 0 points if estimate is < 1000

Part a. Is the duration of Mars' orbit around Sun greater than 1500 days? [High anchor]
a. Yes
b. No

Part b. What do you think the duration of Mars' orbit around Sun is? ___ *days. [687 days]* Scoring for this item:

Score of 1 point if estimate is < 1000 and > 50 and 0 points if estimate is ≥ 1000 or ≤ 50.

CART Scoring
Summed raw scores of 5 to 8 are scored as 3 points
Summed raw scores of 3 and 4 are scored as 2 points
Summed raw score of 2 is scored as 1 point
Summed raw scores of 0 and 1 are scored as 0 points

Subtest: Avoidance of Preference Anomalies Subtest

CART points: 3
Full Form: yes
Short Form: no
Structure: The Avoidance of Preference Anomalies subtest of the CART comprises nine pairs of items. The total raw score on this subtest ranges from 0 to 10 because the omission bias item has a maximum of 2 points. The nine items on this subtest assess: the certainty effect in decision making; outcome bias; the undue weighting of the word "free" effect; omission bias; fairness reversals (two items) and "less-is-more" effects (three items).
Presentation: Items are presented in two widely separated blocks with all a versions in block 1 and all b versions in block 2.

Appendix

Sample Item Pair
Part a. Data indicate that 786 of every 10,000 tourists in a particular country are the victims of a serious crime. Rate the following statement: "It is extremely dangerous for tourists to travel in that country."
1 = Disagree strongly
2 = Disagree moderately
3 = Disagree slightly
4 = Agree slightly
5 = Agree moderately
6 = Agree strongly

Part b. Data indicate that 8.14% of the tourists in a particular country are the victims of a serious crime. Rate the following statement: "It is extremely dangerous for tourists to travel in that country."
1 = Disagree strongly
2 = Disagree moderately
3 = Disagree slightly
4 = Agree slightly
5 = Agree moderately
6 = Agree strongly

Scoring for this item:

Scored 1 if b minus a rating was ≥ 0
Scored 0 if b minus a rating was < 0

CART Scoring
Summed raw scores of 7 to 10 are scored as 3 points
Summed raw scores of 6 are scored as 2 points
Summed raw scores of 4 and 5 are scored as 1 point
Summed raw scores of 0 to 3 are scored as 0 points

Subtest: Argument Evaluation

CART points: 5
Full Form: yes
Short Form: no

Structure: The subject completes two separate sections: one assessing the degree of agreement with 23 focal propositions and another section in which the subject evaluates arguments relevant to the 23 propositions.

Presentation: The two parts of the subtest are separated by several other subtests.

Instructions and Sample Items

Part 1: Prior Opinion Instructions: Please indicate your degree of agreement or disagreement with the following beliefs.

15. Students should have a stronger voice than the general public in setting university policies.
A=Strongly Disagree; B=Disagree; C=Agree; D=Strongly Agree

21. Smoking should be banned in all enclosed public places.
A=Strongly Disagree; B=Disagree; C=Agree; D=Strongly Agree

Part 2: Evaluation Instructions: We are interested in your ability to evaluate counterarguments. First, you will be presented with a belief held by an individual named Dale. Following this, you will be presented with Dale's premise or justification for holding this particular belief. A Critic will then offer a counterargument to Dale's justification for the belief. (We will assume that the Critic's statement is factually correct.) Finally, Dale will offer a rebuttal to the Critic's counterargument. (We will assume that Dale's rebuttal is also factually correct.) You are to evaluate the *strength of Dale's rebuttal to the Critic's counterargument*, regardless of your feeling about the original belief or Dale's premise.

15. Dale's belief: Students should have a stronger voice than the general public in setting university policies.

Dale's premise or justification for belief: Because students are the ones who must ultimately pay the costs of running the university through tuition, they should have a stronger voice in setting university policies.

Critic's counterargument: Tuition covers less than one-half the cost of an education at most state universities (assume statement factually correct), so the taxpayers should have a stronger say in the policies.

Dale's rebuttal to Critic's counterargument: Because it is the students who are directly influenced by university policies (assume statement factually correct), they are the ones who should have the stronger voice.
[Experts' Rating = Very Weak]

Appendix

Indicate the strength of Dale's rebuttal to the **Critic's counterargument**:

1=Very Weak; 2=Weak; 3=Strong; 4=Very Strong

21. Dale's belief: Smoking should be banned in all enclosed public places.

Dale's premise or justification for belief: Smoking should be banned in all enclosed public places because even second-hand smoke poses a significant health risk to nonsmokers.

Critic's counterargument: Since many smokers already refrain from smoking in places where their second-hand smoke poses a health risk to others (assume statement factually correct), it is unnecessary to severely restrict smoking locations.

Dale's rebuttal to Critic's counterargument: While it may be true that many smokers are considerate, it is equally true that many smokers are not so considerate (assume statement factually correct). Banning smoking would be an effective way to ensure that many of us won't be subjected to the risks posed by second-hand smoke.

[Experts' Rating = Very Strong]

Indicate the strength of Dale's rebuttal to the **Critic's counterargument**:

1=Very Weak; 2=Weak; 3=Strong; 4=Very Strong

CART Scoring

A separate regression should be run for each participant with the twenty-three argument evaluation ratings as the criterion variable. The two predictor variables are the experts' rating of the twenty-three items and the subject's prior agreement with the twenty-three propositions. The regression for each subject results in two beta weights: one for the experts' rating and one for the subject's prior agreement. Only the former is scored for CART points. The beta weights for the experts' rating are scored as follows:

Beta weights for experts' rating ≥ 0.450 are scored as 5 points.
Beta weights for experts' rating < 0.450 and ≥ 0.350 are scored as 4 points.
Beta weights for experts' rating < 0.350 and ≥ 0.250 are scored as 3 points.
Beta weights for experts' rating < 0.250 and ≥ 0.150 are scored as 2 points.
Beta weights for experts' rating < 0.150 and > 0 are scored as 1 point.
Beta weights for experts' rating ≤ 0 are scored as 0 points.

Subtest: Knowledge Calibration

CART points: 6
Full Form: yes
Short Form: no
Structure: Part 1 consists of thirty-six knowledge calibration items using the two-choice method. It is followed by an aggregate estimate of performance on the thirty-six items. Part 2 consists of fifteen knowledge calibration items using the confidence interval method. It is followed by an aggregate estimate of performance on the fifteen items.
Presentation: The two parts are run consecutively.

Sample Items
Part 1: Two-choice Probability Estimation

General Knowledge Questions Directions: In this task we would like you to answer a series of questions on a variety of different topics. Please indicate which answer is correct and indicate how certain you are of your answer on the probability scale below:
[asterisk indicates the correct answer]
The Open Door Policy required that

*a. No nation could claim exclusive trading rights in China.
b. Reporters must be allowed to observe the effects of the Chinese Cultural Revolution.

100% chance that I answered correctly (I am certain)
90% chance that I answered correctly
80% chance that I answered correctly
70% chance that I answered correctly
60% chance that I answered correctly
50% chance that I answered correctly (I was just guessing)

Aggregate Estimation Question Of the 36 questions that you just answered, how many do you think you answered correctly? Remember, on multiple-choice items such as these, someone just guessing would expect to get 18 correct just by chance. How many do you think you answered correctly? _____

Appendix

Part 2: Confidence Interval Method

Estimation Task For each of the following items, your task is to make an estimation of some number, like "How old was President Obama when he was first elected?" However, instead of estimating an exact number, we ask that you give a range, such that you think there is a 90% chance that the correct answer lies somewhere in the range. That is, provide a low and high guess such that you are 90% sure that the correct answer falls between the two. In other words, give a range such that you would expect to be wrong only about one out of ten times.

[correct answer in brackets]

I am 90% confident that Elvis Presley's age at the time of his death was somewhere between ___ years and ___ years. [42]

Aggregate Estimation Question Out of all 15 fill-in-blank questions you just answered, for how many of the 15 questions do you think the answer will turn out to be within the interval you gave? _____

CART Scoring

Part 1, two-choice, item-by-item calibration:

Calculate the mean percentage confidence judgment minus the mean percentage correct.

Difference scores of 2% or less are scored as 2

Difference scores > 2% and < 10% are scored as 1

Difference scores > 10% are scored as 0

Part 1, two-choice, aggregate calibration:

Calculate the aggregate estimate minus the number correct.

Difference scores of 0 or less are scored as 1

Difference scores > 0 are scored as 0

Part 2, interval, item-by-item calibration:

Number of hits ≥ 9 are scored as 2

Number of hits ≥ 6 and ≤ 8 are scored as 1

Number of hits < 6 are scored as 0

Part 2, interval, aggregate calibration:

Calculate the aggregate estimate minus the number of hits.

Difference scores of 0 or less are scored as 1

Difference scores > 0 are scored as 0

CART score = total number of points on the two sections of part 1 and the two sections of part 2

Subtest: Rational Temporal Discounting

CART points: 7
Full Form: yes
Short Form: no
Structure: Part 1 of this subtest consists of two fixed-sequence titrations. The first is an ascending sequence involving a constant delayed reward of $100 in three months. Only those choices (twenty in total) involving immediate amounts of $90 and lower are scored. The second is a descending sequence involving a constant delayed reward of $2,000 in one year. Only those choices (eighteen in total) involving immediate amounts of $1600 and lower are scored.

Part 2 of the Temporal Discounting subtest is a thirty-one-item measure that asks participants to indicate the strength of their preference for either a smaller amount of money sooner or a larger amount of money later. The amounts and delays vary from item to item rather than being sequenced as in the titration method. Participants respond on a continuous scale rather than simply picking the smaller or larger amount. In twenty-six of the thirty-one items the delayed larger amount corresponded to a substantial annualized increase in value, which, on a simple interest basis would have resulted in value increases of between 44.7% and 346.7% if earned annually. Five of the items were filler items where the delayed larger amount corresponded to only relatively low percentage increases in value. These fillers were included so as to reduce response bias and demand characteristics favoring the large delayed reward. The filler items do not enter into the raw score.

Presentation: The two parts are run consecutively.

Appendix

Sample Items
Part 1

Temporal Discounting Staircase Increasing Imagine that you are offered a choice between receiving a specific amount of money now or $100 in 3 months. Indicate your preference for each of the following:

1) $1 now or $100 in 3 months
2) $2.50 now or $100 in 3 months
3) $5 now or $100 in 3 months
4) $7.50 now or $100 in 3 months
5) $10 now or $100 in 3 months
6) $15 now or $100 in 3 months
7) $20 now or $100 in 3 months
8) $25 now or $100 in 3 months
9) $30 now or $100 in 3 months
10) $35 now or $100 in 3 months
11) $40 now or $100 in 3 months
12) $45 now or $100 in 3 months
13) $50 now or $100 in 3 months
14) $60 now or $100 in 3 months
15) $65 now or $100 in 3 months
16) $70 now or $100 in 3 months
17) $75 now or $100 in 3 months
18) $80 now or $100 in 3 months
19) $85 now or $100 in 3 months
20) $90 now or $100 in 3 months
21) $92.50 now or $100 in 3 months
22) $95 now or $100 in 3 months
23) $97.50 now or $100 in 3 months
24) $99 now or $100 in 3 months
25) $99.50 now or $100 in 3 months

Temporal Discounting Staircase Decreasing: Imagine that you are offered a choice between receiving a specific amount of money now or $2,000 in 1 year. Indicate your preference for each of the following:

1) $1,990 now or $2,000 in 1 year
2) $1,980 now or $2,000 in 1 year

3) $1,950 now or $2,000 in 1 year
4) $1,900 now or $2,000 in 1 year
5) $1,850 now or $2,000 in 1 year
6) $1,800 now or $2,000 in 1 year
7) $1,700 now or $2,000 in 1 year
8) $1,600 now or $2,000 in 1 year
9) $1,500 now or $2,000 in 1 year
10) $1,400 now or $2,000 in 1 year
11) $1,300 now or $2,000 in 1 year
12) $1,200 now or $2,000 in 1 year
13) $1,000 now or $2,000 in 1 year
14) $900 now or $2,000 in 1 year
15) $800 now or $2,000 in 1 year
16) $700 now or $2,000 in 1 year
17) $600 now or $2,000 in 1 year
18) $500 now or $2,000 in 1 year
19) $400 now or $2,000 in 1 year
20) $300 now or $2,000 in 1 year
21) $200 now or $2,000 in 1 year
22) $150 now or $2,000 in 1 year
23) $100 now or $2,000 in 1 year
24) $50 now or $2,000 in 1 year
25) $20 now or $2,000 in 1 year

Part 2

Temporal Discounting Mixed Questionnaire   Directions: For the next set of items, imagine that you are offered a choice between receiving a specific amount of money sooner or a larger amount later. Your choice would probably depend on how much greater the later amount is, and how long you would have to wait to get the larger amount. For example, you probably would prefer receiving $500 right now rather than receiving $501 in 12 months. You also would prefer receiving $500 in 1 week rather than receiving $25 right now.

If you had a choice, would you prefer $340 now or $400 in 4 months?

a. Very strongly prefer $340 now
b. Strongly prefer $340 now
c. Prefer $340 now

Appendix

d. Prefer $400 in 4 months
e. Strongly prefer $400 in 4 months
f. Very strongly prefer $400 in 4 months

CART Scoring

Part 1, $100 staircase increasing:

Only the first twenty items are scored, 1 each for choosing the delayed reward.
Summed raw scores of 20 scored as 2 CART points
Summed raw scores of 16–19 scored as 1 CART point
Summed raw scores of 15 or less scored as 0 CART points

Part 1, $2000 staircase decreasing:

Only the last eighteen items are scored, 1 each for choosing the delayed reward.
Summed raw scores of 18 scored as 2 CART points
Summed raw scores of 13–17 scored as 1 CART point
Summed raw scores of 12 or less scored as 0 CART points

Part 2, mixed temporal discounting questionnaire:

Only the twenty-six nonfiller items are scored.
Scale is scored: f = 6, e = 5, d = 4, c = 3, b = 2, a = 1.
Summed raw scores of 97 and above scored as 3 points
Summed raw scores > 86 and < 97 are scored as 2 points
Summed raw scores > 72 and ≤ 86 are scored as 1 point
Summed raw scores ≤ 72 are scored as 0 points

CART score = total number of points on the two sections of part 1 and part 2

Subtest: Probabilistic Numeracy

CART points: 9
Full Form: yes
Short Form: yes
Structure: There are nine items in this subtest. The raw number of correct on this subtest is also the number of points on the CART.
Presentation: Items are presented as a block.

Sample Items

For a person over age 60, the chances of getting shingles at some time in their life is 0.06. Out of 30,000 people, over age 60, about how many of them can be expected to get shingles? _____ [1,800]

Imagine that an unvaccinated person has a 10% chance of getting the flu and that the flu vaccine is 80% effective in preventing the flu. What are the chances that a person who has had the vaccine will still get the flu? _____ [2%]

Subtest: Financial Literacy and Economic Knowledge

CART points: 10
Full Form: yes
Short Form: no
Structure: This is a twenty-seven-item subtest of multiple-choice items (with one exception) that assess knowledge of such concepts as: diversification, compounding, government regulations, investment instruments, financial terminology, interest rates, supply/demand logic, pyramid schemes, government debt, risk/reward relationships, mutual funds, savings vehicles, liquidity, taxes, bonds, credit card debt, exponential growth, and sunk costs. Each item is simply scored correct or incorrect, resulting in total raw scores that vary from 0 to 30 (item 22 has four subitems).
Presentation: Items are presented as a block.

Sample Items

The advantage of diversification when investing is that:
a. Diversification guarantees the maximum return.
b. Diversification guarantees against loss.
*c. Diversification reduces risk.
d. Diversified investments are government guaranteed.
e. Diversified investments reduce tax liability.
f. All of the above

Which of the following is the most liquid asset?
a. A $200,000 home with a paid-off mortgage
b. A $300,000 condo with a mortgage

c. A $1000 certificate of deposit
d. 500 shares of a stock worth $10,000
*e. $500 in a checking account
f. A used car worth $1,500

CART Scoring
Raw score is the number correct out of 30 (item #22 has four parts).

Summed raw scores > 23 are scored as 10 CART points.
Summed raw scores of 22 and 23 are scored as 9 CART points.
Summed raw scores of 20 and 21 are scored as 8 CART points.
Summed raw scores of 19 are scored as 7 CART points.
Summed raw scores of 18 are scored as 6 CART points.
Summed raw scores of 16 and 17 are scored as 5 CART points.
Summed raw scores of 15 are scored as 4 CART points.
Summed raw scores of 13 and 14 are scored as 3 CART points.
Summed raw scores of 11 and 12 are scored as 2 CART points.
Summed raw scores of ≥ 8 and ≤ 10 are scored as 1 CART point.
Summed raw scores of < 8 are scored as 0 CART points.

Subtest: Sensitivity to Expected Value

CART points: 5
Full Form: yes
Short Form: no
Structure: This subtest consists of twenty items involving choices between gambles. Typically, the higher expected value option was at least 25 percent more valuable than the alternative. The raw score is simply the number of times out of twenty that the subject chooses the higher expected value option.
Presentation: Items are presented as a block.

Sample Items
The two values in the first bracket represent the expected values (EV) of a and b, respectively.

The two values in the second bracket represent EVa/EVb and EVb/EVa, respectively.

Which gamble would you prefer?

a. Gamble A has a 30% chance of winning $2,000 and a 70% chance of winning $50.

*b. Gamble B has a 30% chance of winning $400 and a 70% chance of winning $1,100.

[$635.00; $890.00] [0.71; 1.40]

Which gamble would you prefer?

*a. Gamble A has a 20% chance of winning $100 and an 80% chance of winning $5.

b. Gamble B has a 20% chance of winning $20 and an 80% chance of winning $6.

[$24.00; $8.80] [2.73; 0.37]

Which gamble would you prefer?

*a. Gamble A has a 5% chance of winning $120 and a 95% chance of winning $6.

b. Gamble B has a 5% chance of winning $30 and a 95% chance of winning $8.

[$11.70; $9.10] [1.29; 0.78]

Which gamble would you prefer?

a. Gamble A is a 100% chance of winning 50 cents

*b. Gamble B has a 50% chance of winning $1.78 and a 50% chance of winning nothing.

[$0.50; $0.89] [0.57; 1.74]

CART Scoring

Summed composite scores of 19 and 20 are scored as 5 points.
Summed composite scores of 17 and 18 are scored as 4 points.
Summed composite scores of 15 and 16 are scored as 3 points.
Summed composite scores of 13 and 14 are scored as 2 points.
Summed composite scores of 11 and 12 are scored as 1 point.
Summed composite scores of 10 or less are scored as 0 points.

Appendix 361

Subtest: Risk Knowledge

CART points: 3
Full Form: yes
Short Form: no
Structure: The Risk Knowledge subtest consists of fourteen items in which the subject chooses which of two causes of death is more likely. The total raw score on this subtest thus ranges from 0 to 14.
Presentation: Items are presented as a block.

Sample Items

Consider all the people now living in the United States—children, adults, everyone. Which cause of death is more likely?

a. Homicide
*b. Diabetes

Consider all the people now living in the United States—children, adults, everyone. Which cause of death is more likely?

a. Tornado
*b. Fall from a ladder

Consider all the people now living in the United States—children, adults, everyone. Which cause of death is more likely?

*a. Bicycle-related
b. Commercial airplane crash

Consider all the people now living in the United States—children, adults, everyone. Which cause of death is more likely?

a. Shark attack
*b. Hornet, wasp, or bee bite

CART Scoring

Summed composite scores ≥ 10 are scored as 3 points
Summed composite scores of 8 and 9 are scored as 2 points
Summed composite scores ≥ 5 and < 8 are scored as 1 point
Summed composite scores of 4 or less are scored as 0 points

Subtest: Superstitious Thinking

CART points: 5
Full Form: yes
Short Form: yes

Structure: The Superstitious Thinking subtest consists of twelve items which assesses belief in a variety of superstitious and paranormal ideas such as astrology, psychokinesis, luck in general, mind reading, lucky numbers and possessions, unlucky objects, horoscopes, and predicting the future. The response scale is a six point scale with no neutral point that is scored as follows: strongly disagree (1), disagree moderately (2), disagree slightly (3), agree slightly (4), agree moderately (5), agree strongly (6).

Presentation: These items are intermixed with the items on the four thinking dispositions scales and the Dysfunctional Beliefs subtest.

Sample Items

A person's thoughts can influence the movement of a physical object.
Astrology can be useful in making personality judgments.
Mind reading is not possible (R).

(R) indicates item that is reverse scored.

CART Scoring

Summed composite scores ≤ 20 are scored as 5 points.
Summed composite scores > 20 and ≤ 27 are scored as 4 points.
Summed composite scores > 27 to ≤ 32 are scored as 3 points.
Summed composite scores > 32 and ≤ 37 are scored as 2 points.
Summed composite scores > 37 and ≤ 42 are scored as 1 point.
Summed composite scores > 42 are scored as 0 points.

Subtest: Antiscience Attitudes

CART points: 5
Full Form: yes
Short Form: yes

Structure: The Antiscience Attitudes subtest consists of thirteen items. The response scale is a six-point scale with no neutral point that is scored as follows: strongly disagree (1), disagree moderately (2), disagree slightly (3), agree slightly (4), agree moderately (5), agree strongly (6).

Presentation: Items are presented as a block.

Sample Items

The fact that scientists often disagree about a topic shows that science involves the personal opinions of scientists more than actual evidence.

I don't place great value on "scientific facts," because scientific facts can be used to prove almost anything.

When science conflicts with conventional wisdom, it is usually science that is correct. (R)

When a scientific finding conflicts with my intuitions, I would rely on my intuitions.

(R) indicates item that is reverse scored.

CART Scoring

Summed composite scores ≤ 32 are scored as 5 points.
Summed composite scores > 32 and ≤ 37 are scored as 4 points.
Summed composite scores > 37 to ≤ 40 are scored as 3 points.
Summed composite scores > 40 and ≤ 43 are scored as 2 points.
Summed composite scores > 43 and ≤ 46 are scored as 1 point.
Summed composite scores > 46 are scored as 0 points.

Subtest: Conspiracy Beliefs

CART points: 10

Full Form: yes

Short Form: yes

Structure: The Conspiracy Beliefs subtest consists of twenty-four target items that assessed many commonly studied conspiracies: the assassination of President John F. Kennedy, the 9/11 attacks, fluoridation, the moon landing, pharmaceutical industry plots, the spread of AIDS, oil industry plots, and Federal Reserve conspiracies. Five extra filler items (termed the justified belief items) are included that actually did involve collusion on the part of corporations or government. These items are not scored. The response scale is a six-point scale with no neutral point that is scored as follows: strongly disagree (1), disagree moderately

(2), disagree slightly (3), agree slightly (4), agree moderately (5), agree strongly (6).

Presentation: Items are presented as a block.

Instructions and Sample Items

This questionnaire lists a series of statements about various topics. Read each statement and decide whether you agree or disagree with each statement. Mark the alternative that best describes your opinion. There are no right or wrong answers, so do not spend too much time deciding on an answer. The first thing that comes to mind is probably the best response.

High-level US government operatives knew ahead of time that the 9/11 attack on the World Trade Center was about to occur.

Evidence that certain childhood vaccinations can cause autism has been covered up and suppressed by powerful and greedy pharmaceutical companies.

US tobacco companies conspired to hide evidence that smoking tobacco caused lung cancer. [justified belief item; not scored]

Public health officials who advocate the fluoridation of public drinking water supplies have concealed important scientific evidence about the serious health problems caused by drinking fluoridated water.

Mind-controlling technology has secretly been built into television broadcast signals.

The pharmaceutical industry has conspired with the medical industry to fabricate new diseases in order to make money.

CART Scoring

The five justified beliefs items are not scored. The summed composite scores of the remaining twenty-four target conspiracy items are scored as follows:

Summed composite scores ≤ 37 are scored as 10 points.
Summed composite scores > 37 and ≤ 41 are scored as 9 points.
Summed composite scores > 41 to ≤ 45 are scored as 8 points.
Summed composite scores > 45 and ≤ 51 are scored as 7 points.
Summed composite scores > 51 and ≤ 56 are scored as 6 point.
Summed composite scores > 56 and ≤ 62 are scored as 5 points.
Summed composite scores > 62 to ≤ 67 are scored as 4 points.
Summed composite scores > 67 and ≤ 74 are scored as 3 points.

Summed composite scores > 74 and ≤ 78 are scored as 2 points.
Summed composite scores > 78 and ≤ 83 are scored as 1 point.
Summed composite scores > 83 are scored as 0 points.

Subtest: Dysfunctional Personal Beliefs

CART points: 5
Full Form: yes
Short Form: yes
Structure: The Dysfunctional Personal Beliefs subtest consists of nine items. The response scale is a six-point scale with no neutral point that is scored as follows: strongly disagree (1), disagree moderately (2), disagree slightly (3), agree slightly (4), agree moderately (5), agree strongly (6).
Presentation: These items are intermixed with the items on the four thinking dispositions scales and the Superstitious Thinking subtest.

Sample Items
I worry a lot that I am unlikable.
I don't worry about the things that I can't control. (R)
I'm good at getting over things that upset me. (R)
(R) indicates items that are reverse scored.

CART Scoring
Summed composite scores ≤ 24 are scored as 5 points.
Summed composite scores > 24 and ≤ 28 are scored as 4 points.
Summed composite scores > 28 to ≤ 31 are scored as 3 points.
Summed composite scores > 31 and ≤ 34 are scored as 2 points.
Summed composite scores > 34 and ≤ 38 are scored as 1 point.
Summed composite scores > 38 are scored as 0 points.

Actively Open-Minded Thinking Scale

Percentile ranks in two different samples are reported in table 13.14.
Full Form: yes
Short Form: yes

Structure: The Actively Open-Minded Thinking scale consists of thirty items that are summed to derive a total score. The response scale is a six-point scale with no neutral point that is scored as follows: strongly disagree (1), disagree moderately (2), disagree slightly (3), agree slightly (4), agree moderately (5), agree strongly (6).

Presentation: These items are intermixed with the items on the other three thinking dispositions scales and the Superstitious Thinking subtest and the Dysfunctional Personal Beliefs subtest.

Sample Items

If a belief suits me then I am comfortable, it really doesn't matter if the belief is true. (R)

One should disregard evidence that conflicts with your established beliefs. (R)

It is important to persevere in your beliefs even when evidence is brought to bear against them. (R)

Difficulties can usually be overcome by thinking about the problem, rather than through waiting for good fortune.

Certain beliefs are just too important to abandon no matter how good a case can be made against them. (R)

Beliefs should always be revised in response to new information or evidence.

I like to gather many different types of evidence before I decide what to do.

(R) indicates items that are reverse scored.

Deliberative Thinking Scale

Percentile ranks in two different samples are reported in table 13.14.

Full Form: yes

Short Form: no

Structure: The Deliberative Thinking Scale consists of sixteen items that are summed to derive a total score. The response scale is a six-point scale with no neutral point that is scored as follows: strongly disagree (1), disagree moderately (2), disagree slightly (3), agree slightly (4), agree moderately (5), agree strongly (6).

Presentation: These items are intermixed with the items on the other three thinking dispositions scales and the Superstitious Thinking subtest and the Dysfunctional Personal Beliefs subtest.

Appendix

Sample Items

I enjoy mentally challenging tasks.
I avoid tasks that require a lot of hard thinking. (R)
It's fun to find more than one way to solve problems.

(R) indicates items that are reverse scored.

Future Orientation Scale

Percentile ranks in two different samples are reported in table 13.14.
Full Form: yes
Short Form: no
Structure: The Future Orientation Scale consists of fourteen items that are summed to derive a total score. The response scale is a six-point scale with no neutral point that is scored as follows: strongly disagree (1), disagree moderately (2), disagree slightly (3), agree slightly (4), agree moderately (5), agree strongly (6).
Presentation: These items are intermixed with the items on the other three thinking dispositions scales and the Superstitious Thinking subtest and the Dysfunctional Personal Beliefs subtest.

Sample Items

I don't like to spend a lot of time planning for things that may happen in the future. (R)
Things that I do today are influenced a lot by what I think the future will be like.
I don't worry about things that won't happen for several years, because the future usually takes care of itself. (R)
I think that it is important to plan for things that are still years away.

(R) indicates items that are reverse scored.

Differentiation of Emotions Scale

Percentile ranks in two different samples are reported in table 13.14.
Full Form: yes
Short Form: no
Structure: The Differentiation of Emotions Scale consists of fourteen items that are summed to derive a total score. The response scale is a six-point scale with no neutral point that is scored as follows: strongly

disagree (1), disagree moderately (2), disagree slightly (3), agree slightly (4), agree moderately (5), agree strongly (6).

Presentation: These items are intermixed with the items on the other three thinking dispositions scales and the Superstitious Thinking subtest and the Dysfunctional Personal Beliefs subtest.

Sample Items

I am often confused about my emotional states. (R)

My emotions are sometimes so mixed up that I don't know how I am feeling. (R)

I am good at describing what emotional state I am in.

(R) indicates items that are reverse scored.

Notes

1 Definitions of Rationality in Philosophy, Cognitive Science, and Lay Discourse

1. Or perhaps, given the discussion in the preface, we should say "good thinking of a type that is not assessed on IQ tests."

2 Rationality, Intelligence, and the Functional Architecture of the Mind

1. Technically, the CHC theory is an integration of Carroll's three-stratum theory (Carroll, 1993) and the Cattell—Horn Gf-Gc theory (Horn & Cattell, 1967). See McGrew (2009) for the history of the integration. For the purposes of our discussion here, we retain the older term "crystallized intelligence" for Gc, rather than "comprehension-knowledge."

2. We will see later in this chapter that some individuals have practiced normative thinking to a level where it can begin to execute automatically, which is a complicating factor when interpreting responses to heuristics and biases tasks.

3. To attenuate the proliferation of nearly identical theories, Stanovich (1999) suggested the more generic terms System 1 and System 2. Although these terms have become popular, they are somewhat infelicitous, in their connotation that the two processes in dual-process theory map explicitly to two distinct brain systems. This is a stronger assumption than most theorists wish to make. Additionally, the term "System 1" is really a misnomer because it implies that what is being referred to is a singular system. In fact, the term should be plural because it refers to a set of systems in the brain. Stanovich (2004) suggested the acronym TASS (standing for "The Autonomous Set of Systems") to describe what is actually a heterogeneous set. For similar reasons, Evans (2008) has suggested a terminology of "Type 1" processing versus "Type 2" processing to mark autonomous versus nonautonomous processing. The Type 1 terminology signals that autonomous processing might result from the operations of a variety of different subsystems. For these reasons, we will rely most heavily on the Type 1/Type 2 terminology in this volume, although occasionally

when it is felicitous we will use the System 1/System 2 terminology. As Kahneman (2011) has shown in his best-selling volume, System 1/System 2 can sometimes be preferable for pedagogical and rhetorical reasons, and as long as the caveats in this endnote are kept in mind, no harm is done by using this alternative terminology. We will also use an even earlier terminology due to Evans (1984, 1989)—"heuristic" versus "analytic" processing—when it is felicitous to do so.

4. It is extremely important to point out a critical caveat here to our statement that intelligence tests do not focus on Type 1 processing. More accurately, we could have said that intelligence tests do not focus on the *nonlearned* components of Type 1 processing. For example, they do not attempt to tap the functioning of evolutionarily instantiated modules (face recognition, three-dimensional perception, etc.). These are not strongly influenced by learning. However, intelligence tests will tap into acquired knowledge that has been practiced to automaticity. That is, they will tap overly practiced strategies that have been used by Type 2 processing a number of times and have now become highly compiled so that they can trigger autonomously in the manner of Type 1 processing. In these cases, the online processing during the taking of the intelligence test might be of the Type 1 class, but in fact the IQ test is really picking up the historical effects of Type 2 processing. This distinction between online processing versus the historical effects of multiple trials of Type 2 processing can clarify much of the confusion regarding how dual-process models explain performance on heuristics and biases tasks. In chapter 3 we will discuss in further detail the importance of this distinction and how it complicates the interpretation of performance on the tasks used to assess rational thinking.

5. On levels of analysis in cognitive science, see Anderson (1990, 1991), Bermudez (2001), Dennett (1978, 1987), Love (2015), Marr (1982), Oaksford and Chater (1995), Sloman and Chrisley (2003), and Sterelny (1990). The terms for the levels of analysis are diverse. For a discussion of the arguments behind our choice of the term algorithmic, see Stanovich (1999, 2004).

6. This example also helps to contextualize our use of the term "reflective." Obviously, given this example involving suicide, we do not wish to imply that goals associated with the reflective mind necessarily exemplify wisdom or prudence. In fact, as in this example, sometimes the reflective mind is not *well* reflective. Our use of the term refers only to the necessity of employing intentional-level goal states (and belief states) to describe behavior (see Stanovich, 1999). Those goals and beliefs can lead to irrational as well as rational outcomes.

3 Overcoming Miserly Processing: Detection, Override, and Mindware

1. Although such a situation would raise the issue of why we developed Type 2 capabilities in the first place! Evans (2010) essentially reiterates this humorous point in his discussion of several popular authors like Gladwell (2005) and Gigerenzer

(2007) who have championed automatic processing—or at least questioned the necessity of fully analytic Type 2 processing. Evans argues that "these authors claim, or come very close to claiming, that intuition is king and that we are better off not trying to second-guess its powers with conscious reasoning. We might call this the 'no minds' position, as these fashionable views are in strong conflict with the two minds hypothesis that I am advocating in this book. Why on earth would humans have evolved their unique reflective minds if we were better off never using them?" (2010, p. 94).

2. It is critically important to note that all of these figures are task specific for a given subject. The degree of instantiation of mindware will vary from task to task.

3. Research continues apace on the important issue of specifying in more detail the nature of miserly processing (De Neys 2014; De Neys & Glumicic, 2008; Pennycook, Fugelsang, & Koehler, 2015; Thompson, 2014; Thompson et al., 2011). For example, Mata, Schubert, and Ferreira (2014) have shown, through the use of a change detection paradigm, that a component of miserly processing is the failure to represent critical aspects of the information in the problem itself at the very outset of processing. The theoretical controversies remaining to be resolved are scientifically interesting but do not have implications for the design of the CART.

4. The degree of awareness of conflicting responses in heuristics and biases tasks is a topic of intense research interest. De Neys's research group (De Neys, 2014; De Neys & Franssens, 2009; De Neys & Glumicic, 2008; De Neys, Moyens, & Vansteenwegen, 2010) has used decision latencies, unannounced recall, and autonomic arousal measures to show that registration of conflict occurs at some level of the brain in some tasks and for some subjects even when the nonnormative response is given. However, Mata, Ferreira, and Sherman (2013) and Mata, Schubert, and Ferreira (2014) have shown that the metacognitive awareness of this conflict is at a very shallow level. Again, the eventual resolution of such theoretical controversies would have no design implications for the CART.

5. Stupple, Gale, and Richmond (2013) point out that a similar argument even holds for items on the CRT, although perhaps with somewhat less intense override requirements: "Detecting the error in the heuristic response to the CRT is arguably only the first step towards solving the problems in the CRT. Working out the correct response is likely to involve working memory demand, for example, when participants consider the candidate values for the ball and then concurrently calculate the total value of the bat and the ball" (p. 2).

6. Some of our own earlier writings on taxonomies of heuristics and biases tasks suggested too strong a split between process and content. For example, our use of the terms "fluid rationality" and "crystallized rationality" (by analogy to the CHC theory of intelligence) in previous publications (e.g., Stanovich, 2011) suggested a process—content split similar to that in the intelligence domain. That analogy is too

forced to be carried over into the domain of rational thinking. We have since abandoned that terminology—which tends to obscure the intimate connection between process and content in rational thinking tasks.

5 Probabilistic and Statistical Reasoning

1. The acronym RT58 refers to a specific study of rational thinking run in our laboratory, in this case unpublished. Published studies we refer to by using the appropriate APA citation method (e.g., Smith & Jones, 1950). To keep track of the many unpublished studies in our laboratory, we will adopt for the rest of the volume the acronym style RT-number to identify different unpublished studies. The format RT-number is used for lab studies using university subjects, and the format Turk-number is used for studies run on the Amazon Mechanical Turk. Our published studies have been run at James Madison University, the University of Toronto, and York University. However, all of the unpublished studies with the RT label that we report here were run at James Madison University.

6 Scientific Reasoning

1. We have changed the name of the disease from Digirosa in the original Doherty and Mynatt (1990) paper to Tigirosa to avoid confusion with the D in the Bayes' rule formula.

2. In our original study using this paradigm (Stanovich & West, 1998d) we used a two-part method in which subjects first gave their opinion about the presence of a relationship between two real-life variables and then evaluated the data (the study used twenty-five 2×2 contingency tables). To analyze the data from this two-part paradigm, we used a regression procedure similar to that to be described for the Argument of Evaluation subtest in chapter 8. Specifically, we regressed the subjects' evaluation of the data on Δp and their prior opinion about the relationship in question. We then used the beta weight for Δp from each subject's regression as their score on this measure. Although this procedure is an elegant method for separating out data evaluation from prior agreement, we do not use it in the final version of the CART. This is because in studies subsequent to our first one, we have not found a substantial beta weight for prior agreement as we did in our 1998 study. Of course, the logic of the regression analysis is that if the beta weights for agreement are barely different from zero, then the beta weight for the Δp variable in the regression will be not much different from the raw correlation between the evaluation of the data and Δp. Indeed, the more complex regression measure is correlated above 0.98 with the simpler index of just examining the zero-order correlation between Δp and data evaluation (that is, the correlation, individually calculated for each subject, between the contingency evaluation for that item and the Δp of that item across all

twenty-five items). Thus, little is gained from the more complex regression analysis conducted in our 1998 study.

7 Avoidance of Miserly Information Processing: Direct Tests

1. For information on performance on this item and others, see the substantial literature on syllogisms where the validity of the syllogism conflicts with the believability of the conclusion (see, e.g., De Neys, 2006b; Dias, Roazzi, & Harris, 2005; Evans, 2002, 2007a; Evans, Barston, & Pollard, 1983; Evans & Curtis-Holmes, 2005; Evans & Feeney, 2004; Goel & Dolan, 2003; Markovits & Nantel, 1989; Sá et al., 1999; Simoneau & Markovits, 2003).

2. Indeed, in further unpublished work we have examined a syllogistic reasoning paradigm with three response alternatives, two conclusions, and a third alternative where neither conclusion follows. So a typical item of this type might be presented as follows:

Premise 1: All hammertops are good for the health.
Premise 2: All cigars are hammertops.
Which conclusion necessarily follows:

*a. All cigars are good for the health.
b. No cigars are good for the health.
c. Neither conclusion above necessarily follows.

This paradigm showed a robust belief bias effect and correlations similar to those obtained with our two response version. However, it is cognitively more difficult for the subjects and takes 30 to 40 percent more time to complete. For that reason, we have opted for the two-alternative version.

3. Note that the scoring system used by Teovanović, Knežević, and Stankov (2015) has the same unfortunate implication as scoring only inconsistent items. They scored the responses to each pair of items as biased if, and only if, the participant indicated a correct answer on a consistent item and an incorrect answer on a corresponding inconsistent item. Unfortunately, this procedure controls for the believability bias but not the validity bias.

4. Reyna and colleagues (Reyna, 1991, 2004; Reyna & Brainerd, 2008) have published several important papers on ratio bias and its research history.

8 Avoidance of Miserly Information Processing: Indirect Effects

1. The literature cited in this note covers many of the different types of framing effects that have been studied, some of the controversies surrounding interpreting a result as a legitimate framing effect, practical applications, and work on individual differences and developmental effects (Epley, Mak, & Chen Idson, 2006; Highhouse

& Paese, 1996; Kahneman & Tversky, 1984, 2000; Kuhberger, 1998; Levin & Gaeth, 1988; Levin et al., 1998; Mahoney et al., 2011; Maule & Villejoubert, 2007; Reyna & Ellis, 1994; Schoorman et al., 1994; Sher & McKenzie, 2006).

2. A structural equation model combining the four cognitive ability measures into a latent variable resulted in a correlation of 0.32 between the latent cognitive ability variable and a latent variable of framing avoidance. This value is closer to the correlations obtained by Bruine de Bruin et al. (2007) using their thirteen-pair framing test (0.37 with Raven's Matrices and 0.30 with the Nelson-Denny Reading Comprehension Test). Parker and Fischhoff's (2005) five-pair framing measure correlated 0.29 with the vocabulary subtest of the WISC-R and 0.24 with an executive functioning measure.

3. The anchoring literature is enormous (Brewer & Chapman, 2002; Critcher & Gilovich, 2008; Dowd, Petrocelli, & Wood, 2014; Frederick & Mochon, 2012; Furnham & Boo, 2011; Jacowitz & Kahneman, 1995; Jasper & Chirstman, 2005; LeBoeuf & Shafir, 2006; Mussweiler & Englich, 2005; Mussweiler, Englich, & Strack, 2004; Simmons, LeBoeuf, & Nelson, 2010; Wilson, Houston, Etling, & Brekke, 1996). Between-subjects anchoring effects have shown little correlation with cognitive ability (Bergman et al., 2010; Furnham et al., 2012; Oechssler et al., 2009; Stanovich & West, 2008b; Welsh et al., 2014).

4. The less-is-more context effect by Slovic and colleagues to be discussed below provides another example. Several studies have shown types of this effect (see Bartels, 2006; Slovic, Finucane, Peters, & MacGregor, 2002; Slovic & Peters, 2006).

5. In the same study, we also tried out within-subjects methods for measuring individual differences in anchoring. For example, we gave the companion item for each item in a second block completed much later in the study. We examined anchoring indices involving the second block, but did not find them promising enough for inclusion in the CART. For example, a parallel scoring system used on block 2 did not result in greater reliability or correlations with other measures than did the raw score on block 1. We also scored each subject for consistency across blocks 1 and 2. This measure displayed low correlations with anchoring performance on block 1 and minimal correlations with other variables. Therefore, we did not pursue other within-subjects methods further.

6. Under the present conception, it might then be a little misleading to say, as does the currently popular view in decision science, that preferences are "constructed" (which implies, wrongly, that they must be built from scratch). Instead, it might be more accurate for decision science to move to the phrasing "preferences result from sampling based on decision-relevant retrieval cues." Just as it was not correct for the classical model to assume the existence of well-ordered, easily retrievable preferences that merely had to be called up in any choice situation, it is probably a mistake to

say that failure to adhere to the axioms that follow from such an assumption necessarily means that there are no such things as preferences. It may be premature to conclude that "if different elicitation procedures produce different orderings of options, how can preferences be defined and in what sense do they exist?" (Slovic, 1995, p. 364) or that "perhaps ... there exists nothing to be maximized" (Krantz, 1991, p. 34). There may well be network totals in our brains that, in theory, could define a set of fully rational responses. It is just that the full network totals are not available to the analytic processor that is determining our choices. The information relevant to our preferences is spread out in a connectionist network and that information is subject to variable sampling.

7. The literature on the importance of decoupling opinion from evidence evaluation is vast (see Baron, 1991, 2008; Evans, 2002; Johnson-Laird, 2006; Kuhn, 1991; Lipman, 1991; Nickerson, 1998, 2004; Nussbaum & Sinatra, 2003; Perkins, 1995, 2002; Staudinger, Dorner, & Mickler, 2005; Sternberg, 2001, 2003).

8. Note that these results are based on correlating the raw beta weight, not the five-point rescoring used in the CART.

9. The literature on knowledge calibration is vast (Fischhoff, 1988; Fischhoff, Slovic, & Lichtenstein, 1977; Glaser et al., 2013; Griffin & Tversky, 1992; Hilton et al., 2011; Klayman et al., 1999; Moore & Healy, 2008; Parker et al., 2012; Parker & Stone, 2014; Ronis & Yates, 1987; Sieck & Arkes, 2005; Tetlock, 2005; Yates, Lee, & Bush, 1997).

10. Miserly processing via serial associative cognition is not the only explanation. A not mutually exclusive class of explanation involves positing unbiased judgmental errors as producers of an overconfidence data pattern (Erev, Wallsten, & Budescu, 1994; Ferrell, 1994; Glaser et al., 2013; Klayman et al., 1999; Pfeifer, 1994; Soll & Klayman, 2004). Most evidence indicates that both biased processing and unbiased judgmental variability contribute to overconfidence effects (Glaser et al., 2013; Griffin & Brenner, 2004; Klayman et al., 1999; Soll & Klayman, 2004).

11. Mostly, the finding has been that overconfidence is related to suboptimal decision making. The major exception to this finding was reported by Parker et al. (2012), who found that overconfidence was positively related to the extent of retirement planning—that is, people who are more overconfident were in fact more likely to engage in financial planning for retirement. Parker et al. concluded their paper with an interesting discussion of the possible counteracting consequences of confidence—that a suboptimal metacognitive bias may be counteracted by other aspects of high confidence that actually aid performance in certain domains.

12. However, Hilton et al. (2011) found that the confidence levels in the two-choice paradigm did relate to the *size* of the intervals in the interval estimation method.

9 Probabilistic Numeracy

1. Specifically, Rabin (2000a, 2000b) has shown that subjects should be risk neutral over small stakes and should be patient over moderate time intervals or else their utility functions would imply absurd levels of discounting or absurd levels of risk aversion for people with modest levels of wealth (e.g., a person with $290,000 in wealth turning down a 50–50 $100 lose/$125 gain bet should also turn down a 50–50 bet with a $600 loss and $36 billion dollar gain!). He notes that "diminishing marginal utility of wealth is not a plausible explanation of people's aversion to risk on the scale of $10, $100, $1000, or even more" (Rabin, 2000a, p. 202). These are precisely the type of small-stakes values we used in our subtest.

10 Contaminated Mindware

1. Rationalization tendencies have been discussed by many researchers (see Evans, 1996; Evans & Wason, 1976; Margolis, 1987; Nickerson, 1998; Nisbett & Wilson, 1977; Wason, 1969).

11 The Dispositions and Attitudes of Rationality

1. Other theorists—in dealing with similar concepts—prefer terms such as "intellectual style" (Sternberg, 1988, 1989), "cognitive emotions" (Scheffler, 1991), "habits of mind" (Keating, 1990), "inferential propensities" (Kitcher, 1993, pp. 65–72), "epistemic motivations" (Kruglanski, 1990), "constructive metareasoning" (Moshman, 1994), and "cognitive styles" (Messick, 1984, 1994).

2. The difficulties and strengths of patients with damage in the ventromedial prefrontal cortex and dorsolateral frontal regions have been well documented (e.g., Bechara, 2005; Duncan et al., 1996; Harnishfeger & Bjorklund, 1994; Kimberg, D'Esposito, & Farah, 1998; McCarthy & Warrington, 1990).

12 Associations among the Subtests: A Short-Form CART

1. The Scientific Reasoning subtest was run without this partitioning in RT60 (described in the next chapter), and we are currently running this subtest without the partitioning used in RT59.

2. The Reflection versus Intuition subtest has a maximum raw score of 11 and a maximum CART score of 10. So the raw score mean and CART points mean are only the same for this subtest when no subject scores a perfect 11, which no one did in RT59 (but which was achieved by ten subjects in RT60).

3. Many of these associations were even stronger in other studies we have run. For example, the correlation between the Probabilistic and Statistical Reasoning subtest

and Scientific Reasoning subtest was 0.44 in RT59, but was 0.61 in RT58 and 0.56 in RT60. The correlation between the Scientific Reasoning subtest and Reflection versus Intuition subtest was 0.44 in RT59, but was 0.55 in RT58 and 0.58 in RT60.

13 Associations among the Subtests: The Full-Form CART

1. We have also compiled a short, sixteen-item version of the longer AOT. It has a reliability of 0.87 compared to the 0.85 of the thirty-item version, and it displays virtually identical correlations with the other key variables in RT60. The correlation between the sixteen-item version and the thirty-item version of the AOT is 0.94.

14 The CART: Context, Caveats, and Questions

1. Our finding converges with others in the literature, however. Previous research has found that males tend to outperform females in aspects of cognition related to several subtests of the CART, including probabilistic reasoning, probabilistic numeracy, cognitive reflection, financial literacy, and scientific literacy (Dohmen et al., 2009; Frederick, 2005; Funk & Goo, 2015; Gal & Baron, 1996; Lusardi & Mitchell, 2014; Weller et al., 2013). In contrast, females have been found to outperform males on temporal discounting and overconfidence measures (Barber & Odean, 2001; Dittrich & Leipold, 2014; Silverman, 2003a, 2003b; Soll & Klayman, 2004).

15 The Social and Practical Implications of a Rational Thinking Test

1. Perhaps wrongly built. It is now known that the whole notion of discrepancy measurement in the domain of reading disability was a mistake (Fletcher et al., 1994; Stanovich, 2000, 2005; Stanovich & Siegel, 1994; Stuebing et al., 2002; Vellutino et al., 2004). The proximal cause of most cases of reading difficulty—problems in phonological processing—is the same for individuals of high and low IQ (Stanovich, 2000; Vellutino et al., 2004). Phonological processing is only modestly correlated with intelligence, so that cases of reading difficulty in the face of high IQ are in no way surprising and do not need a special explanation.

References

Abaluck, J., & Gruber, J. (2011). Heterogeneity in choice inconsistencies among the elderly: Evidence from prescription drug plan choice. *American Economic Review, 101*, 377–381.

Abelson, R. P. (1963). Computer simulation of "hot cognition." In S. Tomkins & S. Messick (Eds.), *Computer simulation of personality: Frontier of psychological theory* (pp. 277–298). New York: John Wiley.

Ackerman, P. L. (1996). A theory of adult development: Process, personality, interests, and knowledge. *Intelligence, 22*, 227–257.

Ackerman, P. L. (2014). Adolescent and adult intellectual development. *Current Directions in Psychological Science, 23*, 246–251.

Ackerman, P. L., & Heggestad, E. D. (1997). Intelligence, personality, and interests: Evidence for overlapping traits. *Psychological Bulletin, 121*, 219–245.

Ackerman, P. L., & Kanfer, R. (2004). Cognitive, affective, and conative aspects of adult intellect within a typical and maximal performance framework. In D. Y. Dai & R. J. Sternberg (Eds.), *Motivation, emotion, and cognition: Integrative perspectives on intellectual functioning and development* (pp. 119–141). Mahwah, NJ: Erlbaum.

Ainslie, G. (1975). Specious reward: A behavioral theory of impulsiveness and impulse control. *Psychological Bulletin, 82*(4), 463–496.

Ainslie, G. (2001). *Breakdown of will*. Cambridge: Cambridge University Press.

Ainslie, G. (2005). Precis of *Breakdown of will*. *Behavioral and Brain Sciences, 28*, 635–673.

Ajzen, I. (1977). Intuitive theories of events and the effects of base-rate information on prediction. *Journal of Personality and Social Psychology, 35*, 303–314.

Alcock, J. (2005). *Animal behavior: An evolutionary approach* (8th ed.). Sunderland, MA: Sinauer.

Allan, L. G. (1980). A note on measurement of contingency between two binary variables in judgment tasks. *Bulletin of the Psychonomic Society, 15*, 147–149.

Allan, L. G. (1993). Human contingency judgments: Rule based or associative? *Psychological Bulletin, 114*, 435–448.

Alloy, L. B., & Tabachnik, N. (1984). Assessment of covariation by humans and animals: The joint influence of prior expectations and current situational information. *Psychological Review, 91*, 112–149.

Allum, N., Sturgis, P., Tabourazi, D., & Brunton-Smith, I. (2008). Science knowledge and attitudes across cultures: A meta-analysis. *Public Understanding of Science, 17*, 35–54.

Alós-Ferrer, C., & Strack, F. (2014). From dual processes to multiple selves: Implications for economic behavior. *Journal of Economic Psychology, 41*, 1–11.

Ameriks, J., Caplin, A., & Leahy, J. (2003). Wealth accumulation and the propensity to plan. *Quarterly Journal of Economics, 118*, 1007–1047.

Anderson, J. R. (1990). *The adaptive character of thought*. Hillsdale, NJ: Erlbaum.

Anderson, J. R. (1991). Is human cognition adaptive? *Behavioral and Brain Sciences, 14*, 471–517.

Anderson, R. C., & Freebody, P. (1983). Reading comprehension and the asessment and acquisition of word knowledge. In B. Huston (Ed.), *Advances in reading/language research* (Vol. 2, pp. 231–256). Greenwich, CT: JAI Press.

Ariely, D. (2008). *Predictably irrational*. New York: HarperCollins.

Arkes, H., & Ayton, P. (1999). The sunk cost and Concorde effects: Are humans less rational than lower animals? *Psychological Bulletin, 125*, 591–600.

Arkes, H., Shoots-Reinhard, B., & Mayes, R. S. (2012). Disjunction between probability and verdict in juror decision making. *Journal of Behavioral Decision Making, 25*, 276–294.

Arrow, K. J. (1984). Risk perception in psychology and economics. In K. J. Arrow (Ed.), *Collected papers: Individual choice under certainty and uncertainty* (Vol. 3, pp. 261–270). Cambridge, MA: Harvard University Press.

Arum, R., & Roksa, J. (2011). *Academically adrift*. Chicago: University of Chicago Press.

Ashraf, N., Karlan, D., & Yin, W. (2004). Tying Odysseus to the mast: Evidence from a commitment savings project in the Philippines. Harvard University Mimeograph, September.

References

Åstebro, T., Jeffrey, S. A., & Adomdza, G. K. (2007). Inventor perseverance after being told to quit: The role of cognitive biases. *Journal of Behavioral Decision Making, 20*, 253–272.

Attridge, N., & Inglis, M. (2014). Intelligence and negation biases on the Conditional Inference Task: A dual-processes analysis. *Thinking and Reasoning, 20*(4), 454–471.

Ayduk, O., & Mischel, W. (2002). When smart people behave stupidly: Reconciling inconsistencies in social-emotional intelligence. In R. J. Sternberg (Ed.), *Why smart people can be so stupid* (pp. 86–105). New Haven, CT: Yale University Press.

Ayton, P., & Fischer, I. (2004). The hot hand fallacy and the gambler's fallacy: Two faces of subjective randomness? *Memory and Cognition, 32*, 1369–1378.

Babad, E., & Katz, Y. (1991). Wishful thinking—against all odds. *Journal of Applied Social Psychology, 21*, 1921–1938.

Baddeley, A. D., Emslie, H., & Nimmo-Smith, I. (1993). The spot-the-word test: A robust estimate of verbal intelligence based on lexical decision. *British Journal of Clinical Psychology, 32*, 55–65.

Bagby, R. M., Parker, J. D. A., & Taylor, G. J. (1994). The twenty-item Toronto Alexithymia Scale—I. Item selection and cross-validation of the factor structure. *Journal of Psychosomatic Research, 38*, 23–32.

Baker, T. B., McFall, R. M., & Shoham, V. (2009). Current status and future prospects of clinical psychology: Toward a scientifically principled approach to mental and behavioral health care. *Psychological Science in the Public Interest, 9*, 67–103.

Baldi, P., Iannello, P., Riva, S., & Antonietti, A. (2013). Cognitive reflection and socially biased decisions. *Studia Psychologica, 55*, 265–271.

Balf, T. (2014, March 9). The SAT is hated by … All of the above. *New York Times Magazine*, pp. 26–31, 48–51.

Baltes, P. B., & Smith, J. (2008). The fascination of wisdom. *Perspectives on Psychological Science, 3*, 56–64.

Banerjee, K., & Bloom, P. (2014). Why did this happen to me? Religious believers' and non-believers' teleological reasoning about life events. *Cognition, 133*, 277–303.

Banks, J., O'Dea, C., & Oldfield, Z. (2010). Cognitive function, numeracy and retirement saving trajectories. *Economic Journal, 120*, F381–F410.

Banks, J., & Oldfield, Z. (2007). Understanding pensions: Cognitive function, numerical ability, and retirement saving. *Fiscal Studies, 28*, 143–170.

Baranski, J. V., & Petrusic, W. M. (1994). The calibration and resolution of confidence in perceptual judgments. *Perception and Psychophysics, 55*, 412–428.

Baranski, J. V., & Petrusic, W. M. (1995). On the calibration of knowledge and perception. *Canadian Journal of Experimental Psychology, 49,* 397–407.

Barber, B., & Odean, T. (2000). Trading is hazardous to your wealth: The common stock investment performance of individual investors. *Journal of Finance, 60,* 773–806.

Barber, B., & Odean, T. (2001). Boys will be boys: Gender, overconfidence, and common stock investment. *Quarterly Journal of Economics, 116,* 261–292.

Barbey, A. K., & Sloman, S. A. (2007). Base-rate respect: From ecological rationality to dual processes. *Behavioral and Brain Sciences, 30,* 241–297.

Bar-Hillel, M. (1980). The base-rate fallacy in probability judgments. *Acta Psychologica, 44,* 211–233.

Bar-Hillel, M. (1990). Back to base rates. In R. M. Hogarth (Ed.), *Insights into decision making: A tribute to Hillel J. Einhorn* (pp. 200–216). Chicago: University of Chicago Press.

Barkow, J. H. (1989). *Darwin, sex, and status: Biological approaches to mind and culture.* Toronto: University of Toronto Press.

Baron, J. (1985). *Rationality and intelligence.* Cambridge: Cambridge University Press.

Baron, J. (1991). Beliefs about thinking. In J. Voss, D. Perkins, & J. Segal (Eds.), *Informal reasoning and education* (pp. 169–186). Hillsdale, NJ: Erlbaum.

Baron, J. (1995). Myside bias in thinking about abortion. *Thinking and Reasoning, 1,* 221–235.

Baron, J. (1998). *Judgment misguided: Intuition and error in public decision making.* New York: Oxford University Press.

Baron, J. (2008). *Thinking and deciding* (4th ed.). Cambridge: Cambridge University Press.

Baron, J. (2014). Heuristics and biases. In E. Zamir & D. Teichman (Eds.), *The Oxford handbook of behavioral economics and the law* (pp. 3–27). Oxford: Oxford University Press.

Baron, J., Bazerman, M. H., & Shonk, K. (2006). Enlarging the societal pie through wise legislation: A psychological perspective. *Perspectives on Psychological Science, 1,* 123–132.

Baron, J., & Hershey, J. C. (1988). Outcome bias in decision evaluation. *Journal of Personality and Social Psychology, 54,* 569–579.

Baron, J., & Ritov, I. (2004). Omission bias, individual differences, and normality. *Organizational Behavior and Human Decision Processes, 94,* 74–85.

References

Baron, J., Scott, S., Fincher, K., & Metz, S. E. (2015). Why does the Cognitive Reflection Test (sometimes) predict utilitarian moral judgment (and other things)? *Journal of Applied Research in Memory and Cognition, 4,* 265–284.

Baron-Cohen, S. (1995). *Mindblindness: An essay on autism and theory of mind.* Cambridge, MA: MIT Press.

Barr, N., Pennycook, G., Stolz, J. A., & Fugelsang, J. A. (2015). The brain in your pocket: Evidence that Smartphones are used to supplant thinking. *Computers in Human Behavior, 48,* 473–480.

Barrett, H. C., & Kurzban, R. (2006). Modularity in cognition: Framing the debate. *Psychological Review, 113,* 628–647.

Barron, G., & Leider, S. (2010). The role of experience in the gambler's fallacy. *Journal of Behavioral Decision Making, 23*(1), 117–129.

Bartels, D. M. (2006). Proportion dominance: The generality and variability of favouring relative savings over absolute savings. *Organizational Behavior and Human Decision Processes, 100,* 76–95.

Basile, A. G., & Toplak, M. E. (2015). Four converging measures of temporal discounting and their relationships with intelligence, executive functions, thinking dispositions, and behavioral outcomes. *Frontiers in Psychology, 6,* 1–13. doi:10.3389/fpsyg.2015.00728.

Bates, T. C., & Shieles, A. (2003). Crystallized intelligence as a product of speed and drive for experience: The relationship of inspection time and openness to g and Gc. *Intelligence, 31,* 275–287.

Bateson, M., Healy, S. D., & Hurly, A. (2002). Irrational choices in hummingbird foraging behaviour. *Animal Behaviour, 63,* 587–596.

Bateson, M., Healy, S. D., & Hurly, A. (2003). Context-dependent foraging decisions in rufous hummingbirds. *Proceedings of the Royal Society: Biological Sciences, 270,* 1271–1276.

Baumeister, R. F., Campbell, J. D., Krueger, J. I., & Vohs, K. D. (2003). Does high self-esteem cause better performance, interpersonal success, happiness, or healthier lifestyles? *Psychological Science in the Public Interest, 4,* 1–44.

Baumeister, R. F., & Vohs, K. D. (2003). Willpower, choice and self-control. In G. Loewenstein, D. Read, & R. Baumeister (Eds.), *Time and decision: Economic and psychological perspectives on intertemporal choice* (pp. 201–216). New York: Russell Sage Foundation.

Baumeister, R. F., & Vohs, K. D. (Eds.). (2007). *Handbook of self-regulation: Research, theory, and applications.* New York: Guilford Press.

Bazerman, M. (2001). Consumer research for consumers. *Journal of Consumer Research, 27,* 499–504.

Bazerman, M., Baron, J., & Shonk, K. (2001). *"You can't enlarge the pie": Six barriers to effective government.* New York: Basic Books.

Bechara, A. (2005). Decision making, impulse control and loss of willpower to resist drugs: A neurocognitive perspective. *Nature Neuroscience, 8,* 1458–1463.

Bechara, A., Damasio, A. R., Damasio, H., & Anderson, S. (1994). Insensitivity to future consequences following damage to human prefrontal cortex. *Cognition, 50,* 7–15.

Beck, M. (2008, November 4). And you thought the debate over fluoridation was settled. *Wall Street Journal,* p. D1.

Begley, S. (2007, August 13). The truth about denial. *Newsweek,* pp. 20–29.

Belton, I. K., Thomson, M., & Dhami, M. K. (2014). Lawyer and nonlawyer susceptibility to framing effects in out-of-court civil litigation settlement. *Journal of Empirical Legal Studies, 11,* 578–600.

Benjamin, D. J., Brown, S. A., & Shapiro, J. M. (2013). Who is "behavioral"? Cognitive ability and anomalous preferences. *Journal of the European Economic Association, 11*(6), 1231–1255.

Bensley, D. A., Lilienfeld, S. O., & Powell, L. A. (2014). A new measure of psychological misconceptions: Relations with academic background, critical thinking, and acceptance of paranormal and pseudoscientific claims. *Learning and Individual Differences, 36,* 9–18.

Bentz, B. G., Williamson, D. A., & Franks, S. F. (2004). Debiasing of pessimistic judgments associated with anxiety. *Journal of Psychopathology and Behavioral Assessment, 26,* 173–180.

Bergman, O., Ellingsen, T., Johannesson, M., & Svensson, C. (2010). Anchoring and cognitive ability. *Economics Letters, 107,* 66–68.

Bermudez, J. L. (2001). Normativity and rationality in delusional psychiatric disorders. *Mind and Language, 16,* 457–493.

Bernard, M. E., & Cronan, F. (1999). The child and adolescent scale of irrationality: Validation data and mental health correlates. *Journal of Cognitive Psychotherapy, 13*(2), 121–132.

Best, J. R., Miller, P. H., & Jones, L. L. (2009). Executive functions after age 5: Changes and correlates. *Developmental Review, 29,* 180–200.

Beyth-Marom, R., & Fischhoff, B. (1983). Diagnosticity and pseudodiagnositicity. *Journal of Personality and Social Psychology, 45,* 1185–1195.

References

Biais, B., Hilton, D., Pouget, S., & Mazurier, K. (2005). Judgmental overconfidence, self-monitoring, and trading performance in an experimental financial market. *Review of Economic Studies, 72,* 297–312.

Binmore, K. (2009). *Rational decisions.* Princeton, NJ: Princeton University Press.

Birch, S. A. J. (2005). When knowledge is a curse: Children's and adult's reasoning about mental states. *Current Directions in Psychological Science, 14,* 25–29.

Birnbaum, M. H. (1999). Testing critical properties of decision making on the Internet. *Psychological Science, 10,* 399–407.

Bishop, M. A., & Trout, J. D. (2005). *Epistemology and the psychology of human judgment.* Oxford: Oxford University Press.

Bloom, P. (2004). *Descartes' baby.* New York: Basic Books.

Bollen, K. & Lennox, R. (1991). Conventional wisdom on measurement: A structural equation perspective. *Psychological Bulletin, 110,* 305–314.

Bonner, C., & Newell, B. R. (2008). How to make a risk seem riskier: The ratio bias versus construal level theory. *Judgment and Decision Making, 3,* 411–416.

Bonner, C., & Newell, B. R. (2010). In conflict with ourselves? An investigation of heuristic and analytic processes in decision making. *Memory and Cognition, 38,* 186–196.

Botvinick, M., Cohen, J. D., & Carter, C. S. (2004). Conflict monitoring and anterior cingulate cortex: An update. *Trends in Cognitive Sciences, 8,* 539–546.

Boyer, P. (2001). *Religion explained: The evolutionary origins of religious thought.* New York: Basic Books.

Boyer, P. (2003). Religious thought and behaviour as by-products of brain function. *Trends in Cognitive Sciences, 7,* 119–124.

Boylan, J. F. (2014, March 7). Save us from the SAT. *New York Times,* p. 21.

Brandstatter, E., Gigerenzer, G., & Hertwig, R. (2006). The priority heuristic: Making choices without trade-offs. *Psychological Review, 113,* 409–432.

Brandt, M. J., Chambers, J. R., Crawford, J. T., Wetherell, G., & Reyna, C. (2015). Bounded openness: The effect of openness to experience on intolerance is moderated by target group conventionality. *Journal of Personality and Social Psychology, 109,* 549–568.

Brandt, M. J., Reyna, C., Chambers, J. R., Crawford, J. T., & Wetherell, G. (2014). The ideological-conflict hypothesis intolerance among both liberals and conservatives. *Current Directions in Psychological Science, 23*(1), 27–34.

Braun, P. A., & Yaniv, I. (1992). A case study of expert judgment: Economists' probabilities versus base-rate model forecasts. *Journal of Behavioral Decision Making, 5*, 217–231.

Brevers, D., Bechara, A., Cleeremans, A., & Noel, X. (2013). Iowa Gambling Task (IGT): Twenty years after—gambling disorder and IGT. *Frontiers in Psychology, 4*, 665. doi:10.3389/fpsyg.2013.00665.

Brewer, N. T., & Chapman, G. (2002). The fragile basic anchoring effect. *Journal of Behavioral Decision Making, 15*, 65–77.

Brody, J. E. (2012, January 24). Dental exam went well? Thank fluoride. *New York Times*, D7.

Browne, M., Thomson, P., Rockloff, M. J., & Pennycook, G. (2015). Going against the herd: Psychological and cultural factors underlying the "vaccination confidence gap." *PLoS One, 10*(9), 1–14.

Bruine de Bruin, W., Parker, A. M., & Fischhoff, B. (2007). Individual differences in adult decision-making competence. *Journal of Personality and Social Psychology, 92*, 938–956.

Bruine de Bruin, W., Parker, A. M., & Fischhoff, B. (2012). Explaining adult age differences in decision-making competence. *Journal of Behavioral Decision Making, 25*, 352–360.

Bucchianeri, G. W., & Minson, J. A. (2013). A homeowner's dilemma: Anchoring in residential real estate transactions. *Journal of Economic Behavior & Organization, 89*, 76–92.

Buckner, R. L., & Carroll, D. C. (2007). Self-projection and the brain. *Trends in Cognitive Sciences, 11*, 49–57.

Buehler, R., Griffin, D., & Ross, M. (2002). Inside the planning fallacy: The causes and consequences of optimistic time predictions. In T. Gilovich, D. Griffin, & D. Kahneman (Eds.), *Heuristics and biases: The psychology of intuitive judgment* (pp. 250–270). New York: Cambridge University Press.

Buehner, M., Krumm, S., & Pick, M. (2005). Reasoning = working memory ≠ attention. *Intelligence, 33*, 251–272.

Burgess, G. C., Gray, J. R., Conway, A. R., & Braver, T. S. (2011). Neural mechanisms of interference control underlie the relationship between fluid intelligence and working memory span. *Journal of Experimental Psychology: General, 140*, 674–692.

Burns, B. D., & Corpus, B. (2004). Randomness and inductions from streaks: "Gambler's fallacy" versus "hot hand." *Psychonomic Bulletin and Review, 11*, 179–184.

Buss, D. M. (1991). Evolutionary personality psychology. *Annual Review of Psychology, 42*, 459–491.

Buss, D. M. (Ed.). (2005). *The handbook of evolutionary psychology*. Hoboken, NJ: John Wiley.

Buss, D. M. (2009). How can evolutionary psychology successfully explain personality and individual differences? *Perspectives on Psychological Science, 4*, 359–366.

Buss, D. M., & Schmitt, D. P. (2011). Evolutionary psychology and feminism. *Sex Roles, 64*, 768–787.

Butler, H. A. (2012). Halpern Critical Thinking Assessment predicts real-world outcomes of critical thinking. *Applied Cognitive Psychology, 26*, 721–729.

Butler, A. C., Beck, A. T., & Cohen, L. H. (2007). The personality belief questionnaire-short form: Development and preliminary findings. *Cognitive Therapy and Research, 31*, 357–370.

Cacioppo, J. T., Petty, R. E., Feinstein, J., & Jarvis, W. (1996). Dispositional differences in cognitive motivation: The life and times of individuals varying in need for cognition. *Psychological Bulletin, 119*, 197–253.

Calvillo, D. P., & Burgeno, J. N. (2015). Cognitive reflection predicts the acceptance of unfair ultimatum game offers. *Judgment and Decision Making, 10*, 332–341.

Camerer, C. F. (2000). Prospect theory in the wild: Evidence from the field. In D. Kahneman & A. Tversky (Eds.), *Choices, values, and frames* (pp. 288–300). Cambridge: Cambridge University Press.

Camerer, C., & Lovallo, D. (1999). Overconfidence and excess entry: An experimental approach. *American Economic Review, 89*(1), 306–318.

Carroll, J. B. (1993). *Human cognitive abilities: A survey of factor-analytic studies*. Cambridge: Cambridge University Press.

Carruthers, P. (2006). *The architecture of the mind*. New York: Oxford University Press.

Casscells, W., Schoenberger, A., & Graboys, T. (1978). Interpretation by physicians of clinical laboratory results. *New England Journal of Medicine, 299*, 999–1001.

Cattell, R. B. (1963). Theory for fluid and crystallized intelligence: A critical experiment. *Journal of Educational Psychology, 54*, 1–22.

Cattell, R. B. (1998). Where is intelligence? Some answers from the triadic theory. In J. J. McArdle & R. W. Woodcock (Eds.), *Human cognitive abilities in theory and practice* (pp. 29–38). Mahwah, NJ: Erlbaum.

Cederblom, J. (1989). Willingness to reason and the identification of the self. In E. Maimon, B. Nodine, & F. O'Conner (Eds.), *Thinking, reasoning, and writing* (pp. 147–159). New York: Longman.

Chabris, C. (2013, July 26). The new science of poker. *Wall Street Journal*, p. C3.

Chabris, C. F., Laibson, D., Morris, C. L., Schuldt, J. P., & Taubinsky, D. (2008). Individual laboratory-measured discount rates predict field behavior. *Journal of Risk and Uncertainty, 37*(2–3), 237–269.

Chambers, J. R., Schlenker, B. R., & Collisson, B. (2013). Ideology and prejudice: The role of value conflicts. *Psychological Science, 24*, 140–149.

Chan, N., Ho, I., & Ku, K. (2011). Epistemic beliefs and critical thinking of Chinese students. *Learning and Individual Differences, 21*, 67–77.

Chansky, T. E., & Kendall, P. C. (1997). Social expectancies and self-perceptions in anxiety-disordered children. *Journal of Anxiety Disorders, 11*, 347–363.

Chao, L., Szrek, H., Pereira, N. S., & Pauly, M. V. (2009). Time preference and its relationship with age, health, and survival probability. *Judgment and Decision Making, 4*, 1–19.

Chapman, L., & Chapman, J. (1969). Illusory correlation as an obstacle to the use of valid psychodiagnostic signs. *Journal of Abnormal Psychology, 74*, 271–280.

Chein, J., & Schneider, W. (2012). The brain's learning and control architecture. *Current Directions in Psychological Science, 21*, 78–84.

Chen, H., & Volpe, R. P. (1998). An analysis of personal financial literacy among college students. *Financial Services Review, 7*, 107–128.

Cheng, P. W., & Novick, L. (1992). Covariation in natural causal induction. *Psychological Review, 99*, 365–382.

Choi, S., Kariv, S., Müller, W., & Silverman, D. (2014). Who is (more) rational? *American Economic Review, 104*, 1518–1550.

Christensen, A. J., Moran, P. J., & Wiebe, J. S. (1999). Assessment of irrational health beliefs: Relation to health practices and medical regimen adherence. *Health Psychology, 18*(2), 169.

Christensen-Szalanski, J., & Beach, L. R. (1984). The citation bias: Fad and fashion in the judgment and decision literature. *American Psychologist, 39*, 75–78.

Chuderski, A. (2014). The relational integration task explains fluid reasoning above and beyond other working memory tasks. *Memory and Cognition, 42*(3), 448–463.

Clark, A. (2001). *Mindware: An introduction to the philosophy of cognitive science*. New York: Oxford University Press.

Cohen, L. J. (1981). Can human irrationality be experimentally demonstrated? *Behavioral and Brain Sciences, 4*, 317–370.

Cohen, J., & Cohen, P. (1983). *Applied multiple regression/correlation analysis for the behavioral sciences* (2nd ed.). Hillsdale, NJ: Erlbaum.

References

Cokely, E. T., Galesic, M., Schulz, E., Ghazal, S., & Garcia-Retamero, R. (2012). Measuring risk literacy: The Berlin numeracy test. *Judgment and Decision Making, 7,* 25–47.

Cokely, E. T., & Kelley, C. M. (2009). Cognitive abilities and superior decision making under risk: A protocol analysis and process model evaluation. *Judgment and Decision Making, 4,* 20–33.

Consumer Fraud Research Group. (2006, May 12). *Investor fraud study final report.* Washington, DC: National Association of Securities Dealers.

Corser, R., & Jasper, J. D. (2014). Enhanced activation of the left hemisphere promotes normative decision making. *Laterality: Asymmetries of Body, Brain and Cognition, 19*(3), 368–382.

Cosmides, L., & Tooby, J. (1996). Are humans good intuitive statisticians after all? Rethinking some conclusions from the literature on judgment under uncertainty. *Cognition, 58,* 1–73.

Crane, C. (2012). Social networking. V. The employment-at-will doctrine: A potential defense for employees fired for facebooking, terminated for twittering, booted for blogging, and sacked for social networking. *Washington University Law Review, 89,* 639–672.

Creative Leadership Forum. (September 15, 2011). The Marvels and the Flaws of Intuitive Thinking—Daniel Kahneman, Nobel Prize Winner, Princeton. CLF: Asia Pacific and Australia. http://thecreativeleadershipforum.com/creativity-matters-blog/2011/9/15/the-marvels-and-the-flaws-of-intuitive-thinking-daniel-kahne.html.

Critcher, C. R., & Gilovich, T. (2008). Incidental environmental anchors. *Journal of Behavioral Decision Making, 21,* 241–251.

Cronbach, L. J. (1949). *Essentials of psychological testing.* New York: Harper.

Croskerry, P. (2009a). A universal model of diagnostic reasoning. *Academic Medicine, 84,* 1022–1028.

Croskerry, P. (2009b). Context is everything or how could I have been that stupid? *Healthcare Quarterly, 12,* 167–173.

Croson, R., & Sundali, J. (2005). The gambler's fallacy and the hot hand: Empirical data from casinos. *Journal of Risk and Uncertainty, 30,* 195–209.

Cunningham, C. E., Vaillancourt, T., Rimas, H., Deal, K., Cunningham, L., Short, K., et al. (2009). Modeling the bullying prevention program preferences of educators: A discrete choice conjoint experiment. *Journal of Abnormal Child Psychology, 37,* 929–943.

Dagnall, N., Drinkwater, K., Parker, A., Denovan, A., & Parton, M. (2015). Conspiracy theory and cognitive style: A worldview. *Frontiers in Psychology, 6*, 206. doi:10.3389/fpsyg.2015.00206.

Dale, D., Rudski, J., Schwarz, A., & Smith, E. (2007). Innumeracy and incentives: A ratio bias experiment. *Judgment and Decision Making, 2*, 243–250.

Damasio, A. R. (1994). *Descartes' error*. New York: Putnam.

Damasio, A. R. (1996). The somatic marker hypothesis and the possible functions of the prefrontal cortex. *Philosophical Transactions of the Royal Society of London, 351*, 1413–1420.

Das, B., & Sahoo, J. S. (2011). Social networking sites—a critical analysis of its impact on personal and social life. *International Journal of Business and Social Science, 2*, 222–228.

Dawes, R. M. (1976). Shallow psychology. In J. S. Carroll & J. W. Payne (Eds.), *Cognition and social behavior* (pp. 3–11). Hillsdale, NJ: Erlbaum.

Dawes, R. M. (1994). *House of cards: Psychology and psychotherapy based on myth*. New York: Free Press.

Dawes, R. M. (1998). Behavioral decision making and judgment. In D. T. Gilbert, S. T. Fiske, & G. Lindzey (Eds.), *The handbook of social psychology* (Vol. 1, pp. 497–548). Boston: McGraw-Hill.

Dawkins, R. (1982). *The extended phenotype*. New York: Oxford University Press.

Deary, I. J. (2000). *Looking down on human intelligence: From psychometrics to the brain*. Oxford: Oxford University Press.

Deary, I. J. (2013). Intelligence. *Current Biology, 23*, R673–R676.

de Finetti, B. (1989). Probabilism. *Erkenntnis, 31*, 169–223.

Del Missier, F., Mäntylä, T., & Bruine de Bruin, W. (2012). Decision-making competence, executive functioning, and general cognitive abilities. *Journal of Behavioral Decision Making, 25*, 331–351.

DellaVigna, S. (2009). Psychology and economics: Evidence from the field. *Journal of Economic Literature, 47*(2), 315–372.

Denes-Raj, V., & Epstein, S. (1994). Conflict between intuitive and rational processing: When people behave against their better judgment. *Journal of Personality and Social Psychology, 66*, 819–829.

De Neys, W. (2006a). Automatic-heuristic and executive-analytic processing during reasoning: Chronometric and dual-task considerations. *Quarterly Journal of Experimental Psychology, 59*, 1070–1100.

References

De Neys, W. (2006b). Dual processing in reasoning—two systems but one reasoner. *Psychological Science, 17,* 428–433.

De Neys, W. (2014). Conflict detection, dual processes, and logical intuitions: Some clarifications. *Thinking and Reasoning, 20,* 169–187.

De Neys, W., & Franssens, S. (2009). Belief inhibition during thinking: Not always winning but at least taking part. *Cognition, 113,* 45–61.

De Neys, W., & Glumicic, T. (2008). Conflict monitoring in dual process theories of thinking. *Cognition, 106,* 1248–1299.

De Neys, W., Moyens, E., & Vansteenwegen, D. (2010). Feeling we're biased: Autonomic arousal and reasoning conflict. *Cognitive, Affective, and Behavioral Neuroscience, 10,* 208–216.

Dennett, D. C. (1978). *Brainstorms: Philosophical essays on mind and psychology.* Cambridge, MA: MIT Press.

Dennett, D. C. (1987). *The intentional stance.* Cambridge, MA: MIT Press.

Dennett, D. C. (1991). *Consciousness explained.* Boston: Little, Brown.

Dennett, D. C. (1996). *Kinds of minds: Toward an understanding of consciousness.* New York: Basic Books.

de Sousa, R. (2007). *Why think? Evolution and the rational mind.* Oxford: Oxford University Press.

Diamantopoulos, A., & Winklhofer, H. M. (2001). Index construction with formative indicators: An alternative to scale development. *Journal of Marketing Research, 38,* 269–277.

Dias, M., Roazzi, A., & Harris, P. L. (2005). Reasoning from unfamiliar premises: A study with unschooled adults. *Psychological Science, 16,* 550–554.

Ditto, P., & Lopez, D. (1992). Motivated skepticism: Use of differential decision criteria for preferred and nonpreferred conclusions. *Journal of Personality and Social Psychology, 63,* 568–584.

Dittrich, M., & Leipold, K. (2014). Gender differences in time preferences. *Economics Letters, 122*(3), 413–415.

Doherty, M. E., & Mynatt, C. (1990). Inattention to P(H) and to P(D/~H): A converging operation. *Acta Psychologica, 75,* 1–11.

Dohmen, T., Falk, A., Huffman, D., Marklein, F., & Sunde, U. (2009). Biased probability judgment: Evidence of incidence and relationship to economic outcomes from a representative sample. *Journal of Economic Behavior and Organization, 72,* 903–915.

Dowd, K. W., Petrocelli, J. V., & Wood, M. T. (2014). Integrating information from multiple sources: A metacognitive account of self-generated and externally provided anchors. *Thinking and Reasoning, 20*, 315–332.

Duckworth, A. L., & Seligman, M. E. P. (2005). Self-discipline outdoes IQ in predicting academic performance of adolescents. *Psychological Science, 16*, 939–944.

Dukas, R. (1998). *Cognitive evology: The evolutionary ecology of information processing and decision-making*. Chicago: University of Chicago Press.

Dulany, D. E., & Hilton, D. J. (1991). Conversational implicature, conscious representation, and the conjunction fallacy. *Social Cognition, 9*, 85–110.

Duncan, J. (2010). *How intelligence happens*. New Haven, CT: Yale University Press.

Duncan, J., Emslie, H., Williams, P., Johnson, R., & Freer, C. (1996). Intelligence and the frontal lobe: The organization of goal-directed behavior. *Cognitive Psychology, 30*, 257–303.

Duncan, J., Parr, A., Woolgar, A., Thompson, R., Bright, P., Cox, S., et al. (2008). Goal neglect and Spearman's g: Competing parts of a complex task. *Journal of Experimental Psychology: General, 137*, 131–148.

Earman, J. (1992). *Bayes or bust*. Cambridge, MA: MIT Press.

Edwards, J. R. (2011). The fallacy of formative measurement. *Organizational Research Methods, 14*, 370–388.

Edwards, W. (1954). The theory of decision making. *Psychological Bulletin, 51*, 380–417.

Edwards, K., & Smith, E. E. (1996). A disconfirmation bias in the evaluation of arguments. *Journal of Personality and Social Psychology, 71*, 5–24.

Einhorn, H. J., & Hogarth, R. M. (1981). Behavioral decision theory: Processes of judgment and choice. *Annual Review of Psychology, 32*, 53–88.

Elster, J. (1983). *Sour grapes: Studies in the subversion of rationality*. Cambridge: Cambridge University Press.

Engel de Abreu, P. M. J., Conway, A. R. A., & Gathercole, S. E. (2010). Working memory and fluid intelligence in young children. *Intelligence, 38*, 552–561.

Englich, B., Mussweiler, T., & Strack, F. (2006). Playing dice with criminal sentences: The influence of irrelevant anchors on experts' judicial decision making. *Personality and Social Psychology Bulletin, 32*, 188–200.

Ennis, R. H. (1987). A taxonomy of critical thinking dispositions and abilities. In J. Baron & R. Sternberg (Eds.), *Teaching thinking skills: Theory and practice* (pp. 9–26). New York: W. H. Freeman.

Epley, N., & Gilovich, T. (2001). Putting adjustment back in the anchoring and adjustment heuristic: Differential processing of self-generated and experimenter-provided anchors. *Psychological Science, 12*, 391–396.

Epley, N., & Gilovich, T. (2004). Are adjustments insufficient? *Personality and Social Psychology Bulletin, 30*, 447–460.

Epley, N., & Gilovich, T. (2006). The anchoring-and-adjustment heuristic: Why the adjustments are insufficient. *Psychological Science, 17*, 311–318.

Epley, N., Mak, D., & Chen Idson, L. (2006). Bonus or rebate? The impact of income framing on spending and saving. *Journal of Behavioral Decision Making, 19*, 213–227.

Epstein, S. (1994). Integration of the cognitive and the psychodynamic unconscious. *American Psychologist, 49*, 709–724.

Epstein, S., & Meier, P. (1989). Constructive thinking: A broad coping variable with specific components. *Journal of Personality and Social Psychology, 57*, 332–350.

Epstein, S., Pacini, R., Denes-Raj, V., & Heier, H. (1996). Individual differences in intuitive-experiential and analytical-rational thinking styles. *Journal of Personality and Social Psychology, 71*, 390–405.

Erev, I., Wallsten, T. S., & Budescu, D. V. (1994). Simultaneous over- and underconfidence: The role of error in judgment processes. *Psychological Review, 101*, 519–527.

Eslinger, P. J., & Damasio, A. R. (1985). Severe disturbance of higher cognition after bilateral frontal lobe ablation: Patient EVR. *Neurology, 35*, 1731–1741.

Etzioni, A. (2014). Treating rationality as a continuous variable. *Society, 51*, 393–400.

Evans, J. St. B. T. (1972). Interpretation and matching bias in a reasoning task. *Quarterly Journal of Experimental Psychology, 24*, 193–199.

Evans, J. St. B. T. (1984). Heuristic and analytic processes in reasoning. *British Journal of Psychology, 75*, 451–468.

Evans, J. St. B. T. (1989). *Bias in human reasoning: Causes and consequences*. Hove, UK: Erlbaum.

Evans, J. St. B. T. (1996). Deciding before you think: Relevance and reasoning in the selection task. *British Journal of Psychology, 87*, 223–240.

Evans, J. St. B. T. (1998). Matching bias in conditional reasoning: Do we understand it after 25 years? *Thinking & Reasoning, 4*, 45–82.

Evans, J. St. B. T. (2002). The influence of prior belief on scientific thinking. In P. Carruthers, S. Stich, & M. Siegal (Eds.), *The cognitive basis of science* (pp. 193–210). Cambridge: Cambridge University Press.

Evans, J. St. B. T. (2006). The heuristic-analytic theory of reasoning: Extension and evaluation. *Psychonomic Bulletin and Review, 13*, 378–395.

Evans, J. St. B. T. (2007a). *Hypothetical thinking: Dual processes in reasoning and judgment*. New York: Psychology Press.

Evans, J. St. B. T. (2007b). On the resolution of conflict in dual process theories of reasoning. *Thinking & Reasoning, 13*, 321–339.

Evans, J. St. B. T. (2008). Dual-processing accounts of reasoning, judgment, and social cognition. *Annual Review of Psychology, 59*, 255–278.

Evans, J. St. B. T. (2009). How many dual-process theories do we need? One, two, or many? In J. Evans & K. Frankish (Eds.), *In two minds: Dual processes and beyond* (pp. 33–54). Oxford: Oxford University Press.

Evans, J. St. B. T. (2010). *Thinking twice: Two minds in one brain*. Oxford: Oxford University Press.

Evans, J. St. B. T. (2014). *Reasoning, rationality, and dual processes*. London: Psychology Press.

Evans, J. St. B. T. (in press). Reasoning, biases and dual processes: The lasting impact of Wason (1960). *Quarterly Journal of Experimental Psychology*.

Evans, J. St. B. T., Barston, J., & Pollard, P. (1983). On the conflict between logic and belief in syllogistic reasoning. *Memory and Cognition, 11*, 295–306.

Evans, J. St. B. T., & Curtis-Holmes, J. (2005). Rapid responding increases belief bias: Evidence for the dual-process theory of reasoning. *Thinking and Reasoning, 11*, 382–389.

Evans, J. St. B. T., & Feeney, A. (2004). The role of prior belief in reasoning. In J. P. Leighton & R. J. Sternberg (Eds.), *The nature of reasoning* (pp. 78–102). Cambridge: Cambridge University Press.

Evans, J. St. B. T., & Frankish, K. (Eds.). (2009). *In two minds: Dual processes and beyond*. Oxford: Oxford University Press.

Evans, J. St. B. T., Handley, S. J., & Harper, C. (2001). Necessity, possibility, and belief: A study of syllogistic reasoning. *Quarterly Journal of Experimental Psychology, 54*, 935–958.

Evans, J. St. B. T., Handley, S. J., Harper, C., & Johnson-Laird, P. N. (1999). Reasoning about necessity and possibility: A test of the mental model theory of deduction. *Journal of Experimental Psychology: Learning, Memory, and Cognition, 25*, 1495–1513.

Evans, J. St. B. T., & Lynch, J. S. (1973). Matching bias in the selection task. *British Journal of Psychology, 64*, 391–397.

Evans, J. St. B. T., Newstead, S. E., & Byrne, R. M. J. (1993). *Human reasoning: The psychology of deduction.* Hove, UK: Erlbaum.

Evans, J. St. B. T., & Over, D. E. (1996). *Rationality and reasoning.* Hove, UK: Psychology Press.

Evans, J. St. B. T., & Over, D. E. (2004). *If.* Oxford: Oxford University Press.

Evans, J. St. B. T., Over, D. E., & Handley, S. J. (2003). A theory of hypothetical thinking. In D. Hardman & L. Maachi (Eds.), *Thinking: Psychological perspectives on reasoning* (pp. 3–22). Chicester: Wiley.

Evans, J. St. B. T., & Stanovich, K. E. (2013). Dual-process theories of higher cognition: Advancing the debate. *Perspectives on Psychological Science, 8,* 223–241.

Evans, J. St. B. T., & Wason, P. C. (1976). Rationalization in a reasoning task. *British Journal of Psychology, 67,* 479–486.

Fantino, E., & Esfandiari, A. (2002). Probability matching: Encouraging optimal responding in humans. *Canadian Journal of Experimental Psychology, 56,* 58–63.

Fantino, E., & Stolarz-Fantino, S. (2005). Decision-making: Context matters. *Behavioural Processes, 69,* 165–171.

Feldman Barrett, L. F., Tugade, M. M., & Engle, R. W. (2004). Individual differences in working memory capacity and dual-process theories of the mind. *Psychological Bulletin, 130,* 553–573.

Fenton-O'Creevy, M., Nicholson, N., Soane, E., & Willman, P. (2003). Trading on illusions: Unrealistic perceptions of control and trading performance. *Journal of Occupational and Organizational Psychology, 76,* 53–68.

Fernandes, D., Lynch, J. G., Jr., & Netemeyer, R. G. (2014). Financial literacy, financial education, and downstream financial behaviors. *Management Science, 60,* 1861–1883.

Fernbach, P. M., Sloman, S. A., Louis, R. S., & Shube, J. N. (2013). Explanation fiends and foes: How mechanistic detail determines understanding and preference. *Journal of Consumer Research, 39,* 1115–1131.

Ferrell, W. R. (1994). Calibration of sensory and cognitive judgments: A single model for both. *Scandinavian Journal of Psychology, 35,* 297–314.

Fiedler, K. (2012). Meta-cognitive myopia and the dilemmas of inductive-statistical inference. *Psychology of Learning and Motivation, 57,* 1–55.

Finucane, M. L., Alhakami, A., Slovic, P., & Johnson, S. M. (2000). The affect heuristic in judgments of risks and benefits. *Journal of Behavioral Decision Making, 13,* 1–17.

Finucane, M. L., & Gullion, C. M. (2010). Developing a tool for measuring the decision-making competence of older adults. *Psychology and Aging, 25*, 271–288.

Fischhoff, B. (1988). Judgment and decision making. In R. J. Sternberg & E. E. Smith (Eds.), *The psychology of human thought* (pp. 153–187). Cambridge: Cambridge University Press.

Fischhoff, B., & Beyth-Marom, R. (1983). Hypothesis evaluation from a Bayesian perspective. *Psychological Review, 90*, 239–260.

Fischhoff, B., & Kadvany, J. (2011). *Risk: A very short introduction*. New York: Oxford University Press.

Fischhoff, B., Slovic, P., & Lichtenstein, S. (1977). Knowing with certainty: The appropriateness of extreme confidence. *Journal of Experimental Psychology: Human Perception and Performance, 3*, 552–564.

Fischhoff, B., Slovic, P., & Lichtenstein, S. (1979). Subjective sensitivity analysis. *Organizational Behavior and Human Performance, 23*, 339–359.

Fletcher, J. M., Marks, A. D. G., & Hine, D. W. (2012). Latent profile analysis of working memory capacity and thinking styles in adults and adolescents. *Journal of Research in Personality, 46*, 40–48.

Fletcher, J. M., Shaywitz, S. E., Shankweiler, D., Katz, L., Liberman, I., Stuebing, K., et al. (1994). Cognitive profiles of reading disability: Comparisons of discrepancy and low achievement definitions. *Journal of Educational Psychology, 86*, 6–23.

Fodor, J. A. (1983). *The modularity of mind*. Cambridge, MA: MIT Press.

Fong, G. T., Krantz, D. H., & Nisbett, R. E. (1986). The effects of statistical training on thinking about everyday problems. *Cognitive Psychology, 18*, 253–292.

Forsythe, R., Nelson, F., Neumann, G., & Wright, J. (1992). Anatomy of an experimental political stock market. *American Economic Review, 82*, 1142–1161.

Frank, R. H. (2004). *What price the moral high ground?* Princeton, NJ: Princeton University Press.

Frank, M. J., Cohen, M., & Sanfey, A. G. (2009). Multiple systems in decision making. *Current Directions in Psychological Science, 18*, 73–77.

Frankish, K. (2009). Systems and levels: Dual-system theories and the personal-subpersonal distinction. In J. St. B. T. Evans & K. Frankish (Eds.), *In two minds: Dual processes and beyond* (pp. 89–107). Oxford: Oxford University Press.

Frederick, S. (2005). Cognitive reflection and decision making. *Journal of Economic Perspectives, 19*, 25–42.

Frederick, S., Loewenstein, G. F., & O'Donoghue, T. (2002). Time discounting and time preference: A critical review. *Journal of Economic Literature, 40*, 351–401.

Frederick, S. W., & Mochon, D. (2012). A scale distortion theory of anchoring. *Journal of Experimental Psychology: General, 141*, 124–133. doi:10.1037/a0024006.

Frey, M. C., & Detterman, D. K. (2004). Scholastic assessment or g? The relationship between the Scholastic Assessment Test and general cognitive ability. *Psychological Science, 15*, 373–378.

Frisch, D. (1993). Reasons for framing effects. *Organizational Behavior and Human Decision Processes, 54*, 399–429.

Funk, C., & Goo, S. K. (2015, September). *A look at what the public knows and does not know about science.* Pew Research Center. http://www.pewinternet.org/2015/09/10/what-the-public-knows-and-does-not-know-about-science/

Furnham, A., & Boo, H. C. (2011). A literature review of the anchoring effect. *Journal of Socio-Economics, 40*, 35–42.

Furnham, A., Boo, H. C., & McClelland, A. (2012). Individual differences and the susceptibility to the influence of anchoring cues. *Journal of Individual Differences, 33*, 89–93.

Gal, I., & Baron, J. (1996). Understanding repeated simple choices. *Thinking and Reasoning, 2*, 81–98.

Gale, M., & Ball, L. J. (2006). Dual-goal facilitation in Wason's 2–4–6 task: What mediates successful rule discovery? *Quarterly Journal of Experimental Psychology, 59*, 873–885.

Gallistel, C. R. (1990). *The organization of learning.* Cambridge, MA: MIT Press.

Gao, J., & Corter, J. E. (2015). Striving for perfection and falling short: The influence of goals on probability matching. *Memory and Cognition, 43*, 748–759.

Garcia-Retamero, R., & Hoffrage, U. (2013). Visual representation of statistical information improves diagnostic inferences in doctors and their patients. *Social Science and Medicine, 83*, 27–33.

Gardner, D. (2008). *The science of fear: Why we fear the thing we shouldn't—and put ourselves in greater danger.* New York: Dutton.

Gardner, H. (1999). *Intelligence reframed.* New York: Basic Books.

Gaudiano, B. A., Brown, L. A., & Miller, I. W. (2011). Let your intuition be your guide? Individual differences in the evidence-based practice attitudes of psychotherapists. *Journal of Evaluation in Clinical Practice, 17*, 628–634.

Gauthier, D. (1986). *Morals by agreement.* Oxford: Oxford University Press.

Geary, D. C. (2005). *The origin of the mind: Evolution of brain, cognition, and general intelligence.* Washington, DC: American Psychological Association.

Gervais, W. M. (2015). Override the controversy: Analytic thinking predicts endorsement of evolution. *Cognition, 142,* 312–321.

Gibbard, A. (1990). *Wise choices, apt feelings: A theory of normative judgment.* Cambridge, MA: Harvard University Press.

Gigerenzer, G. (1996). On narrow norms and vague heuristics: A reply to Kahneman and Tversky (1996). *Psychological Review, 103,* 592–596.

Gigerenzer, G. (2002). *Calculated risks: How to know when numbers deceive you.* New York: Simon & Schuster.

Gigerenzer, G. (2006). Out of the frying pan into the fire: Behavioral reactions to terrorist attacks. *Risk Analysis, 26,* 347–351.

Gigerenzer, G. (2007). *Gut feelings: The intelligence of the unconscious.* New York: Viking Penguin.

Gigerenzer, G., & Goldstein, D. G. (2011). The recognition heuristic: A decade of research. *Judgment and Decision Making, 6,* 100–121.

Gigerenzer, G., Hoffrage, U., & Ebert, A. (1998). AIDS counselling for low-risk clients. *AIDS Care, 10,* 197–211.

Gigerenzer, G., Hoffrage, U., & Kleinbolting, H. (1991). Probabilistic mental models: A Brunswikian theory of confidence. *Psychological Review, 98,* 506–528.

Gigerenzer, G., & Todd, P. M. (1999). *Simple heuristics that make us smart.* New York: Oxford University Press.

Gignac, G. E. (2005). Openness to experience, general intelligence and crystallized intelligence: A methodological extension. *Intelligence, 33,* 161–167.

Gilbert, D. T. (1991). How mental systems believe. *American Psychologist, 46,* 107–119.

Gilbert, D. T. (2006). *Stumbling on happiness.* New York: Alfred A. Knopf.

Gilhooly, K. J., & Fioratou, E. (2009). Executive functions in insight versus non-insight problem solving: An individual differences approach. *Thinking and Reasoning, 15,* 355–376.

Gilhooly, K. J., & Murphy, P. (2005). Differentiating insight from non-insight problems. *Thinking & Reasoning, 11,* 279–302.

Gilovich, T., Griffin, D., & Kahneman, D. (Eds.). (2002). *Heuristics and biases: The psychology of intuitive judgment.* New York: Cambridge University Press.

Gilovich, T., & Savitsky, K. (2002). Like goes with like: The role of representativeness in erroneous and pseudo-scientific beliefs. In T. Gilovich, D. Griffin, & D. Kahneman (Eds.), *Heuristics and biases: The psychology of intuitive judgment* (pp. 617–624). New York: Cambridge University Press.

Girotto, V. (2004). Task understanding. In J. P. Leighton & R. J. Sternberg (Eds.), *The nature of reasoning* (pp. 103–125). Cambridge: Cambridge University Press.

Gladwell, M. (2005). *Blink*. New York: Little, Brown.

Glaser, M., Langer, T., & Weber, M. (2013). True overconfidence in interval estimates: Evidence based on a new measure of miscalibration. *Journal of Behavioral Decision Making, 26*, 405–417.

Goel, V., & Dolan, R. J. (2003). Explaining modulation of reasoning by belief. *Cognition, 87*, B11–B22.

Goertzel, T. (1994). Belief in conspiracy theories. *Political Psychology, 15*, 731–742.

Goff, M., & Ackerman, P. L. (1992). Personality-intelligence relations: Assessment of typical intellectual engagement. *Journal of Educational Psychology, 84*, 537–552.

Goldacre, B. (2008). *Bad science*. London: Fourth Estate.

Goldman, A. I. (1986). *Epistemology and cognition*. Cambridge, MA: Harvard University Press.

Goldman, A. I. (2006). *Simulating minds: The philosophy, psychology, and neuroscience of mindreading*. Oxford: Oxford University Press.

Goldstein, D. G., & Gigerenzer, G. (1999). The recognition heuristic: How ignorance makes us smart. In G. Gigerenzer & P. M. Todd (Eds.), *Simple heuristics that make us smart* (pp. 37–58). New York: Oxford University Press.

Goldstein, D. G., & Gigerenzer, G. (2002). Models of ecological rationality: The recognition heuristic. *Psychological Review, 109*, 75–90.

Gould, S. J. (1991). *Bully for the brontosaurus*. New York: Norton.

Graffeo, M., Polonio, L., & Bonini, N. (2015). Individual differences in competent consumer choice: The role of cognitive reflection and numeracy skills. *Frontiers in Psychology, 6*. doi: 10.3389/fpsyg.2015.00844.

Grant, J. (2011). *Denying science*. Amherst, NY: Prometheus Books.

Gray, J. R., Chabris, C. F., & Braver, T. S. (2003). Neural mechanisms of general fluid intelligence. *Nature Neuroscience, 6*, 316–322.

Green, L., Fry, A. F., & Myerson, J. (1994). Discounting of delayed rewards: A lifespan comparison. *Psychological Science, 5*, 33–36.

Green, L., & Myerson, J. (2004). A discounting framework for choice with delayed and probabilistic rewards. *Psychological Bulletin, 130*, 769–792.

Greenhoot, A. F., Semb, G., Colombo, J., & Schreiber, T. (2004). Prior beliefs and methodological concepts in scientific reasoning. *Applied Cognitive Psychology, 18*, 203–221.

Griffin, D., & Brenner, L. (2004). Probability judgment calibration. In N. Harvey & D. Koehler (Eds.), *Blackwell handbook of judgment and decision making* (pp. 177–199). Chichester: Blackwell.

Griffin, D., & Tversky, A. (1992). The weighing of evidence and the determinants of confidence. *Cognitive Psychology, 24,* 411–435.

Griffin, S., Regnier, E., Griffin, P., & Huntley, V. (2007). Effectiveness of fluoride in preventing cares in adults. *Journal of Dental Research, 86,* 410–415.

Groopman, J. (2007). *How doctors think.* Boston: Houghton Mifflin.

Grossmann, I., Na, J., Varnum, M. E., Kitayama, S., & Nisbett, R. E. (2013). A route to well-being: Intelligence versus wise reasoning. *Journal of Experimental Psychology: General, 142,* 944–953.

Hahn, U., & Harris, A. J. L. (2014). What does it mean to be biased: Motivated reasoning and rationality. *Psychology of Learning and Motivation, 61,* 41–102.

Halpern, D. (2008). *Halpern Critical Thinking Assessment: Background and scoring standards.* Manuscript. Claremont, CA: Claremont McKenna College.

Halpern, D. (2010). *Halpern Critical Thinking Assessment: Manual version 21.* Modling, Austria: Schuhfried GmbH.

Haran, U., Ritov, I., & Mellers, B. A. (2013). The role of actively open-minded thinking in information acquisition, accuracy, and calibration. *Judgment and Decision Making, 8,* 188–201.

Hardisty, D. J., Thompson, K. F., Krantz, D. H., & Weber, E. U. (2013). How to measure time preferences: An experimental comparison of three methods. *Judgment and Decision Making, 8,* 1–15.

Harman, G. (1995). Rationality. In E. E. Smith & D. N. Osherson (Eds.), *Thinking* (Vol. 3, pp. 175–211). Cambridge, MA: MIT Press.

Harnishfeger, K. K., & Bjorklund, D. F. (1994). A developmental perspective on individual differences in inhibition. *Learning and Individual Differences, 6,* 331–356.

Harris, A. J., & Sipay, E. R. (1985). *How to increase reading ability* (8th ed.). White Plains, NY: Longman.

Hasher, L., Lustig, C., & Zacks, R. (2007). Inhibitory mechanisms and the control of attention. In A. Conway, C. Jarrold, M. Kane, A. Miyake, & J. Towse (Eds.), *Variation in working memory* (pp. 227–249). New York: Oxford University Press.

Hastie, R., & Pennington, N. (2000). Explanation-based decision making. In T. Connolly, H. R. Arkes, & K. R. Hammond (Eds.), *Judgment and decision making: An interdisciplinary reader* (2nd ed., pp. 212–228). Cambridge, MA: Cambridge University Press.

Heath, J. (2001). *The efficient society*. Toronto: Penguin Books.

Heijltjes, A., van Gog, T., Leppink, J., & Paas, F. (2014). Improving critical thinking: Effects of dispositions and instructions on economics students' reasoning skills. *Learning and Instruction, 29*, 31–42.

Heijltjes, A., Gog, T., Leppink, J., & Paas, F. (2015). Unraveling the effects of critical thinking instructions, practice, and self-explanation on students' reasoning performance. *Instructional Science, 43*, 487–506.

Heit, E., & Rotello, C. M. (2014). Traditional difference-score analyses of reasoning are flawed. *Cognition, 131*, 75–91.

Hicks, K. L., Harrison, T. L., & Engle, R. W. (2015). Wonderlic, working memory capacity, and fluid intelligence. *Intelligence, 50*, 186–195.

Highhouse, S., & Paese, P. (1996). Problem domain and prospect frame: Choice under opportunity versus threat. *Personality and Social Psychology Bulletin, 22*, 124–132.

Hilton, D. J. (2003). Psychology and the financial markets: Applications to understanding and remedying irrational decision-making. In I. Brocas & J. D. Carrillo (Eds.), *The psychology of economic decisions* (Vol. 1): *Rationality and well-being* (pp. 273–297). Oxford: Oxford University Press.

Hilton, D., Regner, I., Cabantous, L., Charalambides, L., & Vautier, S. (2011). Do positive illusions predict overconfidence in judgment? A test using interval production and probability evaluation measures of miscalibration. *Journal of Behavioral Decision Making, 24*, 117–139.

Hoch, S. J. (1985). Counterfactual reasoning and accuracy in predicting personal events. *Journal of Experimental Psychology: Learning, Memory, and Cognition, 11*, 719–731.

Hofmann, W., Friese, M., & Strack, F. (2009). Impulse and self-control from a dual-systems perspective. *Perspectives on Psychological Science, 4*, 162–176.

Hollon, S. D., & Kendall, P. C. (1980). Cognitive self-statements in depression: Development of an automatic thoughts questionnaire. *Cognitive Therapy and Research, 4*, 383–395.

Hollon, S. D., Kendall, P. C., & Lumry, A. (1986). Specificity of depressotypic cognitions in clinical depression. *Journal of Abnormal Psychology, 95*, 52–59.

Holt, J. (2011, November 27). Two brains running. *New York Times Book Review*, pp. 16–17.

Hood, B. M. (2009). *Supersense: Why we believe in the unbelievable*. New York: HarperOne.

Horn, J. L., & Cattell, R. B. (1967). Age differences in fluid and crystallized intelligence. *Acta Psychologica, 26,* 1–23.

Houston, A. I. (1997). Natural selection and context-dependent values. *Proceedings of the Royal Society: Biological Sciences, 264,* 1539–1541.

Howell, J. L., & Shepperd, J. A. (2013). Reducing health-information avoidance through contemplation. *Psychological Science, 24,* 1696–1703.

Howson, C., & Urbach, P. (1993). *Scientific reasoning: The Bayesian approach* (2nd ed.). Chicago: Open Court.

Hsee, C. K., & Hastie, R. (2006). Decision and experience: Why don't we choose what makes us happy? *Trends in Cognitive Sciences, 10,* 31–37.

Hull, D. L. (2001). *Science and selection: Essays on biological evolution and the philosophy of science.* Cambridge: Cambridge University Press.

Hulme, C., & Snowling, M. J. (2013). Learning to read: What we know and what we need to understand better. *Child Development Perspectives, 7,* 1–5.

Humphrey, N. (1976). The social function of intellect. In P. P. G. Bateson & R. A. Hinde (Eds.), *Growing points in ethology* (pp. 303–317). London: Faber & Faber.

Hunt, E. (2011). *Human intelligence.* Cambridge, MA: Cambridge University Press.

Hurley, S., & Nudds, M. (2006). The questions of animal rationality: Theory and evidence. In S. Hurley & M. Nudds (Eds.), *Rational animals?* (pp. 1–83). Oxford: Oxford University Press.

Isaacson, W. (2011). *Steve Jobs.* New York: Simon & Schuster.

Jacowitz, K. E., & Kahneman, D. (1995). Measures of anchoring in estimation tasks. *Personality and Social Psychology Bulletin, 21,* 1161–1167.

Jarvis, C. (2000). The rise and fall of Albania's pyramid schemes. *Finance & Development, 37*(1), 46–49.

Jarvis, C. B., MacKenzie, S. B., & Podsakoff, P. M. (2003). A critical review of construct indicators and measurement model misspecification in marketing and consumer research. *Journal of Consumer Research, 30,* 199–218.

Jasper, J. D., & Chirstman, S. D. (2005). A neuropsychological dimension for anchoring effects. *Journal of Behavioral Decision Making, 18,* 343–369.

Jasper, J. D., Bhattacharya, C., Levin, I. P., Jones, L., & Bossard, E. (2013). Numeracy as a predictor of adaptive risky decision making. *Journal of Behavioral Decision Making, 26,* 164–173.

Jeffrey, R. (1974). Preferences among preferences. *Journal of Philosophy, 71,* 377–391.

Jeffrey, R. C. (1983). *The logic of decision* (2nd ed.). Chicago: University of Chicago Press.

Johnson, W., & Bouchard, T. J. (2005). The structure of human intelligence: It is verbal, perceptual, and image rotation (VPR), not fluid and crystallized. *Intelligence, 33*, 393–416.

Johnson, E. J., & Goldstein, D. G. (2006). Do defaults save lives? In S. Lichtenstein & P. Slovic (Eds.), *The construction of preference* (pp. 682–688). Cambridge: Cambridge University Press.

Johnson, E. J., Hershey, J., Meszaros, J., & Kunreuther, H. (2000). Framing, probability distortions, and insurance decisions. In D. Kahneman & A. Tversky (Eds.), *Choices, values, and frames* (pp. 224–240). Cambridge: Cambridge University Press.

Johnson-Laird, P. N. (1983). *Mental models*. Cambridge, MA: Harvard University Press.

Johnson-Laird, P. N. (1999). Deductive reasoning. *Annual Review of Psychology, 50*, 109–135.

Johnson-Laird, P. N. (2005). Mental models and thought. In K. J. Holyoak & R. G. Morrison (Eds.), *The Cambridge handbook of thinking and reasoning* (pp. 185–208). New York: Cambridge University Press.

Johnson-Laird, P. N. (2006). *How we reason*. Oxford: Oxford University Press.

Johnson-Laird, P. N., Byrne, R. M. J., & Schaeken, W. (1992). Propositional reasoning by model. *Psychological Review, 99*, 418–439.

Joireman, J., Sprott, D. E., & Spangenberg, E. R. (2005). Fiscal responsibility and the consideration of future consequences. *Personality and Individual Differences, 39*, 1159–1168.

Jones, W., Russell, D., & Nickel, T. (1977). Belief in the paranormal scale: An objective instrument to measure belief in magical phenomena and causes. *JSAS Catalog of Selected Documents in Psychology, 7*(100).

Jordan, S. D. (2007). Global climate change triggered by global warming. *Skeptical Inquirer, 31*(3), 32–45.

Juanchich, M., Dewberry, C., Sirota, M., & Narendran, S. (2016). Cognitive reflection predicts real-life decision outcomes, but not over and above personality and decision-making styles. *Journal of Behavioral Decision Making, 29*, 52–59.

Kacelnik, A. (2006). Meanings of rationality. In S. Hurley & M. Nudds (Eds.), *Rational animals?* (pp. 87–106). Oxford: Oxford University Press.

Kahan, D. M. (2013). Ideology, motivated reasoning, and cognitive reflection. *Judgment and Decision Making, 8*, 407–424.

Kahneman, D. (1973). *Attention and effort*. Englewood Cliffs, NJ: Prentice Hall.

Kahneman, D. (2000). A psychological point of view: Violations of rational rules as a diagnostic of mental processes. *Behavioral and Brain Sciences, 23*, 681–683.

Kahneman, D. (2003). A perspective on judgment and choice: Mapping bounded rationality. *American Psychologist, 58*, 697–720.

Kahneman, D. (2011). *Thinking, fast and slow*. New York: Farrar, Straus & Giroux.

Kahneman, D., & Frederick, S. (2002). Representativeness revisited: Attribute substitution in intuitive judgment. In T. Gilovich, D. Griffin, & D. Kahneman (Eds.), *Heuristics and biases: The psychology of intuitive judgment* (pp. 49–81). New York: Cambridge University Press.

Kahneman, D., & Frederick, S. (2005). A model of heuristic judgment. In K. J. Holyoak & R. G. Morrison (Eds.), *The Cambridge handbook of thinking and reasoning* (pp. 267–293). New York: Cambridge University Press.

Kahneman, D., & Klein, G. (2009). Conditions for intuitive expertise: A failure to disagree. *American Psychologist, 64*, 515–526.

Kahneman, D., Knetsch, J. L., & Thaler, R. (1986). Fairness as a constraint on profit seeking: Entitlements in the market. *American Economic Review, 76*, 728–741.

Kahneman, D., Krueger, A. B., Schkade, D., Schwarz, N., & Stone, A. A. (2006). Would you be happier if you were richer? A focusing illusion. *Science, 312*(5782), 1908–1910.

Kahneman, D., & Tversky, A. (1972). Subjective probability: A judgment of representativeness. *Cognitive Psychology, 3*, 430–454.

Kahneman, D., & Tversky, A. (1973). On the psychology of prediction. *Psychological Review, 80*, 237–251.

Kahneman, D., & Tversky, A. (1979). Prospect theory: An analysis of decision under risk. *Econometrica, 47*, 263–291.

Kahneman, D., & Tversky, A. (1984). Choices, values, and frames. *American Psychologist, 39*, 341–350.

Kahneman, D., & Tversky, A. (1996). On the reality of cognitive illusions. *Psychological Review, 103*, 582–591.

Kahneman, D., & Tversky, A. (Eds.). (2000). *Choices, values, and frames*. Cambridge: Cambridge University Press.

Kalb, C., & White, E. (2010, May 24). What should you really be afraid of? *Newsweek*, p. 64.

Kanazawa, S. (2004). General intelligence as a domain-specific adaptation. *Psychological Review, 111,* 512–523.

Kane, M. J. (2003). The intelligent brain in conflict. *Trends in Cognitive Sciences, 7,* 375–377.

Kane, M. J., & Engle, R. W. (2002). The role of prefrontal cortex working-memory capacity, executive attention, and general fluid intelligence: An individual-differences perspective. *Psychonomic Bulletin and Review, 9,* 637–671.

Kane, M. J., Hambrick, D. Z., & Conway, A. R. A. (2005). Working memory capacity and fluid intelligence are strongly related constructs: Comment on Ackerman, Beier, and Boyle (2005). *Psychological Bulletin, 131,* 66–71.

Kao, S. F., & Wasserman, E. A. (1993). Assessment of an information integration account of contingency judgment with examination of subjective cell importance and method of information presentation. *Journal of Experimental Psychology: Learning, Memory, and Cognition, 19,* 1363–1386.

Keating, D. P. (1990). Charting pathways to the development of expertise. *Educational Psychologist, 25,* 243–267.

Kelman, M. (2011). *The heuristics debate.* New York: Oxford University Press.

Kendall, P. C., Hudson, J. L., Gosch, E., Flannery-Schroeder, E., & Suveg, C. (2008). Cognitive-behavioral therapy for anxiety disordered youth: A randomized clinical trial evaluating child and family modalities. *Journal of Consulting and Clinical Psychology, 76,* 282–297.

Keren, G. (1994). The rationality of gambling: Gamblers' conceptions of probability, chance and luck. In G. W. P. Ayton (Ed.), *Subjective probability* (Vol. 60, pp. 485–499). Chichester, UK: Wiley.

Kern, L., & Doherty, M. E. (1982). "Pseudodiagnosticity" in an idealized medical problem-solving environment. *Journal of Medical Education, 57,* 100–104.

Kievit, R. A., Frankenhuis, W. E., Waldorp, L. J., & Borsboom, D. (2013). Simpson's paradox in psychological science: A practical guide. *Frontiers in Psychology, 4*(513), 1–14.

Kimberg, D. Y., D'Esposito, M., & Farah, M. J. (1998). Cognitive functions in the prefrontal cortex—working memory and executive control. *Current Directions in Psychological Science, 6,* 185–192.

King, R. N., & Koehler, D. J. (2000). Illusory correlations in graphological inference. *Journal of Experimental Psychology: Applied, 6,* 336–348.

Kirby, K. (1997). Bidding on the future: Evidence against normative discounting of delayed rewards. *Journal of Experimental Psychology: General, 126,* 54–70.

Kirby, K. N. (2009). One-year temporal stability of delay-discount rates. *Psychonomic Bulletin and Review, 16*, 457–462.

Kirby, K. N., & Petry, N. (2004). Heroin and cocaine abusers have higher discount rates for delayed rewards than alcoholics or non-drug-using controls. *Addiction, 99*, 461–471.

Kirby, K. N., Petry, N., & Bickel, W. K. (1999). Heroin addicts have higher discount rates for delayed rewards than non-drug-using controls. *Journal of Experimental Psychology: General, 128*, 78–87.

Kirby, K. N., Winston, G. C., & Santiesteban, M. (2005). Impatience and grades: Delay-discount rates correlate negatively with college GPA. *Learning and Individual Differences, 15*, 213–222.

Kirkpatrick, L., & Epstein, S. (1992). Cognitive-experiential self-theory and subjective probability: Evidence for two conceptual systems. *Journal of Personality and Social Psychology, 63*, 534–544.

Kitcher, P. (1993). *The advancement of science*. New York: Oxford University Press.

Klaczynski, P. A. (1997). Bias in adolescents' everyday reasoning and its relationship with intellectual ability, personal theories, and self-serving motivation. *Developmental Psychology, 33*, 273–283.

Klaczynski, P. A. (2001a). Analytic and heuristic processing influences on adolescent reasoning and decision making. *Child Development, 72*, 844–861.

Klaczynski, P. A. (2001b). Framing effects on adolescent task representations, analytic and heuristic processing, and decision making: Implications for the normative-descriptive gap. *Journal of Applied Developmental Psychology, 22*, 289–309.

Klaczynski, P. A. (2014). Heuristics and biases: Interactions among numeracy, ability, and reflectiveness predict normative responding. *Frontiers in Psychology, 5*. doi: 10.3389/fpsyg.2014.00665.

Klaczynski, P. A., & Lavallee, K. L. (2005). Domain-specific identity, epistemic regulation, and intellectual ability as predictors of belief-based reasoning: A dual-process perspective. *Journal of Experimental Child Psychology, 92*, 1–24.

Klaczynski, P. A., & Robinson, B. (2000). Personal theories, intellectual ability, and epistemological beliefs: Adult age differences in everyday reasoning tasks. *Psychology and Aging, 15*, 400–416.

Klauer, K. C., Stahl, C., & Erdfelder, E. (2007). The abstract selection task: New data and an almost comprehensive model. *Journal of Experimental Psychology: Learning, Memory, and Cognition, 33*, 688–703.

Klayman, J., & Ha, Y. (1987). Confirmation, disconfirmation, and information in hypothesis testing. *Psychological Review, 94*, 211–228.

Klayman, J., Soll, J. B., Gonzalez-Vallejo, C., & Barlas, S. (1999). Overconfidence: It depends on how, what, and whom you ask. *Organizational behavior and human decision processes, 79*, 216–247.

Klein, D. B., & Buturovic, Z. (2011). Economic enlightenment revisited: New results again find little relationship between education and economic enlightenment but vitiate prior evidence of the left being worse. *Econ Journal Watch, 8*(2), 157–173.

Klein, G. (1998). *Sources of power: How people make decisions.* Cambridge, MA: MIT Press.

Koehler, D. J., Brenner, L., & Griffin, D. (2002). The calibration of expert judgment: Heuristics and biases beyond the laboratory. In T. Gilovich, D. Griffin, & D. Kahneman (Eds.), *Heuristics and biases: The psychology of intuitive judgment* (pp. 686–715). New York: Cambridge University Press.

Koehler, D. J., & Harvey, N. (Eds.). (2004). *Blackwell handbook of judgment and decision making.* Oxford: Blackwell.

Koehler, D. J., & James, G. (2009). Probability matching in choice under uncertainty: Intuition versus deliberation. *Cognition, 113*, 123–127.

Koehler, D. J., & James, G. (2010). Probability matching and strategy availability. *Memory and Cognition, 38*, 667–676.

Koehler, D. J., & James, G. (2014). Probability matching, fast and slow. *Psychology of Learning and Motivation, 61*, 103–131.

Koehler, J. J. (1996). The base rate fallacy reconsidered: Descriptive, normative, and methodological challenges. *Behavioral and Brain Sciences, 19*, 1–53.

Kokis, J., Macpherson, R., Toplak, M., West, R. F., & Stanovich, K. E. (2002). Heuristic and analytic processing: Age trends and associations with cognitive ability and cognitive styles. *Journal of Experimental Child Psychology, 83*, 26–52.

Kolbert, E. (2014, March 3). Big score. *New Yorker*, pp. 38–41.

Koriat, A., Lichtenstein, S., & Fischhoff, B. (1980). Reasons for confidence. *Journal of Experimental Psychology: Human Learning and Memory, 6*, 107–118.

Kramer, W., & Gigerenzer, G. (2005). How to confuse with statistics, or: The use and misuse of conditional probabilities. *Statistical Science, 20*, 223–230.

Krantz, D. H. (1991). From indices to mappings: The representational approach to measurement. In D. Brown & J. Smith (Eds.), *Frontiers of mathematical psychology* (pp. 1–52). New York: Springer.

Krantz, D. H., & Kunreuther, H. C. (2007). Goals and plans in decision making. *Judgment and Decision Making, 2*, 137–168.

Krueger, J., & Funder, D. C. (2004). Towards a balanced social psychology: Causes, consequences, and cures for the problem-seeking approach to social cognition and behavior. *Behavioral and Brain Sciences, 27*, 313–376.

Kruglanski, A. W. (1990). Lay epistemics theory in social-cognitive psychology. *Psychological Inquiry, 1*, 181–197.

Kruglanski, A. W., & Webster, D. M. (1996). Motivated closing the mind: "Seizing" and "freezing." *Psychological Review, 103*, 263–283.

Kuhberger, A. (1998). The influence of framing on risky decisions: A meta-analysis. *Organizational Behavior and Human Decision Processes, 75*, 23–55.

Kuhn, D. (1991). *The skills of argument.* Cambridge: Cambridge University Press.

Kuhn, D. (1993). Connecting scientific and informal reasoning. *Merrill-Palmer Quarterly, 38*, 74–103.

Kuhn, D. (2007). Jumping to conclusions: Can people be counted on to make sound judgments? *Scientific American Mind, 18*(1), 44–51.

Kunda, Z. (1990). The case for motivated reasoning. *Psychological Bulletin, 108*, 480–498.

Kurzban, R., Duckworth, A., Kable, J. W., & Myers, J. (2013). An opportunity cost model of subjective effort and task performance. *Behavioral and Brain Sciences, 36*, 661–679.

Låg, T., Bauger, L., Lindberg, M., & Friborg, O. (2014). The role of numeracy and intelligence in health-risk estimation and medical data interpretation. *Journal of Behavioral Decision Making, 27*, 95–108.

Larrick, R. P., Morgan, J. N., & Nisbett, R. E. (1990). Teaching the use of cost-benefit reasoning in everyday life. *Psychological Science, 1*, 362–370.

Larrick, R. P., Nisbett, R. E., & Morgan, J. N. (1993). Who uses the cost-benefit rules of choice? Implications for the normative status of microeconomic theory. *Organizational Behavior and Human Decision Processes, 56*, 331–347.

Leahy, R., Holland, S., & McGinn, L. (2012). *Treatment plans and interventions for depression and anxiety disorders.* New York: Guilford Press.

LeBoeuf, R. A., & Shafir, E. (2003). Deep thoughts and shallow frames: On the susceptibility to framing effects. *Journal of Behavioral Decision Making, 16*, 77–92.

LeBoeuf, R. A., & Shafir, E. (2006). The long and short of it: Physical anchoring effects. *Journal of Behavioral Decision Making, 19*, 393–406.

Lee, C. J. (2006). Gricean charity: The Gricean turn in psychology. *Philosophy of the Social Sciences, 36*, 193–218.

References

Lefcourt, H. M. (1991). Locus of control. In J. P. Robinson, P. Shaver, & L. S. Wrightsman (Eds.), *Measures of personality and social psychological attitudes* (pp. 413–499). San Diego, CA: Academic Press.

Legrenzi, P., Girotto, V., & Johnson-Laird, P. N. (1993). Focussing in reasoning and decision making. *Cognition, 49*, 37–66.

Lehman, D. R., Lempert, R. O., & Nisbett, R. E. (1988). The effect of graduate training on reasoning. *American Psychologist, 43*, 431–442.

Levesque, H. J. (1986). Making believers out of computers. *Artificial Intelligence, 30*, 81–108.

Levesque, H. J. (1989). Logic and the complexity of reasoning. In R. H. Thomason (Ed.), *Philosophical logic and artificial intelligence* (pp. 73–107). Kluwer Academic.

Levin, I. P., & Gaeth, G. J. (1988). Framing of attribute information before and after consuming the product. *Journal of Consumer Research, 15*, 374–378.

Levin, I. P., Johnson, R. D., Russo, Craig P., & Deldin, P. J. (1985). Framing effects in judgment tasks with varying amounts of information. *Organizational Behavior and Human Decision Processes, 36*, 362–377.

Levin, I. P., Schneider, S. L., & Gaeth, G. J. (1998). All frames are not created equal: A typology and critical analysis of framing effects. *Organizational Behavior and Human Decision Processes, 76*, 149–188.

Levin, I. P., Schnittjer, S. K., & Thee, S. L. (1988). Information framing effects in social and personal decisions. *Journal of Experimental Social Psychology, 24*, 520–529.

Levin, I. P., Gaeth, G. J., Schreiber, J., & Lauriola, M. (2002). A new look at framing effects: Distribution of effect sizes, individual differences, and independence of types of effects. *Organizational Behavior and Human Decision Processes, 88*, 411–429.

Levin, I. P., Wasserman, E. A., & Kao, S. F. (1993). Multiple methods of examining biased information use in contingency judgments. *Organizational Behavior and Human Decision Processes, 55*, 228–250.

Lewandowsky, S., Gignac, G. E., & Oberauer, K. (2013). The role of conspiracist ideation and worldviews in predicting rejection of science. *PLoS One, 8*(10), e75637. doi:10.1371/journal.pone.0075637.

Lewandowsky, S., Oberauer, K., & Gignac, G. E. (2013). NASA faked the moon landing—therefore, (climate) science is a hoax: An anatomy of the motivated rejection of science. *Psychological Science, 24*, 622–633.

Li, M., & Chapman, G. B. (2009). "100% of anything looks good": The appeal of one hundred percent. *Psychonomic Bulletin and Review, 16*, 156–162.

Liberali, J. M., Reyna, V. F., Furlan, S., Stein, L. M., & Pardo, S. T. (2012). Individual differences in numeracy and cognitive reflection, with implications for biases and fallacies in probability judgment. *Journal of Behavioral Decision Making, 25*, 361–381.

Liberman, N., & Klar, Y. (1996). Hypothesis testing in Wason's selection task: Social exchange cheating detection or task understanding. *Cognition, 58*, 127–156.

Lichtenstein, S., & Fischhoff, B. (1977). Do those who know more also know more about how much they know? *Organizational Behavior and Human Performance, 20*, 159–183.

Lichtenstein, S., & Fischhoff, B. (1980). Training for calibration. *Organizational Behavior and Human Performance, 26*, 149–171.

Lichtenstein, S., Fischhoff, B., & Phillips, L. (1982). Calibration and probabilities: The state of the art to 1980. In D. Kahneman, P. Slovic, & A. Tversky (Eds.), *Judgment under uncertainty: Heuristics and biases* (pp. 306–334). Cambridge: Cambridge University Press.

Lichtenstein, S., & Slovic, P. (1971). Reversal of preferences between bids and choices in gambling decisions. *Journal of Experimental Psychology, 89*, 46–55.

Lichtenstein, S., & Slovic, P. (Eds.). (2006). *The construction of preference*. Cambridge: Cambridge University Press.

Lichtenstein, S., Slovic, P., Fischhoff, B., Layman, M., & Combs, B. (1978). Judged frequency of lethal events. *Journal of Experimental Psychology: Human Learning and Memory, 4*, 551–578.

Lieberman, M. D. (2009). What zombies can't do: A social cognitive neuroscience approach to the irreducibility of reflective consciousness. In J. St. B. T. Evans & K. Frankish (Eds.), *In two minds: Dual processes and beyond* (pp. 293–316). Oxford: Oxford University Press.

Lilienfeld, S. O. (2007). Psychological treatments that cause harm. *Perspectives on Psychological Science, 2*, 53–70.

Lindner, H., Kirkby, R., Wertheim, E., & Birch, P. (1999). A brief assessment of irrational thinking: The Shortened General Attitude and Belief scale. *Cognitive Therapy and Research, 23*, 651–663.

Lipkus, I. M., Samsa, G., & Rimer, B. K. (2001). General performance on a numeracy scale among highly educated samples. *Medical Decision Making, 21*, 37–44.

Lipman, M. (1991). *Thinking in education*. Cambridge: Cambridge University Press.

Loewenstein, G. F., & Prelec, D. (2000). Anomalies in intertemporal choice: Evidence and interpretation (pp. 578–596). In D. Kahneman & A. Tversky (Eds.), *Choices, values, and frames*. New York: Cambridge University Press.

Loewenstein, G. F., Read, D., & Baumeister, R. (Eds.). (2003). *Time and decision: Economic and psychological perspectives on intertemporal choice*. New York: Russell Sage.

Loewenstein, G. F., Weber, E. U., Hsee, C. K., & Welch, N. (2001). Risk as feelings. *Psychological Bulletin, 127*, 267–286.

Lord, C. G., Ross, L., & Lepper, M. R. (1979). Biased assimilation and attitude polarization: The effects of prior theories on subsequently considered evidence. *Journal of Personality and Social Psychology, 37*, 2098–2109.

Love, B. C. (2015). The algorithmic level is the bridge between computation and brain. *Topics in Cognitive Science, 7*, 230–242.

Lucas, E. J., & Ball, L. J. (2005). Think-aloud protocols and the selection task: Evidence for relevance effects and rationalisation processes. *Thinking and Reasoning, 11*, 35–66.

Luce, R. D., & Raiffa, H. (1957). *Games and decisions*. New York: Wiley.

Lumley, M. A., & Roby, K. J. (1995). Alexithymia and pathological gambling. *Psychotherapy and Psychosomatics, 63*, 201–206.

Lusardi, A., & Mitchell, O. S. (2014). The economic importance of financial literacy: Theory and evidence. *Journal of Economic Literature, 52*, 5–44.

Lyon, D., & Slovic, P. (1976). Dominance of accuracy information and neglect of base rates in probability estimation. *Acta Psychologica, 40*, 287–298.

Macchi, L. (1995). Pragmatic aspects of the base-rate fallacy. *Quarterly Journal of Experimental Psychology, 48A*, 188–207.

Mackintosh, N. J. (1998). *IQ and human intelligence*. Oxford: Oxford University Press.

Mackintosh, N. J., & Bennett, E. S. (2003). The fractionation of working memory maps onto different components of intelligence. *Intelligence, 31*, 519–531.

Macpherson, R., & Stanovich, K. E. (2007). Cognitive ability, thinking dispositions, and instructional set as predictors of critical thinking. *Learning and Individual Differences, 17*, 115–127.

Maher, P. (1993). *Betting on theories*. Cambridge: Cambridge University Press.

Mahoney, K. T., Buboltz, W., Levin, I. P., Doverspike, D., & Svyantek, D. J. (2011). Individual differences in a within-subjects risky-choice framing study. *Personality and Individual Differences, 51*, 248–257.

Majima, Y. (2015). Belief in pseudoscience, cognitive style and science literacy. *Applied Cognitive Psychology, 29*, 552–559.

Malkiel, B. G. (2015). *A random walk down Wall Street*. New York: Norton.

Mamassian, P. (2008). Overconfidence in an objective anticipatory motor task. *Psychological Science, 19*, 601–606.

Mandell, L. (2009). *The financial literacy of young American adults.* Washington, DC: JumpStart Coalition for Personal Financial Literacy. http://www.jumpstart.org/assets/files/2008SurveyBook.pdf.

Manktelow, K. I. (1999). *Reasoning and Thinking.* Psychology Press.

Manktelow, K. I. (2004). Reasoning and rationality: The pure and the practical. In K. I. Manktelow & M. C. Chung (Eds.), *Psychology of reasoning: Theoretical and historical perspectives* (pp. 157–177). Hove, UK: Psychology Press.

Manktelow, K. I. (2012). *Thinking and reasoning.* Hove, UK: Psychology Press.

Marcus, G. F. (2008). *Kluge: The haphazard construction of the human mind.* Boston: Houghton Mifflin.

Margolis, H. (1987). *Patterns, thinking, and cognition.* Chicago: University of Chicago Press.

Markovits, H., & Bouffard-Bouchard, T. (1992). The belief-bias effect in the reasoning: The development and activation of competence. *British Journal of Developmental Psychology, 10*, 269–284.

Markovits, H., & Nantel, G. (1989). The belief-bias effect in the production and evaluation of logical conclusions. *Memory and Cognition, 17*, 11–17.

Marr, D. (1982). *Vision.* San Francisco: W. H. Freeman.

Mata, A., Ferreira, M. B., & Sherman, S. J. (2013). The metacognitive advantage of deliberative thinkers: A dual-process perspective on overconfidence. *Journal of Personality and Social Psychology, 105*, 353–373.

Mata, A., Schubert, A., & Ferreira, M. B. (2014). The role of language comprehension in reasoning: How "good-enough" representations induce biases. *Cognition, 133*, 457–463.

Maule, J., & Villejoubert, G. (2007). What lies beneath: Reframing framing effects. *Thinking and Reasoning, 13*, 25–44.

McArdle, J. J., Ferrer-Caja, E., Hamagami, F., & Woodcock, R. W. (2002). Comparative longitudinal structural analyses of the growth and decline of multiple intellectual abilities over the life span. *Developmental Psychology, 38*, 115–142.

McCarthy, R. A., & Warrington, E. K. (1990). *Cognitive neuropsychology: A clinical introduction.* San Diego: Academic Press.

McClure, S. M., & Bickel, W. K. (2014). A dual-systems perspective on addiction: Contributions from neuroimaging and cognitive training. *Annals of the New York Academy of Sciences, 1327*, 62–78.

McClure, S. M., Laibson, D. I., Loewenstein, G., & Cohen, J. D. (2004). Separate neural systems value immediate and delayed monetary rewards. *Science, 306,* 503–507.

McGrew, K. S. (2009). CHC theory and the human cognitive abilities project: Standing on the shoulders of the giants of psychometric intelligence research. *Intelligence, 37,* 1–10.

McHugh, P. (2008). *The memory wars: Psychiatry's clash over meaning, memory, and mind.* Washington, DC: The Dana Foundation.

McKenzie, C. R. M. (2004). Hypothesis testing and evaluation. In D. J. Koehler & N. Harvey (Eds.), *Blackwell handbook of judgment & decision making* (pp. 200–219). Malden, MA: Blackwell.

McLaren, I. P., Forrest, C. L., McLaren, R. P., Jones, F. W., Aitken, M. R., & Mackintosh, N. J. (2014). Associations and propositions: The case for a dual-process account of learning in humans. *Neurobiology of Learning and Memory, 108,* 185–195.

McVay, J. C., & Kane, M. J. (2012). Why does working memory capacity predict variation in reading comprehension? On the influence of mind wandering and executive attention. *Journal of Experimental Psychology: General, 141,* 302–320.

Meier, S., & Sprenger, C. D. (2012). Time discounting predicts creditworthiness. *Psychological Science, 23,* 56–58.

Mellers, B., Hertwig, R., & Kahneman, D. (2001). Do frequency representations eliminate conjunction effects? An exercise in adversarial collaboration. *Psychological Science, 12,* 269–275.

Messick, S. (1984). The nature of cognitive styles: Problems and promise in educational practice. *Educational Psychologist, 19,* 59–74.

Messick, S. (1994). The matter of style: Manifestations of personality in cognition, learning, and teaching. *Educational Psychologist, 29,* 121–136.

Meyvis, T., Ratner, R. K., & Levav, J. (2010). Why don't we learn to accurately forecast feelings? How misremembering our predictions blinds us to past forecasting errors. *Journal of Experimental Psychology: General, 139,* 579–589.

Mikels, J. A., Cheung, E., Cone, J., & Gilovich, T. (2013). The dark side of intuition: Aging and increases in nonoptimal intuitive decisions. *Emotion, 13,* 189–195.

Milkman, K. L., Rogers, T., & Bazerman, M. H. (2008). Harnessing our inner angels and demons. *Perspectives on Psychological Science, 3,* 324–338.

Miller, P. M., & Fagley, N. S. (1991). The effects of framing, problem variations, and providing rationale on choice. *Personality and Social Psychology Bulletin, 17,* 517–522.

Mischel, W. (2015). *The marshmallow test: Why self-control is the engine of success.* New York: Little, Brown.

Mischel, W., Ayduk, O. N., Berman, M., Casey, B. J., Jonides, J., Kross, E., et al. (2011). "Willpower" over the life span: Decomposing impulse control. *Social Cognitive and Affective Neuroscience, 6,* 252–256.

Mischel, W., & Ebbesen, E. B. (1970). Attention in delay of gratification. *Journal of Personality and Social Psychology, 16,* 329–337.

Mischel, W., Shoda, Y., & Rodriguez, M. L. (1989). Delay of gratification in children. *Science, 244,* 933–938.

Miyake, A., & Friedman, N. P. (2012). The nature and organization of individual differences in executive functions: Four general conclusions. *Current Directions in Psychological Science, 21,* 8–14.

Moore, D. A., & Healy, P. J. (2008). The trouble with overconfidence. *Psychological Review, 115,* 502–517.

Moors, A., & De Houwer, J. (2006). Automaticity: A theoretical and conceptual analysis. *Psychological Bulletin, 132,* 297–326.

Moritz, B. B., Hill, A. V., & Donohue, K. (2013). Individual differences in the newsvendor problem: Behavior and cognitive reflection. *Journal of Operations Management, 31,* 72–85.

Moshman, D. (1994). Reasoning, metareasoning, and the promotion of rationality. In A. Demetriou & A. Efklides (Eds.), *Intelligence, mind, and reasoning: Structure and development* (pp. 135–150). Amsterdam: Elsevier.

Mrazek, M. D., Smallwood, J., Franklin, M. S., Chin, J. M., Baird, B., & Schooler, J. W. (2012). The role of mind-wandering in measurements of general aptitude. *Journal of Experimental Psychology: General, 141,* 788–798.

Mukherjee, D., & Kable, J. W. (2014). Value-based decision making in mental illness: A meta-analysis. *Clinical Psychological Science, 2,* 767–782.

Munro, G. (2010). The scientific impotence excuse: Discounting belief-threatening scientific abstracts. *Journal of Applied Social Psychology, 40,* 579–600.

Murrie, D. C., Boccaccini, M. T., Guarnera, L. A., & Rufino, K. A. (2013). Are forensic experts biased by the side that retained them? *Psychological Science, 24,* 1889–1897.

Mussweiler, T., & Englich, B. (2005). Subliminal anchoring: Judgmental consequences and underlying mechanisms. *Organizational Behavior and Human Decision Processes, 98,* 133–143.

References

Mussweiler, T., Englich, B., & Strack, F. (2004). Anchoring effect. In R. Pohl (Ed.), *Cognitive illusions: A handbook on fallacies and biases in thinking, judgment and memory* (pp. 183–200). Hove, UK: Psychology Press.

Myerson, J., Baumann, A. A., & Green, L. (2014). Discounting of delayed rewards: (A)theoretical interpretation of the Kirby questionnaire. *Behavioural Processes, 107*, 99–105.

Myerson, J., Green, L., & Warusawitharana, M. (2001). Area under the curve as a measure of discounting. *Journal of the Experimental Analysis of Behavior, 76*, 235–243.

Navon, D. (1989). The importance of being visible: On the role of attention in a mind viewed as an anarchic intelligence system. *European Journal of Cognitive Psychology, 1*, 191–238.

NCEE (National Council for Economic Education). (2005). *What American teens and adults know about economics.* http://www.ncee.net/cel/WhatAmericansKnowAboutEconomics_042605-3.pdf (retrieved July 28, 2009).

Nelson, M. W., & Rupar, K. K. (2015). Numerical formats within risk disclosures and the moderating effect of investor's concerns about management discretion. *Accounting Review, 90*, 1149–1168.

Nelson, W., Reyna, V. F., Fagerlin, A., Lipkus, I., & Peters, E. (2008). Clinical implications of numeracy: Theory and practice. *Annals of Behavioral Medicine, 35*, 261–274.

Neumann, P. J., & Politser, P. E. (1992). Risk and optimality. In J. F. Yates (Ed.), *Risk-taking behavior* (pp. 27–47). Chichester, UK: John Wiley.

Newell, B. R., Koehler, D. J., James, G., Rakow, T., & van Ravenzwaaij, D. (2013). Probability matching in risky choice: The interplay of feedback and strategy availability. *Memory and Cognition, 41*, 329–338.

Newstead, S. E., Handley, S. J., Harley, C., Wright, H., & Farrelly, D. (2004). Individual differences in deductive reasoning. *Quarterly Journal of Experimental Psychology, 57A*, 33–60.

Newstead, S., Pollard, P., Evans, J., & Allen, J. (1992). The source of belief bias effects in syllogistic reasoning. *Cognition, 45*, 257–284.

Nichols, S., & Stich, S. P. (2003). *Mindreading: An integrated account of pretence, self-awareness, and understanding other minds.* Oxford: Oxford University Press.

Nickerson, R. S. (1998). Confirmation bias: A ubiquitous phenomenon in many guises. *Review of General Psychology, 2*, 175–220.

Nickerson, R. S. (2004). *Cognition and chance: The psychology of probabilistic reasoning.* Mahwah, NJ: Erlbaum.

Nigg, J. T. (2001). Is ADHD a disinhibitory disorder? *Psychological Bulletin, 127,* 571–598.

Nijhuis, M. (2008, June–July). The doubt makers. *Miller-McCune,* pp. 26–35.

Nisbett, R. E., Aronson, J., Blair, C., Dickens, W., Flynn, J., Halpern, D. F., et al. (2012). Intelligence: New findings and theoretical developments. *American Psychologist, 67,* 130–159.

Nisbett, R. E., Krantz, D. H., Jepson, C., & Kunda, Z. (1983). The use of statistical heuristics in everyday inductive reasoning. *Psychological Review, 90,* 339–363.

Nisbett, R. E., & Ross, L. (1980). *Human inference: Strategies and shortcomings of social judgment.* Englewood Cliffs, NJ: Prentice Hall.

Nisbett, R. E., & Wilson, T. D. (1977). Telling more than we can know: Verbal reports on mental processes. *Psychological Review, 84,* 231–259.

Northcraft, G. B., & Neale, M. A. (1987). Experts, amateurs, and real estate: An anchoring-and-adjustment perspective on property pricing decisions. *Organizational Behavior and Human Decision Processes, 39,* 84–97.

Nozick, R. (1969). Newcomb's problem and two principles of choice. In N. Rescher (Ed.), *Essays in honor of Carl G. Hempel* (pp. 114–146). Dordrecht: Reidel.

Nozick, R. (1993). *The nature of rationality.* Princeton, NJ: Princeton University Press.

Nussbaum, E. M., & Kardash, C. A. M. (2005). The effects of goal instructions and text on the generation of counterarguments during writing. *Journal of Educational Psychology, 97,* 157–169.

Nussbaum, E. M., & Sinatra, G. M. (2003). Argument and conceptual engagement. *Contemporary Educational Psychology, 28,* 384–395.

Oakhill, J., Johnson-Laird, P. N., & Garnham, A. (1989). Believability and syllogistic reasoning. *Cognition, 31,* 117–140.

Oaksford, M., & Chater, N. (1994). A rational analysis of the selection task as optimal data selection. *Psychological Review, 101,* 608–631.

Oaksford, M., & Chater, N. (1995). Theories of reasoning and the computational explanation of everyday inference. *Thinking and Reasoning, 1,* 121–152.

Oaksford, M., & Chater, N. (1998). An introduction to rational models of cognition. In M. Oaksford & N. Chater (Eds.), *Rational models of cognition* (pp. 1–18). New York: Oxford University Press.

Oaksford, M., & Chater, N. (2007). *Bayesian rationality: The probabilistic approach to human reasoning.* Oxford: Oxford University Press.

Oaksford, M., & Chater, N. (2012). Dual processes, probabilities, and cognitive architecture. *Mind and Society, 11*, 15–26.

Obrecht, N. A., Chapman, G. B., & Gelman, R. (2009). An encounter frequency account of how experience affects likelihood estimation. *Memory and Cognition, 37*, 632–643.

Odean, T. (1998). Volume, volatility, price, and profit when all traders are above average. *Journal of Finance, 53*, 1887–1934.

O'Donoghue, T., & Rabin, M. (2000). The economics of immediate gratification. *Journal of Behavioral Decision Making, 13*, 233–250.

Oechssler, J., Roider, A., & Schmitz, P. W. (2009). Cognitive abilities and behavioral biases. *Journal of Economic Behavior and Organization, 72*, 147–152.

Offit, P. A. (2008). *Autism's false prophets*. New York: Columbia University Press.

Offit, P. A. (2014, September 25). The anti-vaccination epidemic. *Wall Street Journal*, p. A21.

Okan, Y., Garcia-Retamero, R., Cokely, E. T., & Maldonado, A. (2012). Individual differences in graph literacy: Overcoming denominator neglect in risk comprehension. *Journal of Behavioral Decision Making, 25*, 390–401.

Oliver, J. E., & Wood, T. J. (2014). Conspiracy theories and the paranoid style(s) of mass opinion. *American Journal of Political Science, 58*, 952–966.

Orwell, G. (1968). *As I Please, 1943–1945: The Collected Essays, Journalism and Letters of George Orwell* (Vol. 3). S. Orwell & I. Angus (Eds.). New York: Harcourt Brace Jovanovich.

Osherson, D. N. (1995). Probability judgment. In E. E. Smith & D. N. Osherson (Eds.), *Thinking* (Vol. 3, pp. 35–75). Cambridge, MA: MIT Press.

Over, D. E. (2000). Ecological rationality and its heuristics. *Thinking and Reasoning, 6*, 182–192.

Over, D. E. (2002). The rationality of evolutionary psychology. In J. L. Bermudez & A. Millar (Eds.), *Reason and nature: Essays in the theory of rationality* (pp. 187–207). Oxford: Oxford University Press.

Over, D. E. (2004). Rationality and the normative/descriptive distinction. In D. J. Koehler & N. Harvey (Eds.), *Blackwell handbook of judgment and decision making* (pp. 3–18). Malden, MA: Blackwell.

Oyer, B., Gillespie, M., Issah, M., & Fasko, D. (2012). The role of personality in argument evaluation. *Inquiry: Critical Thinking across the Disciplines, 27*, 40–49.

Pacini, R., & Epstein, S. (1999). The relation of rational and experiential information processing styles to personality, basic beliefs, and the ratio-bias phenomenon. *Journal of Personality and Social Psychology, 76,* 972–987.

Palmer, E. C., Gilleen, J., & David, A. S. (2015). The relationship between cognitive insight and depression in psychosis and schizophrenia: A review and meta-analysis. *Schizophrenia Research, 166,* 261–268.

Parker, A. M., & Fischhoff, B. (2005). Decision-making competence: External validation through an individual differences approach. *Journal of Behavioral Decision Making, 18,* 1–27.

Parker, A. M., Bruine de Bruin, W., & Fischhoff, B. (2015). Negative decision outcomes are more common among people with lower decision-making competence: An item-level analysis of the Decision Outcome Inventory (DOI). *Frontiers in Psychology, 6,* 363. doi:10.3389/fpsyg.2015.00363.

Parker, A. M., Bruine de Bruin, W., Yoong, J., & Willis, R. (2012). Inappropriate confidence and retirement planning: Four studies with a national sample. *Journal of Behavioral Decision Making, 25*(4), 382–389.

Parker, A. M., & Stone, E. R. (2014). Identifying the effects of unjustified confidence versus overconfidence: Lessons learned from two analytic methods. *Journal of Behavioral Decision Making, 27,* 134–145.

Pashler, H., McDaniel, M., Rohrer, D., & Bjork, R. (2009). Learning styles: Concepts and evidence. *Psychological Science in the Public Interest, 9,* 105–119.

Paulhus, D. L. (1991). Measurement and control of response bias. In J. P. Robinson, P. Shaver, & L. S. Wrightsman (Eds.), *Measures of personality and social psychological attitudes* (pp. 17–59). San Diego, CA: Academic Press.

Paulos, J. A. (2003). *A mathematician plays the stock market.* New York: Basic Books.

Pennycook, G., Cheyne, J. A., Seli, P., Koehler, D. J., & Fugelsang, J. A. (2012). Analytic cognitive style predicts religious and paranormal belief. *Cognition, 123,* 335–346.

Pennycook, G., Fugelsang, J. A., & Koehler, D. J. (2015). What makes us think? A three-stage dual-process model of analytic engagement. *Cognitive Psychology, 80,* 34–72.

Perkins, D. N. (1985). Postprimary education has little impact on informal reasoning. *Journal of Educational Psychology, 77,* 562–571.

Perkins, D. N. (1995). *Outsmarting IQ: The emerging science of learnable intelligence.* New York: Free Press.

Perkins, D. N. (2002). The engine of folly. In R. J. Sternberg (Ed.), *Why smart people can be so stupid* (pp. 64–85). New Haven, CT: Yale University Press.

Peters, E. (2012). Beyond comprehension: The role of numeracy in judgments and decisions. *Current Directions in Psychological Science, 21*, 31–35.

Peters, E., Vastfjall, D., Slovic, P., Mertz, C. K., Mazzocco, K., & Dickert, S. (2006). Numeracy and decision making. *Psychological Science, 17*, 407–413.

Petry, N. M. (2001). Substance abuse, pathological gambling, and impulsivity. *Drug and Alcohol Dependence, 63*, 29–38.

Petry, N. M. (2005). *Pathological gambling: Etiology, comorbidity, and treatment.* Washington, DC: American Psychological Association.

Petry, N. M., Bickel, W. K., & Arnett, M. (1998). Shortened time horizons and insensitivity to future consequences in heroin addicts. *Addiction, 93*, 729–738.

Petry, N. M., & Casarella, T. (1999). Excessive discounting of delayed rewards in substance abusers with gambling problems. *Drug and Alcohol Dependence, 56*, 25–32.

Pfeifer, P. E. (1994). Are we overconfident in the belief that probability forecasters are overconfident? *Organizational Behavior and Human Decision Processes, 58*, 203–213.

Pinker, S. (1997). *How the mind works.* New York: Norton.

Pinto-Prades, J., Martinez-Perez, J., & Abellán-Perpiñán, J. (2006). The influence of the ratio bias phenomenon on the elicitation of health states utilities. *Judgment and Decision Making, 1*, 118–133.

Poletiek, F. H. (2001). *Hypothesis testing behaviour.* Hove, UK: Psychology Press.

Politzer, G., & Macchi, L. (2000). Reasoning and pragmatics. *Mind and Society, 1*, 73–93.

Pollard, P., & Evans, J. St. B. T. (1987). Content and context effects in reasoning. *American Journal of Psychology, 100*, 41–60.

Polonioli, A. (2015). Stanovich's arguments against the "Adaptive Rationality" Project: An assessment. *Studies in History and Philosophy of Biological and Biomedical Sciences, 40*, 55–62.

Postman, N. (1988). *Conscientious objections.* New York: Vintage Books.

Prado, J., & Noveck, I. A. (2007). Overcoming perceptual features in logical reasoning: A parametric functional magnetic resonance imaging study. *Journal of Cognitive Neuroscience, 19*, 642–657.

Prencipe, A., Kesek, A., Cohen, J., Lamm, C., Lewis, M. D., & Zelazo, P. D. (2006). Development of hot and cool executive function during the transition to adolescence. *Journal of Experimental Child Psychology, 108*, 621–637.

Primi, C., Morsanyi, K., Chiesi, F., Donati, M. A., & Hamilton, J. (2016). The development and testing of a new version of the Cognitive Reflection Test applying item response theory (IRT). *Journal of Behavioral Decision Making*.

Pronin, E. (2006). Perception and misperception of bias in human judgment. *Trends in Cognitive Sciences, 11*, 37–43.

Pronin, E., Lin, D. Y., & Ross, L. (2002). The bias blind spot: Perceptions of bias in self versus others. *Journal of Personality and Social Psychology Bulletin, 28*, 369–381.

Pyszczynski, T., & Greenberg, J. (1987). Toward an integration of cognitive and motivational perspectives on social inference: A biased hypothesis-testing model. In L. Berkowitz (Ed.), *Advances in Experimental Social Psychology* (Vol. 20, pp. 297–340). San Diego: Academic Press.

Rabin, M. (2000a). Diminishing marginal utility of wealth cannot explain risk aversion. In. D. Kahneman & A. Tversky (Eds.), *Choices, values, and frames* (pp. 202–208). New York: Cambridge University Press.

Rabin, M. (2000b). Risk aversion and expected-utility theory: A calibration theorem. *Econometrica, 68*, 1281–129.

Rachlin, H. (2000). *The science of self-control*. Cambridge, MA: Harvard University Press.

Radford, B. (2005). Ringing false alarms: Skepticism and media scares. *Skeptical Inquirer, 29*(2), 34–39.

Radford, B. (2009). Psychic exploits horrific abduction case. *Skeptical Inquirer, 33*(6), 6–7.

Radford, B. (2010). The psychic and the serial killer. *Skeptical Inquirer, 34*(2), 32–37.

Rakow, T., Newell, B. R., & Zougkou, K. (2010). The role of working memory in information acquisition and decision making: Lessons from the binary prediction task. *Quarterly Journal of Experimental Psychology, 63*, 1335–1360.

Rapoport, A., & Chammah, A. (1965). *Prisoner's dilemma*. Ann Arbor, MI: University of Michigan Press.

Real, L. A. (1991). Animal choice behavior and the evolution of cognitive architecture. *Science, 253*, 980–986.

Redelmeier, D. A., & Shafir, E. (1995). Medical decision making in situations that offer multiple alternatives. *Journal of the American Medical Association, 273*, 302–305.

Reimers, S., Maylor, E. A., Stewart, N., & Chater, N. (2009). Associations between a one-shot delay discounting measure and age, income, education and real-world impulsive behavior. *Personality and Individual Differences, 47*, 973–978.

Resnik, M. D. (1987). *Choices: An introduction to decision theory*. Minneapolis: University of Minnesota Press.

Revlin, R., Leirer, V., Yopp, H., & Yopp, R. (1980). The belief bias effect in formal reasoning: The influence of knowledge on logic. *Memory and Cognition, 8,* 584–592.

Reyna, V. F. (1991). Class inclusion, the conjunction fallacy, and other cognitive illusions. *Developmental Review, 11,* 317–336.

Reyna, V. F. (2004). How people make decisions that involve risk. *Current Directions in Psychological Science, 13,* 60–66.

Reyna, V. F., & Brainerd, C. J. (2007). The importance of mathematics in health and human judgment: Numeracy, risk communication, and medical decision making. *Learning and Individual Differences, 17,* 147–159.

Reyna, V. F., & Brainerd, C. J. (2008). Numeracy, ratio bias, and denominator neglect in judgments of risk and probability. *Learning and Individual Differences, 18,* 89–107.

Reyna, V. F., & Brust-Renck, P. G. (2015). A review of theories of numeracy: Psychological mechanisms and implications for medical decision-making. In B. L. Anderson (Ed.), *Numerical reasoning in judgments and decision making about health* (pp. 215–251). Cambridge: Cambridge University Press.

Reyna, V. F., & Ellis, S. (1994). Fuzzy-trace theory and framing effects in children's risky decision making. *Psychological Science, 5,* 275–279.

Reyna, V. F., Nelson, W. L., Han, P. K., & Dieckmann, N. F. (2009). How numeracy influences risk comprehension and medical decision making. *Psychological Bulletin, 135,* 943–973.

Richerson, P. J., & Boyd, R. (2005). *Not by genes alone: How culture transformed human evolution*. Chicago: University of Chicago Press.

Rips, L. J. (1994). *The logic of proof*. Cambridge, MA: MIT Press.

Roberts, M. J., & Newton, E. J. (2001). Inspection times, the change task, and the rapid-response selection task. *Quarterly Journal of Experimental Psychology, 54A,* 1031–1048.

Rodriguez, M. L., Mischel, W., & Shoda, Y. (1989). Cognitive person variables in delay of gratification of older children at risk. *Journal of Personality and Social Psychology, 57,* 358–367.

Rogers, P. (1998). The cognitive psychology of lottery gambling: A theoretical review. *Journal of Gambling Studies, 14,* 111–134.

Rokeach, M. (1960). *The open and closed mind*. New York: Basic Books.

Roney, C. J. R., & Sansone, N. (2015). Explaining the gambler's fallacy: Testing a gestalt explanation versus the "law of small numbers." *Thinking and Reasoning, 21,* 193–205.

Ronan, K. R., Kendall, P. C., & Rowe, M. (1994). Negative affectivity in children: Development and validation of a self-statement questionnaire. *Cognitive Therapy and Research, 18,* 509–528.

Roney, C., & Trick, L. (2009). Sympathetic magic and perceptions of randomness: The hot hand versus the gambler's fallacy. *Thinking and Reasoning, 15,* 197–210.

Ronis, D. L., & Yates, J. F. (1987). Components of probability judgment accuracy: Individual consistency and effects of subject matter and assessment method. *Organizational Behavior and Human Decision Processes, 40,* 193–218.

Ronson, J. (2015). *So you've been publicly shamed.* New York: Riverhead.

Ross, M., Grossmann, I., & Schryer, E. (2014). Contrary to psychological and popular opinion, there is no compelling evidence that older adults are disproportionately victimized by consumer fraud. *Perspectives on Psychological Science, 9,* 427–442.

Roszkowski, M. J., & Snelbecker, G. E. (1990). Effects of "framing" on measures of risk tolerance: Financial planners are not immune. *Journal of Behavioral Economics, 19,* 237–246.

The Royal Swedish Academy of Sciences (2002a). *Advanced information on the Prize in Economic Sciences 2002.* http://nobelprize.org/nobel_prizes/economics/laureates/2002/ecoadv02.pdf (retrieved August 6, 2007).

The Royal Swedish Academy of Sciences. (2002b). *Press release: The Bank of Sweden Prize in Economic Sciences in memory of Alfred Nobel 2002 information for the public.* http://www.nobel.se/economics/laureates/2002/press.html (retrieved August 6, 2007).

Russo, J. E., & Schoemaker, P. J. H. (1992). Managing overconfidence. *Sloan Management Review, 33*(2), 7–17.

Sá, W., Kelley, C., Ho, C., & Stanovich, K. E. (2005). Thinking about personal theories: Individual differences in the coordination of theory and evidence. *Personality and Individual Differences, 38,* 1149–1161.

Sá, W., & Stanovich, K. E. (2001). The domain specificity and generality of mental contamination: Accuracy and projection in judgments of mental content. *British Journal of Psychology, 92,* 281–302.

Sá, W., West, R. F., & Stanovich, K. E. (1999). The domain specificity and generality of belief bias: Searching for a generalizable critical thinking skill. *Journal of Educational Psychology, 91,* 497–510.

Salthouse, T. A., Atkinson, T. M., & Berish, D. E. (2003). Executive functioning as a potential mediator of age-related cognitive decline in normal adults. *Journal of Experimental Psychology: General, 132*, 566–594.

Salthouse, T. A., & Pink, J. E. (2008). Why is working memory related to fluid intelligence? *Psychonomic Bulletin and Review, 15*, 364–371.

Samuels, R. (2005). The complexity of cognition: Tractability arguments for massive modularity. In P. Carruthers, S. Laurence, & S. Stich (Eds.), *The innate mind* (pp. 107–121). Oxford: Oxford University Press.

Samuels, R. (2009). The magical number two, plus or minus: Dual-process theory as a theory of cognitive kinds. In J. St. B. T. Evans & K. Frankish (Eds.), *In two minds: Dual processes and beyond* (pp. 129–146). Oxford: Oxford University Press.

Samuels, R., & Stich, S. P. (2004). Rationality and psychology. In A. R. Mele & P. Rawling (Eds.), *The Oxford handbook of rationality* (pp. 279–300). Oxford: Oxford University Press.

Samuelson, P. L., & Church, I. M. (2015). When cognition turns vicious: Heuristics and biases in light of virtue epistemology. *Philosophical Psychology, 28*, 1095–1113.

Samuelson, W., & Zeckhauser, R. J. (1988). status quo bias in decision making. *Journal of Risk and Uncertainty, 1*, 7–59.

Satz, D., & Ferejohn, J. (1994). Rational choice and social theory. *Journal of Philosophy, 91*, 71–87.

Savage, L. J. (1954). *The foundations of statistics*. New York: Wiley.

Schaller, M., Asp, C. H., Roseil, M. C., & Heim, S. J. (1996). Training in statistical reasoning inhibits the formation of erroneous group stereotypes. *Personality and Social Psychology Bulletin, 22*, 829–844.

Schapira, M. M., Walker, C. M., Cappaert, K. J., Ganschow, P. S., Fletcher, K. E., McGinley, E. L, & Jacobs, E. A. (2012). The Numeracy Understanding in Medicine Instrument: A measure of health numeracy developed using Item Response Theory. *Medical Decision Making, 32*, 851–865.

Scheffler, I. (1991). *In praise of the cognitive emotions*. New York: Routledge.

Schick, F. (1986). Dutch bookies and money pumps. *Journal of Philosophy, 83*, 112–119.

Schlam, T. R., Wilson, N. L., Shoda, Y., Mischel, W., & Ayduk, O. (2013). Preschoolers' delay of gratification predicts their body mass 30 years later. *Journal of Pediatrics, 162*, 90–93.

Schmidt, F. L., & Hunter, J. (2004). General mental ability in the world of work: Occupational attainment and job performance. *Journal of Personality and Social Psychology, 86*, 162–173.

Schneider, S. L. (1992). Framing and conflict: Aspiration level contingency, the status quo, and current theories of risky choice. *Journal of Experimental Psychology: Learning, Memory, and Cognition, 18*, 1040–1057.

Schneider, W., & Chein, J. (2003). Controlled and automatic processing: Behavior, theory, and biological processing. *Cognitive Science, 27*, 525–559.

Schommer, M. (1990). Effects of beliefs about the nature of knowledge on comprehension. *Journal of Educational Psychology, 82*, 498–504.

Schommer, M. (1993). Epistemological development and academic performance among secondary students. *Journal of Educational Psychology, 85*, 406–411.

Schommer-Aikins, M. (2004). Explaining the epistemological belief system: Introducing the embedded systemic model and coordinated research approach. *Educational Psychologist, 39*, 19–30.

Schoorman, F. D., Mayer, R. C., Douglas, C. A., & Hetrick, C. T. (1994). Escalation of commitment and the framing effect: An empirical investigation. *Journal of Applied Social Psychology, 24*, 509–528.

Schuck-Paim, C., & Kacelnik, A. (2002). Rationality in risk-sensitive foraging choices by starlings. *Animal Behaviour, 64*, 869–879.

Schuck-Paim, C., Pompilio, L., & Kacelnik, A. (2004). State-dependent decisions cause apparent violations of rationality in animal choice. *PLoS Biology, 2*(12), 2305–2315.

Schwartz, L. M., Woloshin, S., Black, W. C., & Welch, H. G. (1997). The role of numeracy in understanding the benefit of screening mammography. *Annals of Internal Medicine, 127*, 966–972.

Schwarz, N. (1996). *Cognition and communication: Judgmental biases, research methods, and the logic of conversation.* Mahwah, NJ: Erlbaum.

Scopelliti, I., Morewedge, C. K., McCormick, E., Min, H. L., Lebrecht, S., & Kassam, K. S. (2015). Bias blind spot: Structure, measurement, and consequences. *Management Science, 61*, 2468–2486.

Scott, K. M., de Wit, I., & Deary, I. J. (2006). Spotting books and countries: New approaches to estimating and conceptualizing prior intelligence. *Intelligence, 34*, 429–436.

Senecal, N., Wang, T., Thompson, E., & Kable, J. W. (2012). Normative arguments from experts and peers reduce delay discounting. *Judgment and Decision Making, 7*, 568–589.

Shaffer, R., & Jadwiszczok, A. (2010). Psychic defective: Sylvia Browne's history of failure. *Skeptical Inquirer, 34*(2), 38–42.

Shafir, E. (1993). Intuitions about rationality and cognition. In K. Manktelow & D. Over (Eds.), *Rationality: Psychological and philosophical perspectives* (pp. 260–283). London: Routledge.

Shafir, E. (1994). Uncertainty and the difficulty of thinking through disjunctions. *Cognition, 50*, 403–430.

Shafir, E. (1998). Philosophical intuitions and cognitive mechanisms. In M. R. DePaul & W. Ramsey (Eds.), *Rethinking intuition: The psychology of intuition and its role in philosophical inquiry* (pp. 59–83). Lanham, MD: Rowman & Littlefield.

Shafir, E. (Ed.). (2003). *Preference, belief, and similarity: Selected writings of Amos Tversky*. Cambridge, MA: MIT Press.

Shafir, E., & LeBoeuf, R. A. (2002). Rationality. *Annual Review of Psychology, 53*, 491–517.

Shafir, E., & Tversky, A. (1992). Thinking through uncertainty: Nonconsequential reasoning and choice. *Cognitive Psychology, 24*, 449–474.

Shafir, E., & Tversky, A. (1995). Decision making. In E. E. Smith & D. N. Osherson (Eds.), *Thinking* (Vol. 3, pp. 77–100). Cambridge, MA: MIT Press.

Shamosh, N. A., & Gray, J. R. (2008). Delay discounting and intelligence: A meta-analysis. *Intelligence, 36*, 289–305.

Shanks, D. R. (1995). Is human learning rational? *Quarterly Journal of Experimental Psychology, 48A*, 257–279.

Shanks, D. R., Tunney, R. J., & McCarthy, J. D. (2002). A re-examination of probability matching and rational choice. *Journal of Behavioral Decision Making, 15*, 233–250.

Shapiro, J. M. (2005). Is there a daily discount rate? Evidence from the food stamp nutrition cycle. *Journal of Public Economics, 89*, 303–325.

Sharot, T. (2011). *Optimism bias*. New York: Pantheon.

Shenhav, A., Rand, D. G., & Greene, J. D. (2012). Divine intuition: Cognitive style influences belief in God. *Journal of Experimental Psychology: General, 141*, 423–428.

Shepperd, J., Klein, W., Waters, E., & Weinstein, N. (2013). Taking stock of unrealistic optimism. *Perspectives on Psychological Science, 8*, 395–411.

Shepperd, J. A., Waters, E., Weinstein, N. D., & Klein, W. M. (2015). A primer on unrealistic optimism. *Current Directions in Psychological Science, 24*, 232–237.

Sher, S., & McKenzie, C. R. M. (2006). Information leakage from logically equivalent frames. *Cognition, 101*, 467–494.

Sherman, J. W., Gawronski, B., & Trope, Y. (2014). *Dual-process theories of the social mind*. New York: Guilford.

Shiffrin, R. M., & Schneider, W. (1977). Controlled and automatic human information processing: II. Perceptual learning, automatic attending, and a general theory. *Psychological Review, 84*, 127–190.

Shiloh, S., Salton, E., & Sharabi, D. (2002). Individual differences in rational and intuitive thinking styles as predictors of heuristic responses and framing effects. *Personality and Individual Differences, 32*, 415–429.

Shipley, W., Gruber, C., Martin, T., & Klein, A. (2009). *Shipley-2*. Los Angeles: Western Psychological Services.

Shiv, B., Loewenstein, G., Bechara, A., Damasio, H., & Damasio, A. R. (2005). Investment behavior and the negative side of emotion. *Psychological Science, 16*, 435–439.

Shleifer, A. (2012). Psychologists at the gate: A review of Daniel Kahneman's *Thinking, fast and slow*. *Journal of Economic Literature, 50*, 1080–1091.

Sieck, W. R., & Arkes, H. R. (2005). The recalcitrance of overconfidence and its contribution to decision aid neglect. *Journal of Behavioral Decision Making, 18*, 29–53.

Sieck, W., & Yates, J. F. (1997). Exposition effects on decision making: Choice and confidence in choice. *Organizational Behavior and Human Decision Processes, 70*, 207–219.

Silverman, I. W. (2003a). Gender differences in delay of gratification: A meta-analysis. *Sex Roles, 49*, 451–463.

Silverman, I. W. (2003b). Gender differences in resistance to temptation: Theories and evidence. *Developmental Review, 23*, 219–259.

Simmons, J. P., LeBoeuf, R. A., & Nelson, L. D. (2010). The effect of accuracy motivation on anchoring and adjustment: Do people adjust from provided anchors? *Journal of Personality and Social Psychology, 99*, 917–932.

Simmons, J. P., & Massey, C. (2012). Is optimism real? *Journal of Experimental Psychology: General, 141*, 630–634.

Simon, A. F., Fagley, N. S., & Halleran, J. G. (2004). Decision framing: Moderating effects of individual differences and cognitive processing. *Journal of Behavioral Decision Making, 17*, 77–93.

Simon, H. A. (1955). A behavioral model of rational choice. *Quarterly Journal of Economics, 69*, 99–118.

Simon, H. A. (1956). Rational choice and the structure of the environment. *Psychological Review, 63*, 129–138.

Simoneau, M., & Markovits, H. (2003). Reasoning with premises that are not empirically true: Evidence for the role of inhibition and retrieval. *Developmental Psychology, 39*, 964–975.

Simonson, I. (2008). Will I like a medium pillow? Another look at constructed and inherent preferences. *Journal of Consumer Psychology, 18*, 155–169.

Sinatra, G. M., & Pintrich, P. R. (Eds.). (2003). *Intentional conceptual change.* Mahwah, NJ: Erlbaum.

Singh, K., Spencer, A., & Brennan, D. (2007). Effects of water fluoride exposure at crown completion and maturation on caries of permanent first molars. *Caries Research, 41*, 34–42.

Sivak, M., & Flannagan, M. J. (2003). Flying and driving after the September 11 attacks. *American Scientist, 91*, 6–7.

Skenazy, L. (2009). *Free-range kids.* San Francisco: Jossey-Bass.

Skyrms, B. (1996). *The evolution of the social contract.* Cambridge: Cambridge University Press.

Sloman, A., & Chrisley, R. (2003). Virtual machines and consciousness. *Journal of Consciousness Studies, 10*, 133–172.

Sloman, S. A. (1996). The empirical case for two systems of reasoning. *Psychological Bulletin, 119*, 3–22.

Sloman, S. A., Over, D., Slovak, L., & Stibel, J. M. (2003). Frequency illusions and other fallacies. *Organizational Behavior and Human Decision Processes, 91*, 296–309.

Slovic, P. (1995). The construction of preference. *American Psychologist, 50*, 364–371.

Slovic, P. (2007). "If I look at the mass I will never act": Psychic numbing and genocide. *Judgment and Decision Making, 2*, 79–95.

Slovic, P., Finucane, M. L., Peters, E., & MacGregor, D. G. (2002). The affect heuristic. In T. Gilovich, D. Griffin, & D. Kahneman (Eds.), *Heuristics and biases: The psychology of intuitive judgment* (pp. 397–420). New York: Cambridge University Press.

Slovic, P., Monahan, J., & MacGregor, D. G. (2000). Violence risk assessment and risk communication: The effects of using actual cases, providing instruction, and employing probability versus frequency formats. *Law and Human Behavior, 24*, 271–296.

Slovic, P., & Peters, E. (2006). Risk perception and affect. *Current Directions in Psychological Science, 15*, 322–325.

Slusher, M. P., & Anderson, C. A. (1996). Using causal persuasive arguments to change beliefs and teach new information: The mediating role of explanation availability and evaluation bias in the acceptance of knowledge. *Journal of Educational Psychology, 88*, 110–122.

Smith, E. R., & DeCoster, J. (2000). Dual-process models in social and cognitive psychology: Conceptual integration and links to underlying memory systems. *Personality and Social Psychology Review, 4*, 108–131.

Smith, S. M., & Levin, I. P. (1996). Need for cognition and choice framing effects. *Journal of Behavioral Decision Making, 9*, 283–290.

Smith, A. R., & Windschitl, P. D. (2015). Resisting anchoring effects: The roles of metric and mapping knowledge. *Memory and Cognition, 43*, 1071–1084.

Sniezek, J. A., & Buckley, T. (1991). Confidence depends on level of aggregation. *Journal of Behavioral Decision Making, 4*, 263–272.

Smullyan, R. M. (1978). *What is the name of this book? The riddle of Dracula and other logical puzzles*. Englewood Cliffs, NJ: Prentice Hall.

Soll, J., & Klayman, J. (2004). Overconfidence in interval estimates. *Journal of Experimental Psychology: Learning, Memory, and Cognition, 30*, 299–314.

Solove, D. J. (2007). *The future of reputation: Gossip, rumor, and privacy on the internet*. New Haven: Yale University Press.

Sorge, G. B., Skilling, T. A., & Toplak, M. E. (2015). Intelligence, executive functions, and decision making as predictors of antisocial behavior in an adolescent sample of justice-involved youth and a community comparison group. *Journal of Behavioral Decision Making, 28*, 477–490.

Sperber, D., Cara, F., & Girotto, V. (1995). Relevance theory explains the selection task. *Cognition, 57*, 31–95.

Stanovich, K. E. (1989). Implicit philosophies of mind—the dualism scale and its relation to religiosity and belief in extrasensory perception. *Journal of Psychology, 123*, 5–23.

Stanovich, K. E. (1993). Dysrationalia: A new specific learning disability. *Journal of Learning Disabilities, 26*, 501–515.

Stanovich, K. E. (1999). *Who is rational? Studies of individual differences in reasoning*. Mahwah, NJ: Erlbaum.

Stanovich, K. E. (2000). *Progress in understanding reading: Scientific foundations and new frontiers*. New York: Guilford Press.

Stanovich, K. E. (2003). The fundamental computational biases of human cognition: Heuristics that (sometimes) impair decision making and problem solving. In J. E. Davidson & R. J. Sternberg (Eds.), *The psychology of problem solving* (pp. 291–342). New York: Cambridge University Press.

Stanovich, K. E. (2004). *The robot's rebellion: Finding meaning in the age of Darwin.* Chicago: University of Chicago Press.

Stanovich, K. E. (2005). The future of a mistake: Will discrepancy measurement continue to make the learning disabilities field a pseudoscience? *Learning Disability Quarterly, 28,* 103–106.

Stanovich, K. E. (2008). Higher-order preferences and the Master Rationality Motive. *Thinking and Reasoning, 14,* 111–127.

Stanovich, K. E. (2009). *What intelligence tests miss: The psychology of rational thought.* New Haven, CT: Yale University Press.

Stanovich, K. E. (2010). *Decision making and rationality in the modern world.* New York: Oxford University Press.

Stanovich, K. E. (2011). *Rationality and the reflective mind.* New York: Oxford University Press.

Stanovich, K. E. (2012). On the distinction between rationality and intelligence: Implications for understanding individual differences in reasoning. In K. Holyoak & R. Morrison (Eds.), *The Oxford handbook of thinking and reasoning* (pp. 343–365). New York: Oxford University Press.

Stanovich, K. E. (2013). Why humans are (sometimes) less rational than other animals: Cognitive complexity and the axioms of rational choice. *Thinking and Reasoning, 19,* 1–26.

Stanovich, K. E. (2015). Rational and irrational thought: The thinking that IQ tests miss. *Scientific American Mind Special Collector's Edition, 23*(4), 12–17.

Stanovich, K. E., Cunningham, A. E., & Feeman, D. J. (1984). Intelligence, cognitive skills, and early reading progress. *Reading Research Quarterly, 19,* 278–303.

Stanovich, K. E., Grunewald, M., & West, R. F. (2003). Cost-benefit reasoning in students with multiple secondary school suspensions. *Personality and Individual Differences, 35,* 1061–1072.

Stanovich, K. E., & Siegel, L. S. (1994). Phenotypic performance profile of children with reading disabilities: A regression-based test of the phonological-core variable-difference model. *Journal of Educational Psychology, 86,* 24.

Stanovich, K. E., & Toplak, M. E. (2012). Defining features versus incidental correlates of Type 1 and Type 2 processing. *Mind and Society, 11,* 3–13.

Stanovich, K. E., Toplak, M. E., & West, R. F. (2008). The development of rational thought: A taxonomy of heuristics and biases. *Advances in Child Development and Behavior, 36,* 251–285.

Stanovich, K. E., & West, R. F. (1997). Reasoning independently of prior belief and individual differences in actively open-minded thinking. *Journal of Educational Psychology, 89*, 342–357.

Stanovich, K. E., & West, R. F. (1998a). Cognitive ability and variation in selection task performance. *Thinking and Reasoning, 4*, 193–230.

Stanovich, K. E., & West, R. F. (1998b). Individual differences in framing and conjunction effects. *Thinking and Reasoning, 4*, 289–317.

Stanovich, K. E., & West, R. F. (1998c). Individual differences in rational thought. *Journal of Experimental Psychology: General, 127*, 161–188.

Stanovich, K. E., & West, R. F. (1998d). Who uses base rates and P(D/~H)? An analysis of individual differences. *Memory and Cognition, 26*, 161–179.

Stanovich, K. E., & West, R. F. (1999). Discrepancies between normative and descriptive models of decision making and the understanding/acceptance principle. *Cognitive Psychology, 38*, 349–385.

Stanovich, K. E., & West, R. F. (2000). Individual differences in reasoning: Implications for the rationality debate? *Behavioral and Brain Sciences, 23*, 645–726.

Stanovich, K. E., & West, R. F. (2007). Natural myside bias is independent of cognitive ability. *Thinking and Reasoning, 13*, 225–247.

Stanovich, K. E., & West, R. F. (2008a). On the failure of intelligence to predict myside bias and one-sided bias. *Thinking and Reasoning, 14*, 129–167.

Stanovich, K. E., & West, R. F. (2008b). On the relative independence of thinking biases and cognitive ability. *Journal of Personality and Social Psychology, 94*, 672–695.

Stanovich, K. E., West, R. F., & Harrison, M. R. (1995). Knowledge growth and maintenance across the life span: The role of print exposure. *Developmental Psychology, 31*, 811.

Stanovich, K. E., West, R. F., & Toplak, M. E. (2011a). Intelligence and rationality. In R. J. Sternberg & S. B. Kaufman (Eds.), *Cambridge handbook of intelligence* (pp. 784–826). New York: Cambridge University Press.

Stanovich, K. E., West, R. F., & Toplak, M. E. (2011b). The complexity of developmental predictions from dual process models. *Developmental Review, 31*, 103–118.

Stanovich, K. E., West, R. F., & Toplak, M. E. (2012). Individual differences as essential components of heuristics and biases research. In K. Manktelow, D. Over, & S. Elqayam (Eds.), *The science of reason* (pp. 335–396). New York: Psychology Press.

Stanovich, K. E., West, R. F., & Toplak, M. E. (2013). Myside bias, rational thinking, and intelligence. *Current Directions in Psychological Science, 22*, 259–264.

Statman, M., Thorley, S., & Vorkink, K. (2006). Investor overconfidence and trading volume. *Review of Financial Studies, 19*, 1531–1565.

Staudinger, U. M., Dorner, J., & Mickler, C. (2005). Wisdom and personality. In R. J. Sternberg & J. Jordan (Eds.), *A handbook of wisdom: Psychological perspectives* (pp. 191–219). New York: Cambridge University Press.

Steen, L. A. (1990). Numeracy. *Daedalus, 119*, 211–231.

Stein, E. (1996). *Without good reason: The rationality debate in philosophy and cognitive science*. Oxford: Oxford University Press.

Steinberg, L., Graham, S., O'Brien, L., Woolard, J., Cauffman, E., & Banich, M. (2009). Age differences in future orientation and delay discounting. *Child Development, 80*, 28–44.

Sterelny, K. (1990). *The representational theory of mind: An introduction*. Oxford: Blackwell.

Sterelny, K. (2001). *The evolution of agency and other essays*. Cambridge: Cambridge University Press.

Sterelny, K. (2003). *Thought in a hostile world: The evolution of human cognition*. Malden, MA: Blackwell.

Sternberg, R. J. (1988). *The triarchic mind*. New York: Viking.

Sternberg, R. J. (1989). Domain-generality versus domain-specificity: The life and impending death of a false dichotomy. *Merrill-Palmer Quarterly, 35*, 115–130.

Sternberg, R. J. (2001). Why schools should teach for wisdom: The balance theory of wisdom in educational settings. *Educational Psychologist, 36*, 227–245.

Sternberg, R. J. (Ed.). (2002). *Why smart people can be so stupid*. New Haven, CT: Yale University Press.

Sternberg, R. J. (2003). *Wisdom, intelligence, and creativity synthesized*. Cambridge: Cambridge University Press.

Sternberg, R. J., & Grigorenko, E. L. (1997). Are cognitive styles still in style? *American Psychologist, 52*, 700–712.

Sternberg, R. J., Grigorenko, E. L., & Zhang, L. F. (2008). Styles of learning and thinking matter in instruction and assessment. *Perspectives on Psychological Science, 3*, 486–506.

Stewart, N. (2009). The cost of anchoring on credit-card minimum repayments. *Psychological Science, 20*, 39–41.

Stigler, S. M. (1983). Who discovered Bayes's theorem? *American Statistician, 37*, 290–296.

Stigler, S. M. (1986). *The history of statistics: The measurement of uncertainty before 1900*. Cambridge, MA: Harvard University Press.

Strathman, A., Gleicher, F., Boninger, D. S., & Edwards, C. S. (1994). The consideration of future consequences: Weighing immediate and distant outcomes of behavior. *Journal of Personality and Social Psychology, 66*, 742–752.

Strough, J. N., Karns, T. E., & Schlosnagle, L. (2011). Decision-making heuristics and biases across the life span. *Annals of the New York Academy of Sciences, 1235*, 57–74.

Strough, J., Parker, A. M., & Bruine de Bruin, W. (2015). Understanding life-span developmental changes in decision-making competence. In T. Hess, J. Strough, & C. Löckenhoff (Eds.), *Aging and decision making: Empirical and applied perspectives* (pp. 235–257). London: Elsevier Academic.

Stuebing, K., Fletcher, J. M., LeDoux, J. M., Lyon, G. R., Shaywitz, S. E., & Shaywitz, B. A. (2002). Validity of IQ-discrepancy classification of reading difficulties: A meta-analysis. *American Educational Research Journal, 39*, 469–518.

Stupple, E., Gale, M., & Richmond, C. (2013). Working memory, cognitive miserliness and logic as predictors of performance on the cognitive reflection test. In M. Knauff, M. Pauen, N. Sebanz & I. Wachsmuth (Eds.), *Proceedings of the 35th annual conference of the cognitive science society*. Austin, TX: Cognitive Science Society.

Sundali, J., & Croson, R. (2006). Biases in casino betting: The hot hand and the gambler's fallacy. *Judgment and Decision Making, 1*, 1–12.

Sunstein, C. R. (2002). *Risk and reason: Safety, law, and the environment*. Cambridge: Cambridge University Press.

Sunstein, C. R., & Vermeule, A. (2009). Conspiracy theories: Causes and cures. *Journal of Political Philosophy, 17*, 202–227.

Svedholm, A. M., & Lindeman, M. (2013). The separate roles of the reflective mind and involuntary inhibitory control in gatekeeping paranormal beliefs and the underlying intuitive confusions. *British Journal of Psychology, 104*, 303–319.

Swami, V., Coles, R., Stieger, S., Pietschnig, J., Furnham, A., Rehim, S., et al. (2011). Conspiracist ideation in Britain and Austria: Evidence of a monological belief system and associations between individual psychological differences and real-world and fictitious conspiracy theories. *British Journal of Psychology, 102*, 443–463.

Taber, C. S., & Lodge, M. (2006). Motivated skepticism in the evaluation of political beliefs. *American Journal of Political Science, 50*, 755–769.

Takemura, K. (1992). Effect of decision time on framing of decision: A case of risky choice behavior. *Psychologia, 35*, 180–185.

Takemura, K. (1993). The effect of decision frame and decision justification on risky choice. *Japanese Psychological Research, 35*, 36–40.

Takemura, K. (1994). Influence of elaboration on the framing of decision. *Journal of Psychology, 128*, 33–39.

Taylor, S. E. (1981). The interface of cognitive and social psychology. In J. H. Harvey (Ed.), *Cognition, social behavior, and the environment* (pp. 189–211). Hillsdale, NJ: Erlbaum.

Tentori, K., & Crupi, V. (2012). On the conjunction fallacy and the meaning of and, yet again: A reply to Hertwig, Benz, and Krauss (2008). *Cognition, 122*, 123–134.

Tentori, K., Crupi, V., & Russo, S. (2013). On the determinants of the conjunction fallacy: Probability versus inductive confirmation. *Journal of Experimental Psychology: General, 142*, 235–255.

Teovanović, P., Knežević, G., & Stankov, L. (2015). Individual differences in cognitive biases: Evidence against one-factor theory of rationality. *Intelligence, 50*, 75–86.

Terjesen, M. D., Salhany, J., & Sciutto, M. J. (2009). A psychometric review of measures of irrational beliefs: Implications for psychotherapy. *Journal of Rational-Emotive and Cognitive-Behavior Therapy, 27*, 83–96.

Tetlock, P. E. (2005). *Expert political judgment*. Princeton, NJ: Princeton University Press.

Tetlock, P. E., & Mellers, B. A. (2002). The great rationality debate. *Psychological Science, 13*, 94–99.

Thagard, P. (1992). *Conceptual revolutions*. Princeton, NJ: Princeton University Press.

Thaler, R. H. (1981). Some empirical evidence on dynamic inconsistency. *Economics Letters, 8*, 201–207.

Thaler, R. H. (1992). *The winner's curse: Paradoxes and anomalies of economic life*. New York: Free Press.

Thaler, R. H. (2013, October 5). Financial literacy, beyond the classroom. *New York Times*. http://www.nytimes.com/2013/10/06/business/financial-literacy-beyond-the-classroom.html.

Thaler, R. H. (2015). *Misbehaving: The making of behavioral economics*: New York: Norton.

Thaler, R. H., & Benartzi, S. (2004). Save more tomorrow: Using behavioral economics to increase employee saving. *Journal of Political Economy, 112*, s164–s187.

Thaler, R. H., & Sunstein, C. R. (2008). *Nudge: Improving decisions about health, wealth, and happiness*. New Haven, CT: Yale University Press.

Thompson, S. C. (2004). Illusions of control. In R. Pohl (Ed.), *Cognitive illusions: A handbook on fallacies and biases in thinking, judgment and memory* (pp. 115–126). Hove, UK: Psychology Press.

Thompson, V. A. (2009). Dual-process theories: A metacognitive perspective. In J. Evans & K. Frankish (Eds.), *In two minds: Dual processes and beyond* (pp. 171–195). Oxford: Oxford University Press.

Thompson, V. A. (2014). What intuitions are… and are not. In B. Ross (Ed.), *Psychology of learning and motivation* (Vol. 60, pp. 35–75). New York: Elsevier.

Thompson, V., & Evans, J. St. B. T. (2012). Belief bias in informal reasoning. *Thinking and Reasoning, 18*, 278–310.

Thompson, V. A., & Johnson, S. C. (2014). Conflict, metacognition, and analytic thinking. *Thinking and Reasoning, 20*, 215–244.

Thompson, V. A., & Morsanyi, K. (2012). Analytic thinking: Do you feel like it? *Mind & Society, 11*, 93–105.

Thompson, V. A., Prowse Turner, J. A., & Pennycook, G. (2011). Intuition, reason, and metacognition. *Cognitive Psychology, 63*, 107–140.

Toates, F. (2005). Evolutionary psychology: Towards a more integrative model. *Biology and Philosophy, 20*, 305–328.

Toates, F. (2006). A model of the hierarchy of behavior, cognition, and consciousness. *Consciousness and Cognition, 15*, 75–118.

Tobacyk, J., & Milford, G. (1983). Belief in paranormal phenomena. *Journal of Personality and Social Psychology, 44*, 1029–1037.

Todd, P. M., & Gigerenzer, G. (2000). Precis of Simple Heuristics that Make Us Smart. *Behavioral and Brain Sciences, 23*, 727–780.

Todd, P. M., & Gigerenzer, G. (2007). Environments that make us smart: Ecological rationality. *Current Directions in Psychological Science, 16*, 167–171.

Tomlin, D., Rand, D. G., Ludvig, E. A., & Cohen, J. D. (2015). The evolution and devolution of cognitive control: The costs of deliberation in a competitive world. *Scientific Reports, 5*. doi:10.1038/srep11002.

Toneatto, T. (1999). Cognitive psychopathology of problem gambling. *Substance Use and Misuse, 34*, 1593–1604.

Tooby, J., & Cosmides, L. (1992). The psychological foundations of culture. In J. Barkow, L. Cosmides, & J. Tooby (Eds.), *The adapted mind* (pp. 19–136). New York: Oxford University Press.

Toplak, M. E., Liu, E., Macpherson, R., Toneatto, T., & Stanovich, K. E. (2007). The reasoning skills and thinking dispositions of problem gamblers: A dual-process taxonomy. *Journal of Behavioral Decision Making, 20*, 103–124.

Toplak, M. E., Sorge, G. B., Benoit, A., West, R. F., & Stanovich, K. E. (2010). Decision-making and cognitive abilities: A review of associations between Iowa Gambling Task performance, executive functions, and intelligence. *Clinical Psychology Review, 30*, 562–581.

Toplak, M. E., & Stanovich, K. E. (2002). The domain specificity and generality of disjunctive reasoning: Searching for a generalizable critical thinking skill. *Journal of Educational Psychology, 94*, 197–209.

Toplak, M. E., & Stanovich, K. E. (2003). Associations between myside bias on an informal reasoning task and amount of post-secondary education. *Applied Cognitive Psychology, 17*, 851–860.

Toplak, M. E., West, R. F., & Stanovich, K. E. (2011). The Cognitive Reflection Test as a predictor of performance on heuristics and biases tasks. *Memory and Cognition, 39*, 1275–1289.

Toplak, M. E., West, R. F., & Stanovich, K. E. (2014a). Assessing miserly processing: An expansion of the Cognitive Reflection Test. *Thinking and Reasoning, 20*, 147–168.

Toplak, M. E., West, R. F., & Stanovich, K. E. (2014b). Rational thinking and cognitive sophistication: Development, cognitive abilities, and thinking dispositions. *Developmental Psychology, 50*, 1037–1048.

Toplak, M. E., West, R. F., & Stanovich, K. E. (2016). Associations between rational thinking tasks and real-world correlates in a community sample. Manuscript submitted.

Trout, J. D. (2008). Seduction without cause: Uncovering explanatory neurophilia. *Trends in Cognitive Sciences, 12*(8), 281–282.

Tschirgi, J. E. (1980). Sensible reasoning: A hypothesis about hypotheses. *Child Development, 51*, 1–10.

Tversky, A. (1996). Contrasting rational and psychological principles of choice. In R. Zeckhauser, R. Keeney, & J. Sebenius (Eds.), *Wise choices* (pp. 5–21). Boston: Harvard Business School Press.

Tversky, A., & Edwards, W. (1966). Information versus reward in binary choice. *Journal of Experimental Psychology, 71*, 680–683.

Tversky, A., & Kahneman, D. (1971). Belief in the law of small numbers. *Psychological Bulletin, 76*, 105–110.

Tversky, A., & Kahneman, D. (1974). Judgment under uncertainty: Heuristics and biases. *Science, 185*, 1124–1131.

Tversky, A., & Kahneman, D. (1981). The framing of decisions and the psychology of choice. *Science, 211*, 453–458.

Tversky, A., & Kahneman, D. (1982). Evidential impact of base rates. In D. Kahneman, P. Slovic, & A. Tversky (Eds.), *Judgment under uncertainty: Heuristics and biases* (pp. 153–160). Cambridge: Cambridge University Press.

Tversky, A., & Kahneman, D. (1983). Extensional versus intuitive reasoning: The conjunction fallacy in probability judgment. *Psychological Review, 90*, 293–315.

Tversky, A., & Kahneman, D. (1986). Rational choice and the framing of decisions. *Journal of Business, 59*, 251–278.

Tversky, A., & Thaler, R. H. (1990). Anomalies: Preference reversals. *Journal of Economic Perspectives, 4*, 201–211.

Valentine, D. A. (1998, May 13). *Pyramid schemes*. Presented at the International Monetary Fund Seminar on Current Legal Issues Affecting Central Banks. Washington, DC: IMF. Retreived from http://www.ftc.gov/speeches/other/dvimf16.shtm on August 29, 2007.

Vallone, R., Griffin, D. W., Lin, S., & Ross, L. (1990). Overconfident prediction of future actions and outcomes by self and others. *Journal of Personality and Social Psychology, 58*, 582–592.

Vellutino, F., Fletcher, J. M., Snowling, M., & Scanlon, D. M. (2004). Specific reading disability (dyslexia): What have we learned in the past four decades? *Journal of Child Psychology and Psychiatry, and Allied Disciplines, 45*, 2–40.

von Neumann, J., & Morgenstern, O. (1944). *The theory of games and economic behavior*. Princeton: Princeton University Press.

Voss, J. F., Perkins, D. N., & Segal, J. (1991). *Informal reasoning and education*. Hillsdale, NJ: Erlbaum.

Wagenaar, W. A. (1988). *Paradoxes of gambling behavior*. Hove, UK: Erlbaum.

Wainer, H. (1993). Does spending money on education help? A reaction to the Heritage Foundation and the *Wall Street Journal*. *Educational Researcher, 22*(9), 22–24.

Waksberg, A. J., Smith, A. B., & Burd, M. (2009). Can irrational behaviour maximise fitness? *Behavioral Ecology and Sociobiology, 63*, 461–471.

Wang, L. (2009). Money and fame: Vividness effects in the National Basketball Association. *Journal of Behavioral Decision Making, 22*, 20–44.

Wason, P. C. (1960). On the failure to eliminate hypotheses in a conceptual task. *Quarterly Journal of Experimental Psychology, 12*, 129–140.

Wason, P. C. (1966). Reasoning. In B. Foss (Ed.), *New horizons in psychology* (pp. 135–151). Harmonsworth: Penguin.

Wason, P. C. (1969). Regression in reasoning? *British Journal of Psychology, 60*, 471–480.

Wasserman, E. A., Dorner, W. W., & Kao, S. F. (1990). Contributions of specific cell information to judgments of interevent contingency. *Journal of Experimental Psychology: Learning, Memory, and Cognition, 16*, 509–521.

Watson, G., & Glaser, E. M. (2006). *Watson-Glaser Critical Thinking Appraisal: Short Form Manual.* Orlando, FL: Harcourt.

Weinstein, N. (1980). Unrealistic optimism about future life events. *Journal of Personality and Social Psychology, 39*, 806–820.

Weller, J. A., Ceschi, A., & Randolph, C. (2015). Decision-making competence predicts domain-specific risk attitudes. *Frontiers in Psychology, 6*, 540. doi:10.3389/fpsyg.2015.00540.

Weller, J. A., Dieckmann, N. F., Tusler, M., Mertz, C. K., Burns, W. C., & Peters, E. (2013). Development and testing of an abbreviated numeracy scale: A Rasch analysis approach. *Journal of Behavioral Decision Making, 26*, 198–212.

Wellman, H. M. (2011). Developing a theory of mind. In U. Goswami (Ed.), *The Wiley-Blackwell handbook of childhood cognitive development* (2nd ed., pp. 258–284). Oxford: Wiley-Blackwell.

Wellman, H. M., & Liu, D. (2004). Scaling of theory-of-mind tasks. *Child Development, 75*, 523–541.

Welsh, M. B., Delfabbro, P. H., Burns, N. R., & Begg, S. H. (2014). Individual differences in anchoring: Traits and experience. *Learning and Individual Differences, 29*, 131–140.

West, R. F., Meserve, R. J., & Stanovich, K. E. (2012). Cognitive sophistication does not attenuate the bias blind spot. *Journal of Personality and Social Psychology, 103*, 506–519.

West, R. F., & Stanovich, K. E. (1997). The domain specificity and generality of overconfidence: Individual differences in performance estimation bias. *Psychonomic Bulletin and Review, 4*, 387–392.

West, R. F., & Stanovich, K. E. (2003). Is probability matching smart? Associations between probabilistic choices and cognitive ability. *Memory and Cognition, 31*, 243–251.

West, R. F., Toplak, M. E., & Stanovich, K. E. (2008). Heuristics and biases as measures of critical thinking: Associations with cognitive ability and thinking dispositions. *Journal of Educational Psychology, 100*, 930–941.

Westbrook, A., & Braver, T. S. (2015). Cognitive effort: A neuroeconomic approach. *Cognitive, Affective, and Behavioral Neuroscience, 15*, 395–415.

Westen, D., Blagov, P., Kilts, C., & Hamann, S. (2006). Neural bases of motivated reasoning: An fMRI study of emotional constraints on partisan political judgment in the 2004 U.S. Presidential Election. *Journal of Cognitive Neuroscience, 18*, 1947–1958.

Wilson, T. D., & Gilbert, D. T. (2005). Affective forecasting: Knowing what to want. *Current Directions in Psychological Science, 14*, 131–134.

Wilson, T. D., Houston, C. E., Etling, K. M., & Brekke, N. (1996). A new look at anchoring effects: Basic anchoring and its antecedents. *Journal of Experimental Psychology: General, 125*, 387–402.

Wilson, T. D., Wheatley, T., Meyers, J. M., Gilbert, D. T., & Axsom, D. (2000). Focalism: A source of durability bias in affective forecasting. *Journal of Personality and Social Psychology, 78*, 821–836.

Wolfe, C. R., & Britt, M. A. (2008). The locus of the myside bias in written argumentation. *Thinking & Reasoning, 14*, 1–27.

Wood, J. M., Nezworski, M. T., Lilienfeld, S. O., & Garb, H. N. (2003). *What's wrong with the Rorschach?* San Francisco: Jossey-Bass.

Wood, M. J., Douglas, K. M., & Sutton, R. M. (2012). Dead and alive: Beliefs in contradictory conspiracy theories. *Social Psychological and Personality Science, 3*, 767–773.

Wood, S., Hanoch, Y., Barnes, A., Liu, P. J., Cummings, J., Bhattacharya, C., et al. (2011). Numeracy and Medicare, Part D: The importance of choice and literacy for numbers in optimizing decision making for Medicare's prescription drug program. *Psychology and Aging, 26*, 295–307.

Wu, A., & Weseley, A. J. (2013). The effects of statistical format and population specificity on adolescent perceptions of cell phone use while driving. *Current Psychology, 32*, 32–43.

Xu, J., & Harvey, N. (2014). Carry on winning: The gamblers' fallacy creates hot hand effects in online gambling. *Cognition, 131*, 173–180.

Yamagishi, K. (1997). When a 12.86% mortality is more dangerous than 24.14%: Implications for risk communication. *Applied Cognitive Psychology, 11*, 495–506.

Yates, J. F., Zhu, Y., Ronis, D., Wang, D., Shinotsuka, H., & Toda, M. (1989). Probability judgment accuracy: China, Japan, and the United States. *Organizational Behavior and Human Decision Processes, 43*, 145–171.

Yates, J. F., Lee, J., & Bush, J. G. (1997). General knowledge overconfidence: Cross-national variations, response style, and "reality." *Organizational Behavior and Human Decision Processes, 70*, 87–94.

Yechiam, E., Kanz, J., Bechara, A., Stout, J. C., Busemeyer, J. R., Altmaier, E., et al. (2008). Neurocognitive deficits related to poor decision making in people behind bars. *Psychonomic Bulletin and Review, 15*, 44–51.

Zahler, D., & Zahler, K. A. (1988). *Test your cultural literacy*. New York: Simon & Schuster.

Zeidner, M., & Matthews, G. (2000). Intelligence and personality. In R. J. Sternberg (Ed.), *Handbook of intelligence* (pp. 581–610). New York: Cambridge University Press.

Zikmund-Fisher, B. J., Sarr, B., Fagerlin, A., & Ubel, P. A. (2006). A matter of perspective: Choosing for others differs from choosing for yourself in making treatment decisions. *Journal of General Internal Medicine, 21*, 618–622.

Zimmerman, J., Broder, P. K., Shaughnessy, J. J., & Underwood, B. J. (1977). A recognition test of vocabulary using signal-detection measures, and some correlates of word and nonword recognition. *Intelligence, 1*, 5–31.

Zweig, J. (2007, February). Winning the home run hitter's game. *Money Magazine*, p. 102.

Author Index

Page numbers followed by a "*t*" indicate a table.

Abaluck, J., 304*t*
Abellán-Perpiñán, J., 302*t*
Abelson, R. P., 169
Ackerman, P. L., 16, 24, 27, 210
Adomdza, G. K., 164
Ainslie, G., 51, 169, 170, 171, 306*t*
Ajzen, I., 84
Alcock, J., 40
Alhakami, A., 288
Allan, L. G., 107, 108
Allen, J., 122
Alloy, L. B., 107
Allum, N., 197
Ameriks, J., 311*t*
Anderson, C. A., 155
Anderson, J. R., 370*n*2.5
Anderson, R. C., 239
Anderson, S., 211
Ariely, D., 153, 320
Arkes, H., 18, 305*t*, 375*n*8.9
Arnett, M., 290
Arrow, K. J., 142
Arum, R., 277
Ashraf, N., 171
Åstebro, T., 164
Atkinson, T. M., 22
Attridge, N., 130
Axsom, D., 33
Ayduk, O. N., 170, 306*t*
Ayton, P., 18, 88

Babad, E., 158
Baddeley, A. D., 239
Bagby, R. M., 212
Baker, T. B., 299*t*, 300*t*
Baldi, P., 112
Balf, T., 294, 316
Ball, L. J., 30, 288
Baltes, P. B., 168, 275
Banerjee, K., 199
Banks, J., 180, 183, 307*t*
Bar-Hillel, M., 81, 84
Baranski, J. V., 164
Barber, B., 309*t*, 377*n*14.1
Barbey, A. K., 84
Barkow, J. H., 40
Baron-Cohen, S., 286
Baron, J., xi, 9, 25, 28, 90, 91, 95,
    107, 153, 154, 155, 156, 157, 161,
    188, 208, 209, 210, 227, 297, 308*t*,
    375*n*8.7, 377*n*14.1
Barr, N., 302*t*
Barrett, H. C., 18
Barron, G., 88
Barston, J., 122, 373*n*7.1
Bartels, D. M., 278, 374*n*8.4
Basile, A. G., 171, 306*t*
Bateson, M., 41
Baumann, A. A., 171
Baumeister, R. F., 51, 168, 169, 170,
    301*t*

Bazerman, M. H., 297, 325, 308*t*
Beach, L. R., 182
Bechara, A., 211, 212, 290, 291, 376*n*11.2
Beck, A. T., 204
Beck, M., 310*t*
Begley, S., 301*t*
Belton, I. K., 303*t*
Benartzi, S., 304*t*
Benjamin, D. J., 187, 188, 308*t*
Bennett, E. S., 16
Bensley, D. A., 309*t*
Bentz, B. G., 204
Bergman, O., 374*n*8.3
Berish, D. E., 22
Bermudez, J. L., 41, 370*n*2.5
Bernard, M. E., 204
Best, J. R., 19
Beyth-Marom, R., 98, 100, 154
Biais, B., 164
Bickel, W. K., 17, 290. 306*t*
Binmore, K., 40
Birnbaum, M. H., 187
Bishop, M. A., 5, 275
Bjork, R., 300*t*, 376*n*11.2
Bloom, P., 199
Boccaccini, M. T., 305*t*
Bollen, K., 283, 284
Boninger, D. S., 211
Bonini, N., 302*t*, 307*t*
Bonner, C., 70, 132, 133
Boo, H. C., 374*n*8.3
Borsboom, D., 251
Bouchard, T. J., 16
Bouffard-Bouchard, T., 122
Boyd, R., 32
Boyer, P., 199
Boylan, J. F., 317
Brainerd, C. J., 133, 180, 373*n*7.4
Brandstatter, E., 324
Brandt, M. J., 200, 202
Braun, P. A., 164
Braver, T. S., 22, 50

Brekke, N., 374*n*8.3
Brennan, D., 310*t*
Brenner, L., 164, 298*t*, 375*n*8.10
Brevers, D., 290
Brewer, N. T., 374*n*8.3
Britt, M. A., 154
Broder, P. K., 239
Brody, J. E., 192
Brown, L. A., 310*t*
Brown, S. A., 187, 188, 308*t*
Browne, M., 309*t*
Bruine de Bruin, W., 146, 270, 272, 273, 304*t*, 305*t*, 307*t*
Brust-Renck, P. G., 60, 180
Bucchianeri, G. W., 303*t*
Buckley, T., 166
Budescu, D. V., 375*n*8.10
Buehler, R., 164
Burd, M., 40
Burgeno, J. N., 301*t*
Burgess, G. C., 22
Burns, B. D., 88
Bush, J. G., 163, 375*n*8.9
Buss, D. M., 147, 207
Butler, A. C., 204
Butler, H. A., 274
Buturovic, Z., 184
Byrne, R. M. J., 30, 103, 137

Cacioppo, J. T., 25, 27, 144, 210, 311*t*
Calvillo, D. P., 301*t*
Camerer, C. F., 164, 297, 312, 303*t*
Campbell, J. D., 301*t*
Caplin, A., 311*t*
Cara, F., 33
Carroll, J. B., 16, 369*n*2.1
Carruthers, P., 18
Casarella, T., 171
Casscells, W., 83
Cattell, R. B., 16, 369*n*2.1
Cederblom, J., 209
Chabris, C., 305*t*

Chabris, C. F., 22, 171
Chambers, J. R., 200, 202
Chammah, A., 137
Chan, N., 274
Chansky, T. E., 310t
Chao, L., 306t
Chapman, G. B., 19, 113, 189, 374n8.3
Chapman, J., 300t
Chapman, L., 300t
Chater, N., 41, 103, 104, 279, 370n2.5
Chein, J., 17
Chen, H., 183, 184, 373n8.1
Cheng, P. W., 107
Chirstman, S. D., 374n8.3
Choi, S., 308t
Chrisley, R., 370n2.5
Christensen-Szalanski, J., 182
Christensen, A. J., 204
Chuderski, A., 22
Church, I. M., 209
Clark, A., 34
Cleeremans, A., 290
Cohen, J., 226, 241, 243, 256, 257
Cohen, J. D., 17, 170, 241, 256, 257
Cohen, L. H., 204, 241, 256, 257
Cohen, L. J., 5, 241, 256, 257, 275, 279, 280
Cohen, M., 17, 241, 256, 257
Cohen, P., 226, 241, 243, 256, 257
Cokely, E. T., 112, 177, 180, 181, 275, 313, 307t
Collisson, B., 200
Combs, B., 188
Conway, A. R. A., 22
Corpus, B., 88
Corser, R., 17, 70, 133
Corter, J. E., 90
Cosmides, L., 83, 147, 279, 327
Crane, C., 326
Creative Leadership Forum, 12
Critcher, C. R., 374n8.3
Cronan, F., 204
Cronbach, L. J., 24

Croskerry, P., 300t
Croson, R., 88, 298t
Crupi, V., 86
Cunningham, A. E., 315
Cunningham, C. E., 299t
Curtis-Holmes, J., 373n7.1

Dagnall, N., 198
Dale, D., 134
Damasio, A. R., 211, 212, 291
Damasio, H., 211, 291
Das, B., 326
David, A. S., 34, 100, 107, 204, 294
Dawes, R. M., 7, 50, 142, 153, 299t
Dawkins, R., 40
Deary, I. J., 16, 239, 317
DeCoster, J., 17
de Finetti, B., 154
De Houwer, J., 18
DellaVigna, S., 312
Del Missier, F., 270
Denes-Raj, V., 132, 133, 309t
De Neys, W., 43, 47, 70
Dennett, D. C., 24, 39, 40, 152, 199, 370n2.5
de Sousa, R., 4
D'Esposito, M., 376n11.2
Detterman, D. K., 294
Dewberry, C., 301t
de Wit, I., 239
Dhami, M. K., 303t
Dias, M., 373n7.1
Ditto, P., 155
Dittrich, M., 255, 377n14.1
Doherty, M. E., 99, 107, 300t, 372n6.1
Dohmen, T., 299t, 377n14.1
Dolan, R. J., 373n7.1
Dorner, J., 168, 375n8.7
Dorner, W. W., 102, 107
Douglas, K. M., 198
Dowd, K. W., 374n8.3
Duckworth, A. L., 50, 306t
Dukas, R., 40

Dulany, D. E., 85
Duncan, J., 16, 22, 376n11.2

Earman, J., 154
Ebbesen, E. B., 170
Ebert, A., 299t
Edwards, C. S., 211
Edwards, J. R., 283
Edwards, K., 155
Edwards, W., 7, 91, 155, 211, 283
Einhorn, H. J., 323
Ellis, S., 374n8.1
Elster, J., 281
Emslie, H., 239
Engel de Abreu, P. M. J., 22
Engle, R. W., 16, 22
Englich, B., 148, 374n8.3
Ennis, R. H., 207
Epley, N., 148, 373n8.1
Epstein, S., 42, 132, 133, 309t, 310t
Erdfelder, E., 103, 104
Erev, I., 375n8.10
Esfandiari, A., 90
Eslinger, P. J., 211
Etling, K. M., 374n8.3
Etzioni, A., 5
Evans, J. St. B. T., xi, 9, 17, 18, 20, 29, 30, 31, 33, 39, 42, 53, 103, 104, 115, 122, 129, 152, 154, 161, 273, 288, 336, 369n2.3, 370n2.3, 370n3.1, 371n3.1, 373n7.1, 375n8.7, 376n10.1

Fagley, N. S., 144
Falk, A., 299t
Fantino, E., 40, 90
Farah, M. J., 376n11.2
Feeman, D. J., 315
Feeney, A., 373n7.1
Feinstein, J., 311t
Feldman Barrett, L. F., 22
Fenton-O'Creevy, M., 297, 309t
Ferejohn, J., 41
Fernandes, D., 183

Fernbach, P. M., 112
Ferreira, M. B., 112, 164, 371n3.3, 371n3.4
Ferrell, W. R., 375n8.10
Fiedler, K., 33
Finucane, M. L., 27, 153, 288, 374n8.4
Fioratou, E., 112
Fischer, I., 88
Fischhoff, B., 27, 71, 98, 100, 154, 162, 165, 167, 188, 189, 270, 278, 374n8.2, 375n8.9
Fletcher, J. M., 302t, 377n15.1
Fong, G. T., 84, 195
Forsythe, R., 305t
Frank, M. J., 17
Frank, R. H., 3
Frankenhuis, W. E., 251
Frankish, K., 17, 41
Franks, S. F., 204
Franssens, S., 47, 70, 371n3.4
Frederick, S., 19, 44, 51, 53, 59, 86, 92, 111, 112, 113, 116, 117, 120, 170, 172, 187, 227, 280, 287, 301t, 302t, 374n8.3, 377n14.1
Freebody, P., 239
Frey, M. C., 294
Friedman, N. P., 19
Friese, M., 51
Frisch, D., 144, 153
Fugelsang, J. A., 371n3.3
Funder, D. C., 182
Funk, C., 377n14.1
Furnham, A., 374n8.3

Gaeth, G. J., 144, 374n8.1
Gal, I., 90, 91, 95, 227, 377n14.1
Gale, M., 116, 288, 371n3.5
Gallistel, C. R., 92
Gao, J., 90
Garb, H. N., 301t
Gardner, D., 189
Gardner, H., 318
Garnham, A., 122

Author Index

Gathercole, S. E., 22
Gaudiano, B. A., 310*t*
Gauthier, D., 34
Gawronski, B., 17
Geary, D. C., 16, 317
Gelman, R., 113
Gervais, W. M., 301*t*
Gibbard, A., 42
Gigerenzer, G., 166, 279, 312, 324, 298*t*, 299*t*, 309*t*, 370*n*3.1
Gignac, G. E., 16, 194, 198, 199
Gilbert, D. T., 33, 289
Gilhooly, K. J., 112
Gilleen, J., 204
Gilovich, T., 142, 148, 300*t*, 374*n*8.3
Girotto, V., 33, 85
Gladwell, M., 370*n*3.1
Glaser, E. M., 157, 305*t*
Glaser, M., 168, 375*n*8.9, 375*n*8.10
Gleicher, F., 211
Glumicic, T., 43, 47, 371*n*3.3, 371*n*3.4
Goel, V., 373*n*7.1
Goertzel, T., 194, 198, 199
Goff, M., 210
Goldacre, B., 192
Goldman, A. I., 209, 285, 286
Goldstein, D. G., 304*t*, 312, 324
Goo, S. K., 377*n*14.1
Gould, S. J., 51, 52
Graboys, T., 83
Graffeo, M., 302*t*, 307*t*
Gray, J. R., 22, 171, 175
Green, L., 170, 171
Greenberg, J., 158
Greene, J. D., 112
Greenhoot, A. F., 154
Griffin, D., 142, 162, 164, 166, 298*t*, 375*n*8.9, 375*n*8.10
Griffin, P., 192, 310*t*
Griffin, S., 192, 310*t*
Grigorenko, E. L., 24, 25
Groopman, J., 164, 300*t*
Grossmann, I., 183, 275

Gruber, J., 146, 304*t*
Grunewald, M., 11*t*, 212, 290
Guarnera, L. A., 305*t*
Gullion, C. M., 27

Ha, Y., 22, 288
Hahn, U., 154, 164, 279, 292
Halleran, J. G., 144
Halpern, D., 270, 274, 275
Hambrick, D. Z., 22
Handley, S. J., 33
Haran, U., 210
Hardisty, D. J., 171
Harman, G., 209
Harnishfeger, K. K., 376*n*11.2
Harris, A. J., 315
Harris, A. J. L., 154, 164, 279, 292
Harris, P. L., 373*n*7.1
Harrison, M. R., 239
Harrison, T. L., 22
Harvey, N., 9, 53, 88, 298*t*
Hasher, L., 19
Hastie, R., 18, 289, 302*t*
Healy, P. J., 164, 165, 168, 375*n*8.9
Healy, S. D., 41
Heath, J., 308*t*
Heggestad, E. D., 27
Heier, H., 309*t*
Heijltjes, A., 210
Heit, E., 129
Hershey, J. C., 153, 303*t*
Hertwig, R., 85, 324
Hicks, K. L., 22
Highhouse, S., 144, 373*n*8.1
Hilton, D. J., 18, 85, 163, 164, 166, 167, 297, 312, 303*t*, 305*t*, 375*n*8.9, 375*n*8.12
Hine, D. W., 302*t*
Ho, C., 11*t*, 195
Ho, I., 274
Hoch, S. J., 164
Hoffrage, U., 166, 299*t*
Hofmann, W., 51

Hogarth, R. M., 323
Holland, S. D., 204, 310*t*
Holt, J., 182
Hood, B. M., 199
Horn, J. L., 16, 369*n*2.1
Houston, A. I., 40
Houston, C. E., 374*n*8.3
Howell, J. L., 311*t*
Howson, C., 154
Hsee, C. K., 18, 289, 291
Huffman, D., 299*t*
Hull, D. L., 32
Hulme, C., 295
Humphrey, N., 199
Hunt, E., 16, 317
Hunter, J., 317
Huntley, V., 192, 310*t*
Hurley, S., 3, 40
Hurly, A., 41

Inglis, M., 130
Isaacson, W., 192

Jacowitz, K. E., 150, 374*n*8.3
Jadwiszczok, A., 192
James, G., 91, 92, 112, 143, 152, 275, 280, 372*n*5.1
Jarvis, C. B., 283, 308*t*
Jarvis, W., 311*t*
Jasper, J. D., 17, 70, 133, 180, 187, 374*n*8.3
Jeffrey, R. C., 7, 137
Jeffrey, S. A., 164
Johnson-Laird, P. N., 33, 103, 122, 137, 375*n*8.7
Johnson, E. J., 303*t*, 304*t*
Johnson, S. C., 43, 47
Johnson, S. M., 288
Johnson, W., 16
Joireman, J., 311*t*
Jones, L. L., 19
Jones, W., 195
Jordan, S. D., 301*t*

Juanchich, M., 301*t*

Kable, J. W., 50, 212, 290
Kacelnik, A., 3, 40
Kadvany, J., 188, 189
Kahan, D. M., 200, 202
Kahneman, D., xi, 8, 9, 10, 12, 17, 18, 19, 33, 39, 41, 44, 50, 51, 54, 55, 60, 61, 63, 81, 84, 85, 86, 89, 112, 116, 120, 142, 143, 144, 145, 147, 149, 150, 151, 152, 153, 164, 168, 181, 189, 278, 279, 280, 287, 289, 312, 320, 298*t*, 370*n*2.3, 374*n*8.1, 374*n*8.3
Kalb, C., 189
Kanazawa, S., 27
Kane, M. J., 16, 22
Kanfer, R., 24
Kao, S. F., 102, 107, 108
Kariv, S., 308*t*
Karlan, D., 171
Karns, T. E., 275
Katz, Y., 158
Keating, D. P., 376*n*11.1
Kelley, C., 11*t*, 195
Kelley, C. M., 180, 275, 313
Kelman, M., 5, 279, 280, 312
Kendall, P. C., 204, 310*t*
Keren, G., 52
Kern, L., 300*t*
Kievit, R. A., 251
Kimberg, D. Y., 376*n*11.2
King, R. N., 300*t*
Kirby, K. N., 170, 171, 174, 175, 306*t*
Kirkpatrick, L., 132
Kitcher, P., 376*n*11.1
Klaczynski, P. A., 27, 136, 154, 155, 180, 274, 275, 313
Klauer, K. C., 103, 104
Klayman, J., 164, 166, 167, 168, 288, 375*n*8.9, 375*n*8.10, 377*n*14.1
Klein, A., 146
Klein, D. B., 184
Klein, G., 18, 19, 42

Klein, W. M., 292
Kleinbolting, H., 166
Knetsch, J. L., 153
Koehler, D. J., 9, 81, 84, 91, 92, 112, 143, 152, 275, 280, 298t, 300, 371n3.3
Kokis, J., 10t, 27, 92, 93, 121, 123, 125t, 127t, 135t, 195, 210, 275
Kolbert, E., 317
Koriat, A., 71
Kramer, W., 298t
Krantz, D. H., 84, 151, 171, 187, 375n8.6
Krueger, A. B., 289
Krueger, J. I., 182, 301t
Kruglanski, A. W., 25, 209, 376n11.1
Ku, K., 274
Kuhberger, A., 374n8.1
Kuhn, D., 155, 195, 274, 375n8.7
Kunda, Z., 158
Kunreuther, H. C., 151, 303t
Kurzban, R., 18, 50

Låg, T., 180, 181, 302t, 307t
Laibson, D. I., 17, 170
Langer, T., 168
Larrick, R. P., 183, 307t
Lavallee, K. L., 27, 154, 274
Layman, M., 188
Leahy, J., 311t
Leahy, R., 204
LeBoeuf, R. A., 142, 144, 374n8.3
Lee, C. J., 279
Lee, J. W., 163, 375n8.9
Lefcourt, H. M., 311t
Legrenzi, P., 33
Lehman, D. R., 106
Leider, S., 88
Leipold, K., 255, 377n14.1
Leirer, V., 122
Lempert, R. O., 106
Lennox, R., 283, 284
Lepper, M. R., 155

Levav, J., 289
Levesque, H. J., 138
Levin, I. P., 102, 108, 143, 144, 374n8.1
Lewandowsky, S., 194, 198, 199
Li, M., 19, 189
Liberali, J. M., 112, 134, 136, 180, 181, 270, 275, 313
Lichtenstein, S., 71, 142, 151, 152, 162, 165, 167, 187, 188, 189, 278, 375n8.9
Lieberman, M. D., 17
Lilienfeld, S. O., 299t, 300t, 301t, 309t
Lin, D. Y., 289
Lin, S., 164
Lindeman, M., 161, 209
Lindner, H., 204
Lipkus, I. M., 180, 307t
Lipman, M., 375n8.7
Liu, D., 286
Liu, E., 299t, 301t, 309t, 311t
Lodge, M., 154, 155
Loewenstein, G. F., 17, 51, 168, 169, 170, 174, 291
Lopez, D., 155
Lord, C. G., 155
Lovallo, D., 164
Love, B. C., 370n2.5
Lucas, E. J., 30
Luce, R. D., 7, 40
Lumley, M. A., 311t
Lumry, A., 310t
Lusardi, A., 52, 183, 377n14.1
Lustig, C., 19
Lynch, J. G., 183
Lynch, J. S., 30
Lyon, D., 81

Macchi, L., 81, 86
MacGregor, D. G., 153, 304t, 374n8.4
Mackintosh, N. J., 16, 317

Macpherson, R., 11t, 124t, 125t, 27, 121, 127t, 135t, 195, 210, 274, 299t, 301t, 309t, 311t
Maher, P., 78
Mahoney, K. T., 144, 374n8.1
Majima, Y., 194, 198
Mak, D., 373n8.1
Malkiel, B. G., 299t, 308t
Mamassian, P., 164
Mandell, L., 183, 184
Manktelow, K. I., xi, xii, 6, 7, 9
Mäntylä, T., 270
Marcus, G. F., 40
Margolis, H., 376n10.1
Marklein, F., 299t
Markovits, H., 53, 122, 373n7.1
Marks, A. D. G., 302t
Marr, D., 370n2.5
Martin, T., 146
Martinez-Perez, J., 302t
Massey, C., 292
Mata, A., 112, 164, 168, 371n3.3, 371n3.4
Matthews, G., 27
Maule, J., 374n8.1
Mayes, R. S., 305t
McArdle, J. J., 16
McCarthy, J. D., 91
McCarthy, R. A., 376n11.2
McClure, S. M., 17, 170
McDaniel, M., 300t
McFall, R. M., 299t, 300t
McGinn, L., 204
McGrew, K. S., 16, 369n2.1
McHugh, P., 301t
McKenzie, C. R. M., 154, 374n8.1
McLaren, R. P., 17
McVay, J. C., 22
Meier, P., 310t
Meier, S., 171, 306t
Mellers, B. A., 5, 85, 210
Meserve, R. J., 11t, 114t, 289
Messick, S., 376n11.1

Meszaros, J., 303t
Meyers, J. M., 33
Meyvis, T., 289
Mickler, C., 168, 375n8.7
Mikels, J. A., 134
Milford, G., 195
Milkman, K. L., 297
Miller, I. W., 310t
Miller, P. H., 19
Miller, P. M., 144
Minson, J. A., 303t
Mischel, W., 170, 171, 306t
Mitchell, O. S., 52, 183, 377n14.1
Miyake, A., 19
Mochon, D., 374n8.3
Monahan, J., 304t
Moore, D. A., 164, 165, 168, 375n8.9
Moors, A., 18
Moran, P. J., 204
Morgan, J. N., 183, 307t
Morgenstern, O., 7
Moritz, B. B., 112, 302t
Morsanyi, K., 43
Moshman, D., 376n11.1
Moyens, E., 47, 371n3.4
Mrazek, M. D., 22
Mukherjee, D., 212, 290
Müller, W., 308t
Munro, G., 155
Murphy, P., 112
Murrie, D. C., 305t
Mussweiler, T., 148, 374n8.3
Myers, J., 50
Myerson, J., 170, 171
Mynatt, C., 99, 107, 372n6.1

Nantel, G., 53, 122, 373n7.1
Narendran, S., 301t
Navon, D., 50
NCEE (National Council for Economic Education)., 183, 184
Neale, M. A., 148, 303t
Nelson, F., 305t

Nelson, L. D., 374n8.3
Nelson, M. W., 304t
Nelson, W. L., 180
Netemeyer, R. G., 183
Neumann, G., 305t
Neumann, P. J., 325
Newell, B. R., 70, 91, 92, 132, 133, 134
Newstead, S. E., 30, 103, 122, 129, 130
Newton, E. J., 30
Nezworski, M. T., 301t
Nichols, S., 285, 286
Nickel, T., 195
Nickerson, R. S., 53, 375n8.7, 376n10.1
Nijhuis, M., 301t
Nimmo-Smith, I., 239
Nisbett, R. E., 16, 84, 106, 183, 307t, 376n10.1
Noel, X., 290
Northcraft, G. B., 148, 303t
Noveck, I. A., 17
Novick, L., 107
Nozick, R., 137, 209
Nudds, M., 3, 40
Nussbaum, E. M., 375n8.7

Oakhill, J., 122
Oaksford, M., 41, 103, 104, 279, 370n2.5
Oberauer, K., 194, 198, 199
Obrecht, N. A., 113
Odean, T., 305t, 309t, 377n14.1
O'Donoghue, T., 170
Oechssler, J., 275, 374n8.3
Offit, P. A., 192, 193, 300t, 310t
Okan, Y., 134
Oldfield, Z., 180, 183, 307t
Oliver, J. E., 194, 198, 199, 200, 202
Orwell, G., 194
Osherson, D. N., 78
Over, D. E., xii, 6, 18, 30, 33, 40, 83, 279, 288
Oyer, B., 161

Pacini, R., 132, 309t
Paese, P., 144, 374n8.1
Palmer, E. C., 204
Parker, A. M., 27, 165, 166, 168, 270, 297, 374n8.2, 375n8.11, 375n9.9
Parker, J. D. A., 212
Pashler, H., 300t
Paulhus, D. L., 201, 248
Paulos, J. A., 308t
Pauly, M. V., 306t
Pennington, N., 302t
Pennycook, G., 112, 309t, 371n3.3
Pereira, N. S., 306t
Perkins, D. N., 34, 207, 274, 375n8.7
Peters, E., 19, 60, 153, 180, 287, 288, 374n8.4
Petrocelli, J. V., 374n8.3
Petrusic, W. M., 164
Petry, N. M., 88, 171, 290, 306t
Petty, R. E., 311t
Pfeifer, P. E., 375n8.10
Phillips, L., 167
Pinker, S., 12, 147, 328
Pinto-Prades, J., 302t
Pintrich, P. R., 309t
Poletiek, F. H., 288
Politser, P. E., 325
Politzer, G., 86
Pollard, P., 122, 336, 373n7.1
Polonio, L., 302t, 307t
Polonioli, A., 279
Pompilio, L., 40
Postman, N., 63
Powell, L. A., 309t
Prado, J., 17
Prelec, D., 174
Prencipe, A., 171
Primi, C., 116
Pronin, E., 289, 290
Pyszczynski, T., 158

Rabin, M., 187, 188, 376n9.1
Rachlin, H., 169

Radford, B., 189, 192
Raiffa, H., 7, 40
Rakow, T., 92
Rand, D. G., 112
Rapoport, A., 137
Ratner, R. K., 289
Read, D., 51, 168, 169, 170
Real, L. A., 40
Redelmeier, D. A., 303t
Regnier, E., 192, 310t
Reimers, S., 171
Resnik, M. D., 78
Revlin, R., 122
Reyna, V. F., 60, 133, 180, 306t, 373n7.4, 374n8.1
Richerson, P. J., 32
Richmond, C., 116, 371n3.5
Rimer, B. K., 180
Rips, L. J., 137
Ritov, I., 153, 210
Roazzi, A., 373n7.1
Roberts, M. J., 30
Robinson, B., 154
Roby, K. J., 311t
Rockloff, M. J., 309t
Rodriguez, M. L., 170, 306t
Rogers, P., 52
Rogers, T., 297
Rohrer, D., 300t
Roider, A., 275
Rokeach, M., 209
Roksa, J., 277
Ronan, K. R., 310t
Roney, C. J. R., 53, 88
Ronis, D., 164, 165, 375n8.9
Ronson, J., 326
Ross, L., 155, 164, 289
Ross, M., 164, 183, 289
Roszkowski, M. J., 144
Rotello, C. M., 129
Rowe, M., 310t
Rufino, K. A., 305t
Rupar, K. K., 304t

Russell, D., 195
Russo, J. E., 164, 167
Russo, S., 86

Sá, W., 11t, 106, 121, 123, 124t, 127t 129, 161, 195, 209, 210, 273, 275, 373n7.1
Sahoo, J. S., 326
Salhany, J., 204
Salthouse, T. A., 22
Samsa, G., 180
Samuels, R., 18, 279, 280
Samuelson, P. L., 209
Samuelson, W., 304t
Sanfey, A. G., 17
Sansone, N., 88
Santiesteban, M., 171
Satz, D., 41
Savage, L. J., 7, 40, 137, 153, 187
Savitsky, K., 300t
Schaeken, W., 137
Schapira, M. M., 180, 270
Scheffler, I., 376n11.1
Schick, F., 8, 78
Schkade, D., 289
Schlam, T. R., 306t
Schlenker, B. R., 200
Schlosnagle, L., 275
Schmidt, F. L., 317
Schmitt, D. P., 147
Schmitz, P. W., 275
Schneider, S. L., 144
Schneider, W., 17, 18
Schnittjer, S. K., 144
Schoemaker, P. J. H., 164, 167
Schoenberger, A., 83
Schommer-Aikins, M., 25
Schommer, M., 209
Schoorman, F. D., 144, 374n8.1
Schryer, E., 183
Schubert, A., 371n3.3, 371n3.4
Schuck-Paim, C., 40
Schwartz, L. M., 180

Schwarz, N., 152, 289
Sciutto, M. J., 204
Scopelliti, I., 290
Scott, K. M., 239
Segal, J., 274
Seligman, M. E. P., 306t
Senecal, N., 172
Shaffer, R., 192
Shafir, E., 41, 52, 137, 142, 143, 144, 151, 152, 278, 280, 281, 303t, 374n8.3
Shamosh, N. A., 171, 175
Shanks, D. R., 91, 102, 108
Shapiro, J. M., 171, 187, 188, 306t, 308t
Sharot, T., 292
Shaughnessy, J. J., 239
Shenhav, A., 112
Shepperd, J. A., 164, 292, 311t
Sher, S., 374n8.1
Sherman, J. W., 17
Sherman, S. J., 112, 164, 371n3.4
Shiffrin, R. M., 18
Shiloh, S., 144
Shipley, W., 146
Shiv, B., 291
Shleifer, A., 312
Shoda, Y., 170, 306t
Shoham, V., 299t, 300t
Shonk, K., 297, 308t
Sieck, W. R., 144, 375n8.9
Siegel, L. S., 377n15.1
Silverman, D., 308t
Silverman, I. W., 255, 377n14.1
Simmons, J. P., 292, 374n8.3
Simon, A. F., 144
Simon, H. A., 50
Simoneau, M., 373n7.1
Simonson, I., 151, 152
Sinatra, G. M., 309t, 375n8.7
Singh, K., 310t
Sipay, E. R., 315
Sirota, M., 301t
Sivak, M., 309t

Skenazy, L., 189
Skilling, T. A., 305t
Skyrms, B., 40
Sloman, A., 370n2.5
Sloman, S. A., 51, 83, 84, 93
Slovak, L., 83
Slovic, P., 19, 81, 142, 151, 152, 153, 162, 187, 188, 189, 278, 287, 288, 304t, 374n8.4, 375n8.6, 375n8.9
Slusher, M. P., 155
Smith, A. B., 40, 151
Smith, E. E., 155
Smith, E. R., 17
Smith, J., 168, 275
Smith, S. M., 144
Smullyan, R. M., 137
Snelbecker, G. E., 144
Sniezek, J. A., 166
Snowling, M. J., 295
Soll, J. B., 166, 167, 168, 375n8.10, 377n14.1
Solove, D. J., 326
Sorge, G. B., 305t
Spangenberg, E. R., 311t
Spencer, A., 310t
Sperber, D., 33
Sprenger, C. D., 171, 306t
Sprott, D. E., 311t
Stahl, C., 103, 104
Stankov, L., 373n7.3
Statman, M., 305t
Staudinger, U. M., 168, 375n8.7
Stein, E., 5, 275, 279
Steinberg, L., 171
Sterelny, K., 370n2.5
Sternberg, R. J., ix, 24, 25, 168, 318, 375n8.7, 376n11.1
Stewart, N., 148, 303t
Stibel, J. M., 83
Stich, S. P., 279, 280, 286
Stigler, S. M., 79
Stolarz-Fantino, S., 40
Stone, A. A., 289

Stone, E. R., 165, 166, 168, 375n8.9
Strack, F., 17, 51, 148, 374n8.3
Strathman, A., 25, 168, 211
Strough, J., 270, 275
Stuebing, K., 377n15.1
Stupple, E., 116, 371n3.5
Sundali, J., 88, 298t
Sunde, U., 299t
Sunstein, C. R., 188, 194, 198, 297, 307t, 309t, 320
Sutton, R. M., 198
Svedholm, A. M., 161, 209
Swami, V., 198
Szrek, H., 306t

Tabachnik, N., 107
Taber, C. S., 154, 155
Takemura, K., 144
Taylor, S. E., 50
Tentori, K., 86
Terjesen, M. D., 204
Tetlock, P. E., 5, 164, 375n8.9
Thagard, P., 209
Thaler, R. H., 8, 9, 142, 151, 153, 174, 183, 297, 304t, 308t, 312, 320
Thee, S. L., 144
Thompson, K. F., 171
Thompson, V. A., 43, 47, 154, 161, 371n3.3
Thomson, M., 303t
Thomson, P., 309t
Thorley, S., 305t
Toates, F., 17
Tobacyk, J., 195
Todd, P. M., 279, 324
Tomlin, D., 17, 50
Toneatto, T., 52, 299t, 301t, 309t, 311t
Tooby, J., 83, 147, 279, 327
Trick, L., 53
Trope, Y., 17
Trout, J. D., 5, 189, 275
Tschirgi, J. E., 106
Tugade, M. M., 22

Tunney, R. J., 91
Tversky, A., 8, 9, 10, 12, 17, 39, 50, 51, 54, 55, 60, 63, 81, 84, 85, 89, 91, 137, 142, 143, 144, 145, 147, 149, 151, 152, 153, 163, 166, 189, 279, 280, 281, 298t, 312, 374n8.1, 375n8.9

Underwood, B. J., 239
Urbach, P., 154

Valentine, D. A., 308t
Vallone, R., 164
Vansteenwegen, D., 47, 371n3.4
Vellutino, F., 377n15.1
Vermeule, A., 194, 198
Villejoubert, G., 374n8.1
Vohs, K. D., 51, 169, 301t
Volpe, R. P., 183, 184
von Neumann, J., 7
Vorkink, K., 305t
Voss, J. F., 274

Wagenaar, W. A., 52, 88, 299t
Wainer, H., 298t
Waksberg, A. J., 40
Waldorp, L. J., 251
Wallsten, T. S., 375n8.10
Wang, L., 19, 189
Warrington, E. K., 376n11.2
Warusawitharana, M., 170
Wason, P. C., 30, 31, 61, 103, 137, 288, 376n10.1
Wasserman, E. A., 102, 107, 108
Waters, E., 292
Watson, G., 157, 305t
Weber, E. U., 171, 291
Weber, M., 168
Webster, D. M., 25, 209
Weinstein, N., 291, 292
Welch, N., 291
Weller, J. A., 60, 95, 180, 181, 227, 270, 377n14.1

Wellman, H. M., 286
Welsh, M. B., 374n8.3
Weseley, A. J., 304t
Westbrook, A., 50
Westen, D., 155
Wheatley, T., 33
White, E., 189
Wiebe, J. S., 204
Williamson, D. A., 204
Wilson, N. L., 306t
Wilson, T. D., 33, 289, 374n8.3, 376n10.1
Windschitl, P. D., 151
Winklhofer, H. M., 283
Winston, G. C., 171
Wolfe, C. R., 154
Wood, J. M., 301t
Wood, M. J., 198, 374n8.3
Wood, S., 307t
Wood, T. J., 194, 198, 199, 200, 202
Wright, J., 305t
Wu, A., 304t

Xu, J., 53, 88, 298t

Yamagishi, K., 134, 153
Yaniv, I., 164
Yates, J. F., 144, 163, 164, 165, 375n8.9
Yechiam, E., 290
Yin, W., 171
Yopp, H., 122
Yopp, R., 122

Zacks, R., 19
Zahler, D., 165
Zahler, K. A., 165
Zeckhauser, R. J., 304t
Zeidner, M., 27
Zhang, L. F., 24
Zikmund-Fisher, B. J., 153
Zimmerman, J., 239
Zougkou, K., 92
Zweig, J., 182

Subject Index

Affective forecasting, 289–290
Affect substitution, 287–288
Alexithymia, 212
Algorithmic mind, 23–26, 121, 208
Anchoring effects, 55, 147–151
Attribute substitution, 19, 112, 120, 287
Automaticity, 18, 36–37
Autonomous mind, 24

Base rates, 81–85
  causal vs. noncausal, 84–85
Bayes' theorem, 78–79, 98. *See also* Likelihood ratio
Bayesian reasoning, 78–80, 97–103
Between-subjects design, 144, 149–150, 155–156, 288
Bias blind spot, 289–290

CART (Comprehensive Assessment of Rational Thinking)
  assessment settings, 276–278
  CART16, 265–268
  caveats to, 269–296
  coaching effects, 295–296
  context of, 269–296
  cultural context, 292–293
  development of, 92–93, 105–106, 116–119, 123–130, 136–140, 144–145, 150–151, 157–160
  framework for, 63–71
  full-form study (RT60), 234–268
  full-form test, 233–268
    percentile ranks, 261–263
    vs. short-form, 259–261, 264–266
  grounded in heuristics and bias literature, 9–13, 63–64
  and intelligence, correlation with, 293–294
  measuring propensities rather than abilities, 80–81, 85, 87, 90, 97, 219–220, 313–314, 321–322
  multifarious nature of, 284–286
  points allocated to subtests, 71–72, 183, 221, 331
  practical applications, 164, 171, 297–312
  previous studies, 93–95, 106–110, 117–119, 131–132, 136–137, 139–140, 146, 150–151, 153–154, 161–162, 166–168, 171–175, 181, 184–186, 200–202, 213–216
  process and knowledge in, 68–71
  as prototype (beta) test, 269, 292, 297
  psychometric structure, 236–237, 242, 244–245, 252–254, 260, 266–267, 282–286
  residual CART, 259–261
  restriction of range, 276–277
  short-form study (RT59), 220–223
  short-form test, 219–231
    percentile ranks, 231, 262
  subtest classification, 64–66, 68–71
  38-item two-subtest version, 264–268

CART subtests and scales (in order of appearance and matching the structure of table 4.1)
Probabilistic and Statistical Reasoning, 77–95, 262, 264–268
  instructions and sample items, 332–334
Scientific Reasoning, 97–110, 262, 264–268
  instructions and sample items, 335–340
Reflection vs. Intuition Subtest, 111–119
  instructions and sample items, 340
Belief Bias in Syllogistic Reasoning, 119–132
  instructions and sample items, 340–342
Ratio Bias, 132–137
  instructions and sample items, 342–344
Disjunctive Reasoning, 137–140
  instructions and sample items, 344–345
Framing, 142–146, 344–345
  instructions and sample items, 345–347
Anchoring, 147–151
  instructions and sample items, 347–348
Preference Anomalies, 151–154
  instructions and sample items, 348–349
Argument Evaluation Test, 154–162
  instructions and sample items, 349–351
Knowledge Calibration, 162–168
  instructions and sample items, 352–354
Rational Temporal Discounting, 168–175
  instructions and sample items, 354–357
Probabilistic Numeracy, 177–181
  instructions and sample items, 357–358
Financial Literacy and Economic Knowledge, 181–186, 325
  instructions and sample items, 358–359
Sensitivity to Expected Value, 187–199
  instructions and sample items, 359–360
Risk Knowledge, 188–189
  instructions and sample items, 361
Rejection of Superstitious Thinking, 195–196
  instructions and sample items, 362
Rejection of Antiscience Attitudes, 197
  instructions and sample items, 362–363
Rejection of Conspiracy Beliefs, 198–203
  instructions and sample items, 363–365
Avoidance of Dysfunctional Personal Beliefs, 203–205
  instructions and sample items, 365
Actively Open-Minded Thinking Scale, 209–210, 245–249, 314–315
  instructions and sample items, 365–366
Deliberative Thinking Scale, 210–211
  instructions and sample items, 366–367
Future Orientation Scale, 211
  instructions and sample items, 367
Differentiation of Emotions Scale, 211–212
  instructions and sample items, 367–368
Cognitive decoupling, 28–29, 43–44. *See also* Type 2 processing
  and executive functioning, 19, 21
  and fluid intelligence, 22, 26
  belief bias and, 120–121

Subject Index

functions of, 19–22
  sustained, 20
Cognitive misers, 19. *See also* Miserly processing
  three types, 49, 53–55
Cognitive Reflection Test, 53, 59–61, 111–113
  intelligence and, 113–116
"Cold" override, 169–170
Conflict detection, 43–44
  link with mindware, 44–48, 55–57
Conjunction errors, 85–87
Constructed preference theory, 151–152
Control group reasoning, 106–107
Converging evidence, 104–105
Covariation detection, 102–103, 107–109
Critical thinking, 121, 157

Decision making, context effects in, 142–154
Decision-Making Competence Scale, 270–273
Decontextualization, 323–327
Delay of gratification, 168. *See also* CART subtest: Rational Temporal Discounting
Denominator neglect. *See* Ratio bias
Descriptive invariance, 142–143
Dual-process theory, 16–20, 28–32, 34–35
  intelligence in, 21–22
  and knowledge structures, 34–35
  Type 1 and Type 2 processing (*see* Type 1 processing; Type 2 processing)

Falsifiability, 103–104
Focal bias, 30–33. *See also* Serial associative cognition
Formative vs. reflective assessment, 283–284
Four-card selection task, 30–31, 103–104
Framing effects, 54–55, 142–146
Fundamental computational biases, 147, 162, 198–199

Gambler's fallacy, 87–89
Great Rationality Debate, 278–281, 312, 323–328
  Meliorists and Panglossians, 278–279, 323, 327–328

Halpern Critical Thinking Assessment, 274–275
Heuristics and biases tasks, 9–12
  basis of the CART, 9–13, 63–64
  construal uncertainty a strength in, 85, 87, 90, 97
  critics of, 278–281, 312, 321–323
  and IQ tests, missing from, 12–13
  knowledge and process intertwined, 55–57, 68–71
  logic of, 42–49
"Hot" override, 169–170
Hypothesis testing, 103–104
Hypothetical thinking, 20

Ignoring P(D/~H). *See* Likelihood ratio
Impression management, 200–201, 205, 214, 249–250, 294–295
Intelligence
  CHC theory, 16, 284–285, 320
  crystallized, 16, 35
  fluid, 16, 22
  grounded vs. permissive theories, 15–16, 26, 318
  mindware problems and, 192–194
  political correctness and, 316–317
  vs. rationality distinction, 26, 312–314, 315–316
  tests, 60, 119, 121
Intuitive response, 45–46, 112, 119
Iowa Gambling Task, 212, 290–291

"Kinds of minds," 22–25, 35–37, 39–41
Knowledge projection, 193
Knowledge structures. *See* Mindware

Likelihood ratio, 98–103, 154
Linda problem, 85–87

Mental models, 137
Metarepresentation, 20, 29, 39
Mindware, 34–35, 43
  automatized vs. accessed, 36–37
  contaminated, 52–53, 66–67, 191–205, 294–295
  crystallized intelligence and, 35
  gap, 52–53, 66–67, 177–189
  instantiation continuum, 45–49, 57–61
  override and conflict detection, link with, 44–49, 68–71
  vs. process dependence, 55–61, 68–71
  Type 1 processing and, 35–37
  Type 2 processing and, 35–37
Miserly processing, 32–34, 39–57, 64–66, 70, 111, 132, 141
Myside bias, 154–162. *See also* CART subtest; Argument Evaluation Test

Numeracy, probabilistic, 60–61, 177–181

Overconfidence, 162–168. *See* CART subtests and scales, Knowledge Calibration
Override, 19–20, 43–44, 132–133
  algorithmic capacity for, 43
  conflict detection and mindware, link with, 44–49, 55–57

Political attitudes, conspiracy beliefs and, 200, 202
Preattentive processes, 29, 35
Probabilistic reasoning, 8–9, 60–61
Probability calculus, 78–79
Probability matching, 90–92
Pseudoscience, 191–194
  high intelligence and, 192–194

Rational choice theory, 7–9, 152–153
Rationality, 3–6
  axiomatic approach, 7–9, 40–41, 63–64, 142, 152, 187
  categorical vs. continuous, 7–9, 63–64
  cognitive requirements of, 37–38, 208
  definitions of, 4–6
  epistemic, 6–7, 72–73, 77, 163
  evolutionary psychology and, 327
  individual differences, 10–14, 21–29, 41
  instrumental, 6–7, 72–73
  intelligence and, 13–15, 26–28, 35, 293–294, 312–320, 322
  correlation with, 315–316
  metarationality, 282
  missing from IQ tests, 12–13, 312–315, 322
  modernity and, 322–328
  as not subpersonal, 41–42
  as propensity rather than ability, 80–81
  in real-life domains, 164, 171, 297–312
  thin theory vs. broad theory, 281–282
  tripartite model and, 35–37, 314–315
Recognition heuristic, 324–325
Reflective level of processing, 32, 146
Reflective mind, 24–26, 28–30, 121, 208, 314–315, 324–325
Risk/benefit relationship, 287–288

Sample size, 89–90
Scientific reasoning, 60–61
Serial associative cognition, 29–32, 50–51, 54, 143–144, 148, 164
Sex differences, 95, 110, 227–230, 251–258, 293
Simulation, 20, 29
Social desirability. *See* Impression management
Subjects, Amazon Mechanical Turk vs. university, 185–186, 203, 238–241

Subject Index

Syllogistic reasoning
  belief bias in, 53–54, 111–119, 154
  believability bias vs. validity bias, 129–131
System 1, 17. *See also* Type 1 processing
System 2, 17. *See also* Type 2 processing

Taxonomy of thinking errors, 49–53
  contaminated mindware, 52–53
  default to autonomous mind, 49–50
  default to serial associative cognition with a focal bias, 52
  failure of sustained override, 50–52
  mindware gaps, 52
Thinking dispositions, 25–26, 28, 67–68, 121, 207–216, 313–315
  AOT as a global mental attitude, 314–315
  as independent predictors of rational thought, 27
  and intelligence, associations with, 27
  percentiles, 263
  in the process/knowledge dimensions, 69–70
Tripartite model, 22–29, 208
  fleshed out, 28–32
  knowledge structures in, 34–35
  levels of analysis in, 24–26
  and locus of individual differences, 24–26
  rationality and, 35–37, 208
Type 1 processing, 17–20
  autonomy of, 17–18, 21
  benign vs. hostile environments for, 18–19, 42–43, 147, 182–183, 297, 321–328
  mindware and, 35–37
  normative responding and, 37
  overriding, 19–20, 22, 28–29, 43
Type 2 processing, 17–20
Typical vs. optimal performance situations, 24–26, 28, 67–68

Unrealistic optimism, 291–292
Utility maximization, 7–9, 63–64, 77, 137, 142–143

Vividness, 189

Wason 2–4–6 task, 288
Willpower, 169–170
Within-subject designs, individual differences in, 144, 149–150, 155–156, 288
Working memory, 22
WYSIATI tendency, 33, 55